TRAUMA

Clinical and Biological Aspects

TRAUMA

Clinical and Biological Aspects

Edited by

Stacey B. Day

Director, Department of Biomedical Communications and Medical Education
Sloan—Kettering Institute for Cancer Research
New York, New York

PLENUM MEDICAL BOOK COMPANY • New York and London

JJN 4M

Library of Congress Cataloging in Publication Data

Conference on the Pathobiology of Trauma, University of Minnesota, 1973.
  Trauma: clinical and biological aspects.

  Includes bibliographical references and index.
  1. Traumatology—Congresses. I. Day, Stacey B., ed. II. Title. [DNLM: 1. Wounds
and injuries—Congresses. W0700 C745t 1973]
  RD131.C64 1973   617'.21   74-30105
  ISBN 0-306-30834-7

Selected papers from proceedings of a conference on the Pathobiology of Trauma,
held at the University of Minnesota, Summer 1973, under the direction of Stacey
B. Day, M. D., Ph. D., D. Sc., Head, Bell Museum of Pathobiology, University of
Minnesota Medical School, and Ellis S. Benson, M. D., Professor and Chairman,
Departments of Laboratory Medicine and Pathology, University of Minnesota
Medical School.

© 1975 Plenum Publishing Corporation
227 West 17th Street, New York, N. Y. 10011

United Kingdom edition published by Plenum Publishing Company, Ltd.
4a Lower John Street, London, W1R 3PD, England

Plenum Medical Book Company is an imprint of Plenum Publishing Corporation

Printed in the United States of America

FOR OWEN WANGENSTEEN
MASTER SURGEON
IN HIS SEVENTY-FIFTH YEAR

Foreword

Late in summer 1973, the Bell Museum of Pathology of the
University of Minnesota Medical School, held a three day symposium
to discuss some of the biological aspects of trauma disease. The
meeting was intended to be a contribution to the study of trauma
by placing emphasis upon many of the basic biologic issues related
to trauma injuries. It was also hoped to put in focus perspectives
from which constructive interaction between basic research scientists
and practicing clinicians could flow. It is our belief that it is
through such interdisciplinary exchanges that intelligent progress
and new developments will occur. Moreover we believe that the
student body and the practicing physician can equally share this
scientific backdrop, and in a sense must so participate, for in a
technological society trauma injuries face the prospect of an almost
exponential increase in numbers, decade to decade.

With these views in mind we brought together biologists,
biophysicists, biochemists, pathologists, physicians and surgeons,
in an effort to build a new bridge of collaborative understanding
between unrelated disciplines, which, in the past, have characteristically rarely inter-related, one to the other. There were
no hard and fast rules set in the selection of topics. Not
surprisingly, therefore, the program developed into one of diverse
and remarkable scope and breadth. Papers for discussion varied
from the biochemistry of collagen metabolism to features of trauma
characteristic of aviation and air trauma accidents. Of these
many important and stimulating papers it has been possible to select
only a few for publication. Generally speaking these papers have
been chosen as representing a highly competent intermediate balance
from within a broad field of endeavor. It is hoped that these
selections will provide interest for both clinicians as well as
other scientists working in the field of trauma biology. Since
the whole approach to the subject in this symposium was interdisciplinary in nature, any classification in the text is bound to be
arbitrary. Reading the selections in a different order from their
sequential chapter order will not impair their value.

The problem of selecting some papers and omitting others
has been a delicate one, and has made me more than ever sensitive
of the efforts of the distinguished and outstanding faculty who
took time out of pressing schedules to develop intensive viewpoints
and discussions which highlighted the Minneapolis meetings.
I acknowledge with grateful thanks the interaction of all those
who participated in these proceedings.

With respect to the essays contributed in this text, it is my
hope that they will reflect the quality of the symposium as a whole.
Most of the writings presented here are studies by leaders in their
fields. Most contributions develop knowledge of the discipline
as a whole, in as brief a span as possible and towards the goal of
integrating the basic biologic thrust of new knowledge with the
clinical efforts at trauma control. It is from such conceptual
approaches and inter-related efforts that most good, in the
foreseeable future, is likely to come.

September 1974   Stacey B. Day

Contents

BIOLOGICAL PERSPECTIVES OF SOME TRAUMA INJURIES

M. J. Narasimhan and Stacey B. Day

Bell Museum of Pathobiology

University of Minnesota

Some biological aspects of trauma are presented in the following table. This paper presents a brief overview of some of these injuries.

(i) Molecular Dynamics and Biology of Physical Injuries:

Contact Injuries

Compression-Decompression Injuries: Blast Injuries

Vibratory-Sonic Injuries

Thermal Injuries: Heat Stress, Cold Injuries, Cryogenic Stress

Electrical Injuries

Photopic Injuries

Irradiation Trauma

Acceleration-Deceleration and Gravitational Trauma

(ii) Molecular Traumatology and Mechanisms of Chemical and Biological Agents:

Physico-chemical Types: Osmotic and Ionic Agents

Animal Poison Injuries

Plant Poison Injuries

Microorganisms and their Toxins

Hyperbaric Oxygen Injury

Immunological Trauma

Chromosomal and Intraembryonic Injury: Genetic
    Traumatology

(iii) Molecular Metabolism in Trauma and Related Features in
      Molecular Traumatology

Ultra-cellular Response to Trauma

Local and Systemic Metabolism in Trauma

Neurological Features of Trauma

Ectopic Extravasation of Physiological Fluids as
    Traumatants

Withdrawal Causing Molecular Injury

Psychological and Emotional Trauma

Growth-maturation Defects

Hypersusceptibility States in Trauma

Hypo-hyper-responses to Trauma

Concept of "Combined Injuries"

Response to Trauma in the "Gnotobiotic State"

CONCEPT OF MOLECULAR TRAUMATOLOGY

Molecular pathology is a branch of molecular biology. To a
limited degree, pathologic manifestations reflect biological
sequences representing changed dynamic states falling short of
normal or anticipated normal function of the cell as it is generally
supposed to perform at the molecular level of organization. Study
of functioning cell dynamics of the system is crucial to any
interpretation of injury. For practical purposes, the term

"molecular" also indicates that architecture of structure (as well as the physiological expression of function) includes concern with ultra or minute molecular anatomical entities, many of which are of an order of magnitude resoluble only with the electron microscope.

At this order of function, several other biological factors are of concern. For example, in biological terms, motion and function are usually viewed as implying fundamental constants in the dynamic structure of a cell (and, by extension, to a tissue, organ and ultimately to the whole animal body). Other parameters must be included, i.e., concepts of chemical change, spatial orientation (or re-orientation) in dimensional planes (two dimensional or three dimensional orders). Such visualizations encourage understanding of chemical phases, stereochemical physiology, and even physical forces measurable in terms of nucleus, electrons, protons, or as fields of magnetic force providing the basic background of pathology as a branch of molecular science. Indeed, for such entities as RNA and DNA, it is only at the molecular level that they are best open to analysis, not only in conceptual realizations of the "characteristic normal cell" but as possible provident signposts that may point the way to greater understanding of injured or "pathologic" cells or tissues.

Such chemical, physical, and biological alterations lend themselves to two overviews:

(i)   Clinico-pathological

(ii) Chemico-physico, by which analysis of various metabolic pathways that could or would be involved or substantially altered in any trauma might be explained.

TRAUMA

A working definition of trauma should describe a dynamic transfer of energy from one system in a higher state of energy to a physiological system in a lower state of energy, either directly through continuous contact or via a secondary medium, or, in absence of any medium, by secondary induction. The result of this process is discontinuity in the morphological structure of the biological system, with corresponding alteration in function and efficiency of the physiological organization and component parts or whole system receiving the energy(injury). A further factor that may follow as a natural sequel to such transfer of energy is acute (violent) or chronic(gradual, progressive and additive) response of the host system, manifest as the "reaction to injury." Such response is classically recognized as the features of inflammation(rubor, calor, tumor, dolor, and functio lassiae).

Differentiation must be made between molecular trauma and gross trauma.

GROSS TRAUMA

There is a dissolution of tissue with discontinuity and a secondary zone wherein cells are in shock (concept of cellular shock). A third region around this zone of cells in shock demonstrates a state of reactive hyperemia with outpouring of extracellular transudate, lymph and blood. As a result of this phase, cellular and humoral mechanisms are brought to the site of injury and act as a type of irrigation and debridement agent for injured cell fragments and other toxic metabolites produced or liberated from damaged or destroyed cells. Gross trauma is conspicuous by its macroscopic pathology.

Compared to gross trauma, molecular trauma may occur in few cells. Thus, a large residuum of cells remains intact and provides the potential "to recoup" - to take over the functions of the injured cells without any apparent impairment of tissue function. This situation contrasts markedly with the forms of dissolution pathology characteristic of gross pathology. Furthermore, molecular injury may be localized to a single cell or to a particular ultracellular fraction of a cell, as for example, during cellular nucleolar replication (mitosis) where no gross pathology may be observed but in which derivations from characteristic normal patterns of cell behavior serve as a clue to indicate molecular injury. Such a form of molecular injury without gross pathology is seen where cells have received sublethal doses of irradiation, or during the incubation period of intracellular obligate viruses. Injury to the genetic coding of nucleolar material within a cell provides a further illustration of the critical difference between gross and molecular trauma. Thus, in the event of molecular trauma at the level of the genetic coding material, "traumatic" sequellae may be transmitted to future generations of cells, f2, f3, f4 and so on to a theoretically possible f-infinity of generations. In gross trauma, a lesion is restricted to those cells that are delivered of the trauma, and neither their progeny nor generational offspring bear the stigmata of the wound to the initial (ancestral) cells. Similarly, in gross trauma (notwithstanding instances of infection and endotoxin production, or the phasic release of "toxins" conceivably correlated with the release of various H-like substances, including catecholamines, neuramines, vasodilatory amines liberated by injured cells or from their constituent parts) no sustained harmful action at remote sites is usual. In the case of molecular trauma, however, notwithstanding the fact that the extent and degree of injury at the original site may be small, remote effects can be severe and protracted.

Consideration should be given also to the processes of tissue repair and healing following insult by these two different forms of trauma. As a generalization, the process of tissue repair in cases of gross trauma is usually simpler, faster and more spontaneous than in cases of molecular trauma, for, although systemic components do play a significant role, the manifestation of injury requiring healing is characteristically <u>localized</u>. In the case of molecular trauma, spontaneous recovery is the exception rather than the rule for the molecular architecture and associated physico-chemical and physiological functions are more subtly disturbed. Indeed, there may be no repair at all, as usually understood, but progression and propagation of results following injury along quite distinct pathological lines. This type of biological response is seen in certain neoplasms where, following external stress, cells have undergone phenomena of molecular trauma followed by abnormal or post-traumatic aberrant "healing."

TRAUMA AND HEALING

Although one appears to be the antithesis of the other, a clear relationship exists between trauma and healing. Conceptually, one should think of trauma (molecular or gross) not so much as an absolute acute insult occurring over a fractional period of time, but as a dynamic process with a sequential flow of biological events. This concept does not embrace that chronic type of trauma in which a noxious agent persists in contact (directly or indirectly) with a tissue or cell.

This biological dynamism is well seen in acute violent injuries. Following the initial transfer of energy, there is a phase of propagation of energy wave of injury to adjacent and sub-adjacent tissues. In addition, as cells in cellular shock recover, dead cells in the central zone of destruction act as secondary traumatic agents eliciting reactive phenomena that we recognize as inflammation. In the event of subsequent infection, the combination of necrotic cells and superimposed bacteria promote further damage and delay the onset of healing.

It is useful to view healing not so much as a process that begins after a trauma is inflicted but as a biological (evolutionary) phenomenon that may occur pari-passu with stages of the injury. Consider the events that follow injury of a mechanical nature, as from a high velocity missile (bullet or shrapnel). Oozing of blood almost always occurs. Clotting follows with fibrin formation and the laying down of fibrinogen. In this sense, a process of healing has commenced with the induction of a trauma. In the same way, primary spontaneous contraction and retraction of blood vessels following severance by firearm shrapnel, as here described, consti- tutes a similar physiological first step in the healing process. The seeming paradox may involve the foreign body - which, if lodged,

provokes reactive oozing of lymph and tissue fluid, thereby aiding
in washing out and removing injured cells and tissue debris. Yet,
by the same stimulus, hyper-perfusion of the injured site may
contribute to the wider spread of infection. An unusual occurrence
in North America, although a good illustrative case of such dynamics,
is seen following snake bite, in which a localized injury may spread
to adjoining areas, initially through lymphatics, until eventually
a whole body systemic effect is created. Comparable physio-
pathologic sequences may follow in train of spreading factor and
other permeases characteristic of cellulitis and clostridial
gangrene.

CLASSIFICATION OF INJURY: MOLECULAR CONSIDERATIONS

Innumerable classifications of trauma exist, each based on
variable criteria - anatomic site, regional classification, extent
and impairment of physiologicalfunction (as in burn injuries), and
other listings based on clinical symptomatology with or without
clinico-pathological correlations. Wound injuries provide such
overview - classifications may include open or closed wounds,
incisions (contused, penetrating or puncture), coup or contre-coup
or avulsion, etc. Molecular considerations make a classification
based on genesis and type of injury useful, and the remainder of
this chapter will consider a brief overview of some traumas studied
from the perspective of the molecular level of biology.

Mechanical Contact Injuries

Chronometrically physical injuries may be single contact
injury, repetitive contact injury, and sustained contact injury,
which, depending on the angle of contact, may be: (a) end-on
contact, (b) angular or tangential, oblique, (c) an avulsion.

These factors are important in terms of time of duration of
contact as well as nature of angle of contact, for entirely
different lesions accompany variable components of these factors.
One can readily see that the debridement and healing process would
be different as the nature of the mechanical contact is different.
For example, where a tissue suffers an incised wound, molecular
bonding by fibroblast can more easily take place. Such bonding
varies with the line of pull of the musculature at the site and is of
biological importance since collagen deposition is exactly perpen-
dicular to the direction of muscle pull. In an avulsion injury,
on the contrary, where cutaneous tissues may be separated from
underlying muscles, collagen bonding is more difficult since
fibroblast deposition and eventual collagen deposition occur parallel
to the line of cells and not perpendicular-wise, which would be

required in a tangential or rotary avulsion injury.

Compression-Decompression Injuries

Such injuries are commonly caused by sudden transformations
of energy in the medium in which the body may be. A sudden wave of
compression of the molecules in the medium followed by a sudden wave
of rare faction or decompression of the medium initiates and
propagates a "pressure wave." Such a pressure wave is formed of
zones of compressed molecules separated by zones of molecules in
a rarified state. A simple harmonic curve exists in the energy
equilibrium of the system. When this wave of compression hits the
surface or mucous membrane or viscera of the body (solid as in the
case of some abdominal organs) fluid (gas or liquid) as in the
thoracic cavity with lungs or in the cochlea of the inner ear, the
corresponding constituents of the cavity or tissue are compressed
accordingly to their nature - in respect to elasticity, resilience,
compliance and tensile strength. As these variables differ from
organ to organ and from body to body, rates and degrees of compression
may differ. Further, difference in relative rates of inertia of
cell types or of tissues may also contribute to relative differences
between tissue, extra-cellular fluid, and the vessel wall and blood
within such vessel walls. A moment's thought will confirm that,
both in the process of compression and in the process of
decompression, differential rates of movement of different types of
tissue induces shearing stress followed by shearing strain. This
causes tissue rupture and cell damage via trauma. Exposure to
blasts as from bombing may result in injuries (outside of direct
mechanical contact wounds) through such mechanisms as outlined
above. Co-related injuries may include rupture of minute blood
vessels with extravazation and, should exposure be sudden and
severe, internal shock may be correspondingly grave.

Often compression-decompression injuries are first manifest
upon the brain. Normally, the entire cerebral-cerebellar complex
is suspended in the cranial cavity by the falciform ligament
superiorly and by the tentum inferiorly. The former ligament permits
lateral shear strain but opposes antero-posterior movement. The
latter ligament permits anchoring of the brain from a supra-infra
up and down movement. The brain stem itself acts as the fixation
point or fulcrum for this suspended mass. In addition, the neurons
are considered suspended in the extraneural fluid, the foot
processes of the astrocytes being attached to the cerebral capillaries
as secondary fixation points in the suspension of neurons in extra-
neuronal fluid. Any sudden compression-decompression tends, therefore, to impart a vibration to the neurons resulting in the classical
"diffuse neuronal injury" clinically known as concussion. Secondary
vascular rupture enhances the injury.

Mechanisms of action in air, water and other fluid media are
similar. In a liquid, due to the fact that the molecules are
closely situated to each other, inter-molecular binding forces are
greater. Hence, waves of compression-decompression propagate at
a greater and faster rate than in gaseous media. Moreover, dissipation of forces of a blast in air radiate in all directions with a
solid angle of 360 , thus that vector which forms the shortest
distance between source and target (person, organ system, cell or
tissue), is only one of many radii over which the blast might be
effective. In a liquid medium, dissipation of forces is far less
than in gaseous media, so that injuries sustained by underwater
blasts are greater, more severe, and relatively more fatal than
those occurring on land. Significantly, if the fluid is in a closed
system, the individual (cells, tissue) being suspending in such a
medium, ashock force delivered from without (as at the water surface,
from air, etc.) amplifies the shock wave and results in attendant
increase in severity of blast injury. This phenomenon of amplification is one of the physiological features of the fluid dynamics
of the inner ear in which compression waves delivered by the pes
of the stapedius to the inner tympanic membrane undergo amplification.

Concussion, thoraco-abdominal injury, rupture of such membranes
as the tympanum, rupture of hollow viscera, or solid compressible
viscera as liver and spleen are particularly likely to be damaged
by blast trauma. So are small bore vessels of the nature of capillaries which lack the usual muscular reinforcement characteristic
of arterioles or smaller and larger arteries.

As a generalization, the physics expressed in delivering blast
injuries may be correlated clinically depending upon maximum or
peak pressure, moment or impulse, and duration of shock wave.
Some clinical examples of injuries delivered in this way will be
briefly described.

(1)<u>Blast injuries in solid media</u> (such as sustained from armor
plate or steel decks of ships). The initial shock to such a solid
material sets up secondary accelerations that target upon the
body, pathological effects being maximum at site of contact and
radiating with reference to the anatomy of the skeletal structures
involved. For example, combined compound fractures of the tarsal
bones and the distal ends of the tibia and fibular are not uncommon
lesions on shipboard explosions and have given rise to the descrip-
tions "destroyer heel" or "deck heel" for these types of injuries.

(2)Hollow viscera containing air and gas are more vulnerable than
solid viscera due to bursting or shedding effects which occur at
interphases between the two different media - solid and gaseous -
which have a very large difference in density. Biological trauma
is seen when a pressure wave hits such a gas-containing viscera as
the lungs with their alveoli.

When a shock wave passes through a liquid containing gas bubbles, a "spalling" effect or implosion occurs. On the surface of such bubbles, pressure energy is transformed into kinetic energy. As a result, the compressed bubbles develop a pressure tension and bubbles implode, giving rise to strong pressure waves. Such compression-decompression of gas in the intestines may result in perforation. Less critical lesions may be seen - hemorrhagic areas but no perforation.

(3)Characteristically, injuries to the thoracic wall provoke pulmonary destruction due to differential acceleration and inertia between the lungs and the thoracic wall itself. Of long debate has been the question of whether differences occur between injuries caused during inspiration and those resulting during expiration. Some opinion is extant that the most extensive pulmonary injuries occur during phases of inspiration.

(4)Blast injury and embolism are not uncommon concomitants. An initial compression wave of blast followed by decompression has been invoked as cause of increase in the amount of air or gas content in the blood due to compression. During a decompression phase, such gas may be released as air emboli which may lodge characteristically in the arterial side of the circulatory system (coronary arteries, pulmonary arteries, arteries at the base of the brain, for example). Intravascular fat emboli have been reported in brain, lungs, and other organs following blast injury.

Blast Injury and Physiological Events
In the Severely Injured Blast Victim

Well marked patterns or sequellae of events of function and pathological correlation follow blast injury. There may be characteristically a phase of apnoea followed by respiratory distress and death. In moderately severe injuries, hyperapnoeia may occur with a washing out of carbon dioxide and an alkaline arterial pH. Experimental studies indicate that oxygen administration or hyper-oxygenation, by dilating pulmonary vessels, hence contributing to increase of hemorrhage, may have a worsening effect in such patients. Cardiovascular changes may include acute cardiac arrest, bradycardia, and electrocardiographic changes including low voltage flattened QRS complexes, altered T waves, and occasional widening, elevation and notching of the P waves. Extrasystoles and auricular or ventricular fibrillation can occur. At time of blast injury, sharp rise in blood pressure may occur followed by sustained hypertension. In some cases characteristic neurologic aberrations may occur - areflexia, deafness, visual disturbances, disturbances in olfaction and gustation, as well as changes in states of consciousness, memory, and alterations of normal alpha rhythms to adjusted rates of 1 to 7 per second.

Other possible biological features associated with this trauma
may include lowered body temperature, low hemoglobin and reduced
red cell volume, low content of serum potassium, and, in some cases
(bomb victims), a <u>high</u> carbon dioxide content in the blood of blast
casualties.

SONIC, ULTRASONIC AND SUBSONIC TRAUMA

Higher extension of mechanical pressure (compression and
decompression) beyond that described involves possibilities for
physical force to pass into the zone of sound frequencies.
Generally speaking, the human ear functions over a range of sound
from 20 to 20,000 c.p.s. Lower animals have capacities to hear
both above and below values of normal human threshold - bats hear
at frequencies of 150,000 c.p.s., while the range of sensitivity
for the human ear is between 2,000 to 3,000 c.p.s. <u>Intensity of
sound</u> is the primary determinant of sonic trauma. For the human,
a range of convenience for hearing varies between 40-70 decibels
and, at levels of noise present on highways and subways with heavy
traffic (80 - 100 decibels), marked strain on otic endurance is
noted. Levels of 120-130 decibels induce discomfort, and pain
is felt at 140 decibels. Since the decibel scale is a compressed
logarithmic, rather than proportional linear scale, a sound
differing from a preceding phase by 20 decibels is, in fact, ten
times more intense.

Ultrasonic sounds have been shown to have a "blocking" effect
on neuronal tissue and myoneural functions and interneuronal synapses
in the transmission of neuronal impulses. As a result, a state akin
to local neuronal block anesthesia is induced. It is probable
that the mechanism of action relates to molecular volume and attrac-
tion between the molecules. Such factors are known to induce general
anesthesia in the central nervous system. The role of ultrasonic
frequencies in contributing to trauma on cells of systems with
slow turnover rate is of interest. Such sound affects cells that
are newly forming and causes breakdown. Exostotic cells at the
ephyseal ends is a good example. Localized cellular necrosis with
decalcification of tissue will result. Ultrasonic trauma has been
utilized therapeutically in treatment of arthritis of aging and in
prevention of formation of new bone spicules.

THERMAL INJURY

Application or contact with a heated body (or gas) which is
dry may induce <u>burns</u> of varying degrees of severity. Pathophysio-
logical sequellae may be gross and molecular in nature, as in the

denaturing of proteins. Various degrees of systemic involvement
may occur and numerous physiological systems may be integrated in
the overall picture: neural and hormonal systems, involvement
of hypothalamus and adrenal systems, and inter-relationship with
wider neurological and system organ structures. At the molecular
level, such components as vasodilatory and circulatory amines -
kinins, for example, may contribute to the picture. An overall
so-called "umbrella concept" has been postulated by Day which views
the burn wound as a "total body injury."

Contact with wet heat (as in steam scalds) results in greater
transudation of fluid than customarily seen in dry burns. Hemoconcentration, loss of extracellular fluid followed by loss of
intracellular fluid, dehydration and their sequellae - with
possible ultimate renal failure - may occur. In contrast to these
more frequently known forms of thermal trauma, chronic heat stress
presents other unique features. In heat stroke (severe sun stroke),
exposure to dry heat is cardinal. Heat is transmitted through
radiation. There is gradual exhaustion of the body's
thermoregulatory centers, heat cramps and exhaustion. Dehydration
with loss of electrolytes is important. Sodium and potassium loss
are principally involved, and are responsible for the clinical
features - cramps, muscle fatigue, and exhaustion.

ELECTRICAL INJURIES

Electrical energy and electrical discharges produce physical
trauma. Determinants are the amount of current rather than voltage,
the degree of tissue resistance and the time period duration through
which the current flows. The vector or direction through which the
electrical current passes in the body will have bearing upon the
type of lesion or degree of injury involved. For example, passage
of a current between the two upper extremities may not cause a
cardiac arrest since the cardiac respiratory center is uninvolved.
A vector of current, however, that passes from scalp to toe may
provoke changes in the medullo-pontine structures leading to
respiratory or cardiac arrest.

Hemodynamic changes in electrical injury show various patterns.
There may be an early first phase of slight hypotension associated
with cardiac arrest. There could be a first phase of hypotension
followed by a second hypertensive phase lasting a few seconds to
be again followed by a third long lasting hypotensive phase accompanied by veno-capillary vasodilation leading to death. Effects on
the heart may include bradycardia, disturbances in rhythm
(extrasystoles), and ventricular fibrillation. Disturbances in
electrical conductivity of the heart could lead to lengthening of
the PR interval and other associated phenomena leading to block or

dissociation – atrio ventricular block, bundle branch block, etc.
Deviations of the ST segment may occur including inversion of the
T wave.

Changes in the nervous system may be diverse and include
sudden unconsciousness with recovery followed by mental confusion,
delirium states with headaches, convulsion, paraesthesias, and
amnesia in certain cases. Intracranial hypertension and cerebral
edema may occur due to stimulation of the cerebral vascular
regulative centers (angiospastic and angioparalytic) with probable
changes in the permeability and polarity of the neuronal cells
themselves. Following high voltage and high current injury, there
may be areas of necrosis in the brain due to electrocoagulation,
cerebral softening, and punctiform hemorrhages.

Ocular changes occur. Conjunctivitis, retinitis, iridocyclitis
– which might in part be due to light and heat, suggesting a
photopic burn in addition to the electrical phenomenon. Ocular
cataract which occurs on the side of the body affected, even 4-6
months after the electrical accident, is due to the passage of
current with changes in the lens protein, as well as probable
rotation of the triangular lenticular fibers, whose steric relationship to each other, and to the lens capsule, is responsible for the
specific index of light transmission. This, if changed, could
lead to opacities terminating in cataracts. Auditory disturbances
include hypoacousia and deafness as well as labyrinthine and
vestibular hypoexcitability.

Passage of an electrical current through muscle results in
injury related to anatomical, physical and chemical (electrolyte)
properties of muscle whereby it acts as a conductor of electricity.
The resistance of the muscle causes a release of heat within its
substance due to the absorption of current (Joule effect), which
results in thermo-coagulation and "cooking" of the muscle fiber
itself. It is not uncommon to find vessels, nerves, and internal
organs intact (if currents have passed through them), alongside
severely affected striated muscles that have undergone cooking
and thermocoagulation. Secondary effects from electrical burns
are due to shock, with hemoconcentration due to loss of water and
electrolytes. Protein loss with renal shock is due to large amounts
of myoglobin and hemoglobin liberated from muscle (not seen in
thermal injuries), which, along with large amounts of chromoproteins,
lead to acidic urine, renal shutdown, and renal failure.

COLD INJURY

Acute cold injury has both regional localized and wider systemic
effects. At the molecular level cold stress injury results in death
and aseptic necrosis of cells. Usually there is concomitant vaso-

constriction resulting in ischemia and tissue anoxia.

Traumatic Injuries and Cold

Traumatic responses to cold may follow a dry-wet pattern.
Clinically immersion foot syndrome and trench syndrome illustrate
the wet type of injury. Frost bite exemplifies the dry type of
injury. Symptoms, which depend upon duration of exposure, include
edema, ischemia, hyperparasthesia, and ultimately gangrene.
An obvious comparison between these two traumas is reflected in
the observation that immersion foot syndrome and trench foot
syndrome arise over an exposure involving days. Frostbite may
occur in minutes. Wet foot syndromes show a spectrum of
"graduated injury" increasing in severity towards the tip of the
extremity. In frostbite a sharp line of demarcation is common.

Cryogenic Injuries

Basic pathogenesis is due to cooling and supercooling. This
causes formation of ice crystals which segregate from normal solution
in the extracellular and cellular spaces and act as inner foreign
bodies exerting mechanical and biochemical trauma. Crystal size
is approximately a linear function of rate of advancement of the
freezing boundary from the point of contact of the freezing agent
towards the center. Of importance in cryogenic stress is
differentiation between rapid freezing and slow freezing. In rapid
freezing crystal formation is intracellular. In slow freezing
the formation of crystals is primarily in the extracellular spaces.
Biochemical osmotic reactions due to ice crystals occur as the size
of a crystal grows thus creating a hypertonic phase around it. This
induces transudation of water from inside cells to the outside of the
cell thereby contributing to further growth of the crystal itself.
This may occur until all water save matabolic water (i.e., that bound
to protein, sugar, and other macromolecules) has left the cell,
so that a stage is reached at which the cell consists of concentrated
protein and other metabolic solutes sandwiched between ice crystals,
which may in fact, be larger than the original cells themselves.
It has been postulated that, so far as considerations of cryogenic
stress are concerned, two types of cells may be differentiated:
cells as in most soft tissues which collapse on loss of water and
undergo comparatively little real morphologic damage, and other
cells which not only collapse but undergo morphologic damage at all
phases of cellular anatomy including membrane levels. Further,
osmosis resulting in outflow of cellular water results in a hypotonic
solution inside the cell. The degree of concentration is directly
proportional to the rate and lowering of temperature. It has been
suggested that the actual cause of injury is not so much cold itself
as change in solute concentration inside the cell. Thus it has been

proposed that the mechanism of injury both in freezing as well as
in drying - namely a kind of osmotic dehydration - is similar
(Smith, 1930; Lovelock, 1953). A rise in lactic acid concentration
in muscles is noted after subjection to cold stress. Moreover,
the effect of change in electrolyte concentration, especially
sodium chloride, has been attributed not to its osmotic phenomena
so much as to its lyotrophic property. Some enzymes as dehydrogenase are very sensitive to freezing. Other enzymes like catalase
and lipase are more active when they are frozen. Hence supercooling
and freezing is considered to increase the concentration and
activity of these enzymes which are detrimental to the cells themselves. At low temperatures one molecule of carbohydrate binds one
molecule of free amino group of casein and other proteins. In this
way dehydration and freezing might be expected to cause uncontrolled
binding of free amino acid groups harmful to the cell.

In rapid freezing the intracellular formation of ice crystals
is very fast. Initially microcrystals form which grow and coalesce
with each other to form larger crystals which exert mechanical,
chemical, osmotic, and lyotrophic activity on the cell itself. These
mechanisms appear to be basic in the dynamic interplays involved
in cryogenic stress injury.

PHOTO-TRAUMA AND SUN BURNS

As a generalization exposure to light wavelengths of 3000 to
3500 A° produces tanning. Excess sunlight produces severe photobic
trauma. The skin becomes erythematous. Irritation, pruritis and
a sensation of burning may occur. In severe burning bullae and
vesicle formation occur associated frequently with desquamation
and hyperpigmentation. Systemic effects such as shock, chills and
fever may occur, possibly due to absorption of material from the
skin. Histologically, in solar dermatitis, photopic injury is
characterized by pailler and vesicular lesions with liquifactive
degeneration in the basal cell layer and keratotic plugging of
hair follicles. Non-specific inflammation and edema in the dermis
and epidermis may occur with cell infiltration around blood vessels.

A third form of photopic injury is due to photosensitization
accompanied by local cutaneous and systemic effects. Some observers
have indicated photosensitization and photodynamic injury by contact
and through systemic distribution (including the dermis) of sensi-
tizing agents. Essential oils, plants, drugs like quinacrine,
atabrine, sulphonamides, demethylchlortetracycline and diseases
like lupus erythematosus and dermatomyositis in which sunlight
causes an exacerbation of the lesion, probably by stimulating the
cells, are included in this grouping. Also contact dermatitis

with berlock and meadow grass and vitamin deficiencies such as pelle-
gra. In porphyria, especially of the cutanea tarda type, punctiform
hemorrhages and vesicles develop.

Melanin granules form a primary defense line against photopic
and irradiation injury. Blum has considered that the stratum
corneum was a primary protective layer and that melanin is due only
a minor role. However, the greater incidence and frequency of
sunburn in albinoid and relatively non-melanized populations
(particularly Caucasians) and its near absence in dark races –
Negroid, Asian, and African, would suggest a greater protective
action for melanin.

                            IRRADIATION TRAUMA

Ionizing radiation, whether produced by x rays, beta, gamma
or theta rays, has similar effects on man. The mechanisms of
pathogenesis of injury includes:

1) direct effect of radiation upon the tissues
2) effects produced by ionization of water around the area
   by radiation
3) effects due to metabolites from damaged tissue produced
   by secondary injury
4) effects produced by the autonomic and endocrine systems
   secondary to irradiation damage
5) fluid and electrolyte imbalances in the irradiated animal
   or person
6) other ill-defined factors which are noted to occur.

Not unremarkably, since the skin and epidermis form the first
barrier of the defense of the body against irradiation, pathobiologic
changes are seen initially in the skin. Invariably there is a first
wave of erythema (first day) due to a direct action of the irradiation on blood vessels, which may also result in vasoconstriction.
The second wave of erythema (second to third week) is due to action
of irradiation waves on component cells of the blood. There are
varying degrees of modification and damage to the neuromuscular
controls of the peripheral vascular system. The third wave of
erythema (sixth week) is due to tissue damage and release of histamine
and other tissue hormones and metabolites from damaged tissues.
In addition to quite distinct morphologic changes there are also
physiologic changes in irradiated vessels which show low vasoconstriction response to epinephrin and norepinephrin. Among direct
actions of irradiation are decreased blood flow in the vasa vasorum
supply to the blood vessel wall as well as ionization of solutions
in the immediate area of the blood vessels. Increased permeability
in damaged capillary walls is characteristic. In spite of vasoconstriction, most animals suffer an initial phase of hypotension,

thought to be mediated by reflex mechanisms, especially vagal and
extra-vagal reflexes. In higher doses of radiation, vasodilatory
material (VDM) and vasoexcitory material (VEM) are liberated. By
the second day of injury there is a loss of DNA and lipid from the
heart. By the fourth day post trauma in addition to the lineal
decrease in DNA there is also a decrease in the actomyosin and
potassium concentrations in the heart.

Radioactive wastes from peaceful use of atomic energy as well
as resumption of nuclear weapons testing has required consideration
by the molecular pathobiologist. Gamma irradiations including
strontium 90, caesium 137 and iodine 131 included among fission
products liberated in such blasts are of importance. Strontium 90
has a long physical half-life and is rapidly taken into the skeleton
where it may be retained over long periods. Prior to fixation
in bony tissue strontium 90 may circulate in the body for a short
period of time which may, nonetheless, be sufficiently long to
produce lesions in radio-sensitive organs. In addition to atomic
injury effects, strontium 90 poisoning has been observed in luminous
dial painters in watch factories (Czechoslovakia) where radium 226
is also involved. There is alteration in erythropoiesis. Reticulo-
cytosis and hyperplasia of the erythroblast series in the bone
marrow may occur with a rise in the number and frequency of aneuploid
cells in the exposed group. Ytrium 90 and calcium 45 are known to
be carcinogens producing osteogenic tumors. Cesium 137 may produce
chronic effects due to slow release of the isotope. The autonomic
nervous system is involved with hematological changes like leukopenia, especially of the granulocyte series as well as lymphopenia
with eosinopenia (thrombocyte levels remaining normal). Iodine 131
affects primarily the thyroid gland but may also affect other cells,
particularly those in mitotic division. As long as they do not divide
the epithelial cells of the thyroid are extremely radio resistant.
However, the developing thyroid which, as in children, may be actually
proliferating, is much more radiosensitive and can provoke extensive
damage not seen in the adult thyroid gland. Investigations in
Rongelap and Utirik Islands twelve years after exposure to radiation
fallout revealed that several children, especially girls, had
developed localized carcinomatous nodules in the thyroid gland.
Hypothyroidism association with growth retardation improved upon
administration of thyroid hormone. Cerium 144, Ruthenium 106,
Zirconium 95, Sulphur 35, Phosphorus 32, Tritium, and Cobalt 60
may all produce extensive irradiation damage. With the exception
of cobalt 60, used for irradiation therapy in the treatment of
malignancies, the agents noted above have been studied extensively
only in animals.

MOLECULAR INJURY: CHROMOSOMAL INJURY DUE TO IRRADIATION

X rays, gamma, beta, and other forms of actinic irradiation produce injury in chromosomes themselves or by ionizing the fluid around the chromosomes result in chromosomal aberrations. Injury is less severe if cells are not in process of division and such cells may persist unmodified in interphase for several years. Chromosomal aberrations following radiation injury may persist to a greater degree in cells that are actually reproducing, i.e., gonadal cells. As with a malignant cell which begins as a response to injury, chromosomal aberrations via f-1, f-2, f-3 generations may be transmitted to unlimited subsequent generations causing congenital defects and untimely deaths. From the viewpoint of cytogenetics two types of molecular injury predominate:

1) Chromosomal injuries which occur during the pre-DNA synthesis portion of the interphase during which time the chromosome behaves as a single thread. Aberrations at this stage are later replicated when the chromosomes become involved in cell division. They then always involve both the chromatids of the chromosomes as seen in the metaphase.

2) Chromatid type aberrations are induced after the chromosomes have divided into two strands, namely during DNA synthesis and post-DNA synthesis portions of the life cycle. The injury may be at the same level (isochromatid break) or at different levels (non-isochromatid break). Radiations and other agents causing trauma to such cytogenetic systems in which chromosomal discontinuity results encourage situations of recombination between the broken ends, frequently in pathologic forms.

It may be that the actual aberration lies not so much in the break but in abnormal reformation or reconstitution of chromosomal continuity.

It is interesting to consider that normal breaking of the chromatids which are essential for cell mitosis and cell replication is induced both by physical and chemical processes within the cell itself through external control - thus hypothermia retards the rate of normal breakage of chromosome for cell divisions. However, an abnormal form of breakage due to ionizing energy has been considered to be due to a kind of molecular bombardment of energy setting up a resonance injury in chromosomes or chromatids which result in breakage at abnormal levels. Extrapolating the molecular injury in the chromosomal strand or the chromatid strands in terms of stereochemistry we see that the split in the DNA chain results in cleaving of the adenine-thymine hydrogen bonds or the cytosine-guanine hydrogen bonds, namely a cleaving of the "steps" with further breakage in the phosphoric acids, oxygen, sugar "bannister"

in the DNA staircase. The chromosomal break may be eccentric to
the centrosome in both arms of the chromatids leading to fragments
which are shed off giving rise to a deletion phenomena or all
following break at both ends of the chromatid strands, the central
area may fold upon itself giving rise to a ring formation (asymme-
trical interchange) whereby there is transfer of genetic material
across the two folded arms of the ring itself.

Induction is common in tissues which have a high mitotic rate,
i.e., bone marrow, corneal epithelium, and testicular germinal
epithelium, and can be induced in other tissues such as peripheral
lymphocytes and regenerating liver cells.

ACCELERATION-DECELERATION AND GRAVITATIONAL TRAUMA

To some degree the earlier definition of trauma as a dissolution
or discontinuity of membranes in organs of the body requires
augmentation in terms of acceleration-deceleration and gravitational
injury. Physical forces can induce trauma by creating a disequilibrium or an unbalancing between different parts of the body, due to
the fact that individual parts and different tissues in organ systems
have a differential inertia. Thus the same physical force may cause
a disequilibrium or an unequal distribution of fluid segments and
energy systems resulting in trauma. An aircraft pilot when he pulls
out of a dive suffers a g-positive acceleration. On a bomb-diving
run the same pilot suffers a g-negative acceleration which might
amount to 5-6g units within 3-4 seconds. Due to the differential
inertia of the vascular contents and the vascular chambers in blood
vessels, as well as to the fact that the vasomotor center has not
adapted to the rapid accelerative changes, a positive-g stress leads
to sucking off or drawing off of the blood from the cranial cavity
in its relative accumulation in the thoracic abdominal and peripheral extremities. This sudden acceleration cerebral ischemia
results in the "black-out" syndrome due to ischemia of the retinal
vessels and other parts of the cranial contents resulting in temporary blindness and unconsciousness. Concomitantly fluid is extruded
out of the vascular compartment in extravascular spaces whereas
increased capillary pressure and concomitant fragility results
in petechiae and minute hemorrhages along the peripheral extremities.
Negative acceleration or a g-minus stress occurs when the head is
directed outwards as the pilot makes a high velocity turn. Blood
rushes into the cranial cavity very much as in apoplexy, resulting
in the "red-out syndrome." The vessels of the head and neck are
engorged, resulting in cutaneous hemorrhages, severe throbbing pain
and concomitant respiratory embarrassment as the abdominal viscera
is pushed against the thoracic diaphragm.

Sudden deceleration in a driver who has been traveling at a

high speed or velocity in a vehicle results in unconsciousness.
This is due to an impinging of the cranial contents within the
bony skull as well as deliverance of a kind of a shock movement
to the brain itself. Again, this is due to the difference in the
relative inertia of the cranium and its contents. Sudden
unconsciousness due to concussion resulting in the diffuse neuronal
injury syndrome, as well as coup and contre-coup injuries on the
opposite side to the site of initial impact, may occur.

Other detailed discussions of the physiopathology of weightlessness and the effects of gravity on the cell at the ultrastructural level fall beyond the scope of this review.

SELECTED BIBLIOGRAPHY

Barcroft, J.: Blast injuries. Brit. Med. Jour., 1:89, 1941.

Barrow, D. W., Rhoads, H. T.: Blast concussion injury. JAMA, 125:
        900, 1944.

Bender, M. A.: X-ray induced chromosome aberrations in normal
        diploid human tissue culture. Science, 126:974, 1957.

Benzinger, T.: Physiological effects of blast in air and water.
        German Aviation Medicine, World War II, USAF U. S. Govt.
        Printing Office, Washington, D. C., 1950. p. 1225.

Blum, H. F.: The physiological effects of sunlight on man. Physiol.
        Rev., 25:483, 1945.

Burton, R. R., Sluka, S. J., Besxh, E. L., Smith, A. H.: Hematolog-
        ical criteria of chronic acceleration stress and adapta-
        tion. Aerospace Med., 38:1240, 1967.

Chu, E. H. Y., Giles, N. H., Passano, K.: Types and frequencies of
        human chromosome aberrations induced by X-rays. Proc.
        Nat. Acad. Sc., 47:830, 1961.

Clark, S. L., Ward, J. S.: The effects of rapid compression waves
        on animals submerged in water. Surg. Synec. & Obstet.,
        77:403, 1943.

Clemedson, C. J.: An experimental study on air blast injuries. Acta
        Physiol. Scand., Vol. 18 (supplement 61), 1949.

        Correlation between respiratory phase and extent of lung
        damage in air blast injury. Jour. Appl. Physiol., 7:38,
        1954.

        Mechanical response of different parts of a living body to
        a high explosive shock wave impact. Amer. J. Physiol.

        Blast injury. Physiol. Rev. 36:336, 1956.

Clemedson, C. J., Pettersson, H.: Propagation of a high explosive
        air shock wave through different parts of an animal body.
        Amer. J. Physiol., 184-119, 1956.

Cohen, H. G., Biskind, R.: Pathologic aspects of atmospheric blast
        injuries in man. Arch. Pathol., 42:12, 1946.

Day, Stacey B., Macmillan, B. G., Altemeier, W. A.: Curling's
        Ulcer: An Experiment of Nature. Charles C. Thomas, 1972.

DeSaga, H.: War experiences concerning blast effects on man. German Aviation Medicine, World War II, USAF U. S. Govt. Printing Office, Washington, D. C., 1950. p. 1274.

Fischer, H.: Electrical burns. Proc. Intern. Symp. Elec. Accidents. INS-CIS-EDF, Geneva, 1964.

Frucht, A. H.: Die Schallgeschwindigkeit in menschlichen und tierischen Geweben. Ztschr. für gesamte Expr. Med., 120: 526, 1953.

Fuhrman, F. A., Crismon, J. M.: Studies on gangrene following cold injury: A method for producing gangrene by means of controlled injury by cold. Jour. Clin. Invest., 26: 229, 1947.

Goodall, M. C.: Sympatho-adrenal response to gravitational stress. Jour. Clin. Invest., 41:197, 1962.

Greaves, F. C., Draeger, R. H., Brines, O. A., Shaver, J. S., Corey, E. L.: An experimental study of underwater concussion. U. S. Naval Medical Bulletin, 41:339, 1943.

Hadfield, G., Ross, J. M., Swain, R. H. A., Drury-White, J. M., Jordon, A.: Blast from high explosives. Lancet, 2:478, 1940.

Heard, B. E.: The histological appearances of some normal tissues at low temperature. Brit. Jour. Surg., 42:430, 1955.

Hemingway, A.: Physiological effects of heat and cold. Ann. Rev. Physiol., 7:163, 1945.

Hertzman, A. B., Roth, L. W.: The reactions of the digital artery and minute pad arteries to local cold. Amer. Jour. Physiol., 136:680, 1942.

Hyvarinen, J., Pyykko, I., Sundberg, S.: Vibration frequencies and amplitudes in the aetiology of traumatic vasospastic disease. Lancet, 1:791, 1973.

Hooker, D. R.: Physiological effects of air concussion. Amer. Jour. Physiol., 67:219, 1924.

Krohn, P. L., Witterridge, D., Zuckerman, S.: Physiological effects of blast. Lancet, 1:252, 1942.

Large, A., Heinbecker, P.: Nerve degeneration following prolonged cooling of an extremity. Ann. Surg., 120:742, 1944.

Latner, A. L.: Low pressure phase of blast. Lancet, 2:303, 1942.

Lea, C. H., Hannan, R. S.: The effect of activity of water, of pH
          and of temperature on the primary reaction between casein
          and glucose. Biochem. et biophysic. acta., 3:313, 1949.

Leavell, B. S.: Acute heart failure following blast injury. War
          Medicine, 7:162, 1945.

Lewis, T.: Observations upon the reactions of the vessels of the
          human skin to cold. Heart., 15:177, 1929.

Logan, D. D.: Detonation of high explosives in shell and bomb and
          its effects. Brit. Med. Jour., 2:816, 1939.

Lovelock, E. J.: The hemolysis of human red blood cells by freezing
          and thawing. Biochem. et biophysic. acta., 10:414, 1953.

Meryman, H. T.: Mechanisms of freezing in living cells and tissues.
          Science, 124:515, 1956.

Meryman, H. T.: Tissue freezing and local cold injury. Physiol.
          Rev., 37:233, 1957.

Nelson, A.: Late effects of radiation. Taylor & Francis Ltd.,
          London and Van Nostrand Reinhold, New York, 1970.

Offerhaus, L., Dejongh, J. C.: Homeostatic regulation of the circu-
          lation during prolonged gravitational stress ($+G_z$). Aero-
          space Med., 38:468, 1967.

Oyama, J., Platt, W. T.: Carbohydrate metabolism of mice exposed
          to simulated changes in gravity. Amer. Jour. Physiol.,
          207:411, 1964.

Parin, V. V., Yegorov, B. B., Bayeyskiy, R. M.: Physiological
          measurements in space. Principles and methods. ATD
          Press Abstr., 5(98). Lib. of Congress, Washington,
          D. C., 1967.

Pollard, E. C.: Cellular effects of weightlessness. Hypodynamics
          and Hypogravics, p. 109, M. McCally, Editor. Acad.
          Press, New York, 1968.

Puck, T. T.: Relationship between reproductive death and induction
          of chromosome anamolies by X-irradiation of euploid human
          cells in vitro. Proc. Nat. Acad. Sc., 44:772, 1958.

Ravell, S. H.: The accurate estimation of chromatid breakage, and relevance to a new interpretation of chromatid aberrations induced by ionizing radiations. Proc. Roy. Soc. B., 150:562, 1959.

Robb-Smith, A. H. T.: Pulmonary fat embolism. Lancet, 1:135, 1941.

Sax, K.: Chromosome aberrations induced by X-rays. Genetics, 23: 494, 1938.

Schardin, H.: The physical principles of the effects of a detonation. German Aviation Medicine, World War II. USAF U. S. Govt. Printing Office. Washington, D. C., 1950. p. 1207.

Scher, S.: Air embolism from blast. Brit. Med. Jour., 1:797, 1941.

Smith, E. C., Moran, T.: The formation of lactic acid in desiccated amphibian muscles. Proc. Roy. Soc. Lond., 106:122, 1930.

Tumbridge, R. E., Wilson, J. V.: The pathological and clinical findings in blast injury. Quart. Jour. Med., 12:169, 1943.

Ungley, C. C.: Discussion on immersion injuries and vasomotor disorders of the limbs in wartime. Proc. Roy. Soc. Med., 36:518, 1943.

Wunder, C. C., Duling, B., Bengele, H.: Gravity as a biological determinant. Hypodynamics and Hypogravics, M. McCally, Editor. Acad. Press, New York, 1968. p. 1.

Zuckerman, S.: Experimental study of blast injury to the lungs. Lancet, 2:219, 1940.

HOMEOSTASIS AND HETEROSTASIS

Hans Selye

Institut de medecine et de chirurgie experimentales

Universite de Montreal

Claude Bernard[1] was the first to point out clearly that the internal medium of living organisms is not merely a vehicle for carrying nourishment to cells but that " it is the fixity of the 'milieu interieur' which is the condition of free and independent life." W. B. Cannon[2] suggested the designation "homeostasis" for "the coordinated physiological processes which maintain most of the steady states in the organism."

When faced with stressful situations that require systemic adaptation, the organism can respond through three essentially distinct mechanisms: (1) nervous – by conscious planning of defense, innate or conditioned reflexes and autonomic "emergency reactions" (partly mediated through neurohormones); (2) immunologic and phagocytic – antibody formation, activation of the reticuloendothelial system; (3) hormonal – through the syntoxic hormones which permit tolerance of the pathogen without attacking it (e.g., glucocorticoids preventing inflammation without destroying its cause) or catatoxic substances that eliminate the aggressor (e.g., certain steroids and drugs which accelerate the biodegradation of toxicants without inducing tissue resistance to them).[3]

In recent years, the discovery of the important role played by hepatic microsomal enzyme induction in defense against certain toxicants led to the development of a new field of pharmacology designated "xenobiochemistry" and defined as the "biochemistry of foreign organic compounds."[4] This science is concerned with the fate of "xenobiotics" (from the Greek "xenos" and "bios" for "stranger to life"), that is, compounds which are foreign to the metabolic network of the organism.[5]

At first, the term xenobiochemistry appeared to be a particularly suitable label for the new approach, because both the toxicants and the inducers of their biodegradation were foreign to the body (Most of the initial work was performed with barbituates, polycyclic hydrocarbons, insecticides, etc.). Indeed, Brodie and his coworkers,[6] who probably did more for the development of this field than any other group of investigators, came to the conclusion that presumably the defensive enzyme systems "are not essential to the normal economy of the body, but operate primarily against the toxic influences of foreign compoundsthat gain access to the body from the alimentary tract."

The large number of investigators and the fast-growing litera- ture dealing with problems in this field have recently prompted the founding of a special journal under the name Xenobiotics. Yet, nowadays, hardly anyone doubts that neither the substrates nor the inducers of such defensive enzymes need be foreign to the body's economy.[7] Of course, in the case of poisons completely foreign to the organism, any amount in the body is excessive and hence "foreign," whereas normal body constituents become "foreign" only if their concentration greatly exceeds physiologic levels. Thus, various hormones and hormone derivatives can protect against both exogenous and endogenous damaging substances, catatoxically by inducing enzymes capable of accelerating their biodegradation, or syntoxically by increasing tissue resistance to them.[3] Among the most important natural compounds amenable to catatoxic or syntoxic defense are: steroid hormones, bile acids, bile pigments, fat-soluble vitamins, inflammatory mediators (histamine, serotonin, prostaglan- dins), and so forth, all of which are either made within the body or absorbed from the diet as essential physiologic ingredients of living matter.

The salient feature of these adaptive mechanisms is therefore not that they attack only substances foreign to the body but that they establish a new equilibrium between the body and an unusually high level of the potential pathogen, either by destroying the excess (catatoxic action) or by making tissues tolerant to it (syntoxic action).

When such an abnormal equilibrium must be established to protect against potential pathogens, I propose to speak of heterostasis (heteros=other; stasis=fixity) as the establishment of a new steady state by exogenous (pharmacologic) stimulation of adaptive mechanisms through the development and maintenance of dormant defensive tissue reactions. In a sense this would be the counterpart of homeostasis (homeo=like) which has been defined as the maintenance of a normal steady state by means of endogenous (physiologic) responses. In heterostasis, as in homeostasis, the fixity of the "milieu interieur" is not absolute. Both Bernard[1] and Cannon[2] have clearly realized that in order to maintain a state of relative stability the body cannot remain completely inert but must answer each stimulus with an appropriate counter stimulus to maintain equilibrium. However, in homeostasis, this equilibrium is maintained with small fluctuations near the physiologic level, whereas in heterostasis unusual defense reactions are mobilized to permit resistance to unusual aggression. Furthermore, in general, homeostasis, unlike heterostasis, depends upon rapidly developing and vanishing readjustments, although - as in all biologic classifications - transitional types are common.

Of course, both homeostasis and heterostatic reactions, though usually of defensive value, may be worthless or even harmful under certain circumstances. This can be illustrated by the following examples: (1) In predisposed individuals, excessive neuroendocrine "emergency reactions" may precipitate a cardiovascular accident. (2) Antibody formation can result in anaphylaxis or allergy; indeed, even phagocytosis can offer a suitable milieu to certain organisms and protect them against blood-borne antimicrobials (e.g., in a tubercle). (3) Defensive syntoxic hormones can interfere with the encapsulation and destruction of microorganisms by suppressing useful inflammatory or ommunologic reactions. Similarly, by accelerating the biotransformation of relatively innocuous compounds, catatoxic steroids can enhance the production of highly pathogenic metabolites.

The most salient difference between homeostasis and heterostasis is that the former maintains the normal steady state by physiologic reactions, whereas the latter "resets the thermostat" to maintain a higher state of defense by artificial exogenous intervention. Heterostasis can be accomplished by exogenous administration of natural (e.g., hormones) or artificial (e.g., barbituates) substances capable of inducing responses of this kind. Furthermore, the heterostatic reactions may be syntoxic (e.g., corticoids, anti-inflammatory drugs), permitting coexistence with potential pathogens by evoking responses which make our tissues indifferent to certain potentially pathogenic stimuli. Thus, they suppress excessive inflammatory which represent the very essence of some diseases. However, they may also be catatoxic (e.g., some anabolic hormones, barbituates, insectisides) if they enhance drug-metabolizing enzyme activity and thereby accelerate the biodegradation of potential pathogens. The most potent catatoxic substance known to date is pregnenolone-16 -carbonitrile (PCN); this is a close relative of the naturally occurring pregnenolone and yet an artificial compound in that it is produced synthetically by the introduction of a carbonitrile group into its natural congener.

Thus, the concept of heterostasis includes both syntoxic and catatoxic reactions induced by natural or artificial exogenous compounds. On the other hand, it excludes all therapeutic measures

which act directly (or "passively") not by stimulating the body's own adaptive capacities. For example, (1) Antidotes, which act directly upon potential pathogens (e.g., buffers that neutralize acids, chemicals which inactivate poisons by directly combining with them, antimicrobial agents). (2) All forms of substitution therapy (e.g., vitamin C for scurvy, corticoids for Addison's disease, transplanted or artificial kidneys for renal failure) which merely represent chemical or mechanical prostheses.
(3) The ablation of diseased tissues (infected parts, tumors).
(4) Passive "shielding" (e.g., by lead screens against x rays, receptor blockade against drugs, competition for a substrate).

Thus, the types of adaptation, which depend upon active parti-cipation of the body are: homeostatic – (a) syntoxic, (b) catatoxic; heterostatic – (a) syntoxic, (b) catatoxic.

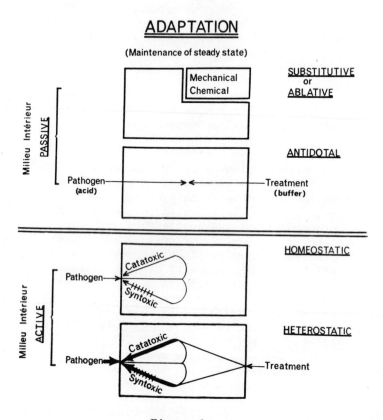

Figure 1

The list comprises all active defense reactions irrespective of their nervous, immunologic, phagocytic, or hormonal mediation. It can be incorporated into a general classification of adaptive mechanisms, whether the "milieu interieur" participates actively or passively, as summarized in figure 1.

REFERENCES

1. C. Bernard. Lecons sur les phenomenes de la vie communs aux animaux et aux vegetaux. 2 vols. Paris: Bailliere, 1878-1879.

2. W.B. Cannon: In A. Pettit (ed.). A Charles Richet: ses amis, ses collegues, ses eleves, p. 91. Paris: Editions Medicales, 1926.

3. H. Selye. Hormones and Resistance. Heidelberg: Springer-Verlag, 1971.

4. R. T. Williams. Detoxification Mechanisms: the metabolism and detoxication of drugs, toxic substances and other organic compounds. New York: Wiley, 1959.

5. H. S. Mason, J. C. North, and M. Vanneste. Fed. Proc., 24: 1172, 1965.

6. B. B. Brodie, J. Axelrod, J. R. Cooper, L. Gaudette, B. N. La Du, C. Mitoma, and S. Udenfriend. Science, 121: 603, 1955.

7. A. H. Conney. Pharmacol. Rev., 19:317, 1967.

COLLAGEN METABOLISM IN WOUND HEALING

Quentin T. Smith

School of Dentistry

University of Minnesota

A wound results in disruption of the normal continuity of body structures. Wound healing is restoration of the continuity of injured tissues. Collagen plays a crucial role in healing of wounds of skin and other tissues. There is general agreement that synthesis of collagen is required for restoration of physical strength of wounds. It is now also widely recognized that clarification of the precise role of collagen in the wound healing process requires not only information on the quantity of collagen present during the healing process, but, in addition, data on the rate at which collagen is synthesized, deposited and resorbed. Collagen biosynthesis includes unique biochemical processes, such as hydroxylation of proline to hydroxproline. Activity of these separate biosynthetic steps, likewise, must be evaluated for a thorough understanding of the function of collagen in wound healing. This review will emphasize collagen metabolism of cutaneous wounds not only because of the clinical and therapeutic significance of the healing of surgical incisions, but also since most experiments on collagen in wound healing have used skin injuries. The data to be summarized have been selected mainly from recent studies on "normal" wound healing. Unfortunately, few facts are available concerning collagen metabolism in healing complications such as keloids.

COURSE OF WOUND HEALING

Clinical and experimental injury to the skin may be produced in several ways: mechanical force, skin incision, skin excision, burns, application of caustic substances, radiation, etc. Incised wounds may be readily inflicted in a uniform manner and, thus, have been the most frequently used experimental wound. The healing of cutaneous wounds produced by various agents, however, involves the same mechanisms occurring in the same sequence.

The principal phases in the course of wound repair have been known for more than a century. After wounding, the space between the tissues becomes filled and sealed with a firm blood clot. Acute inflammation, characterized by vasodilation and appearance of neutrophils ensues. During this same healing phase fibrocytes and fibroblasts in the wound margin swell and present a more embryonal appearance. These fibroblasts have the capacity of migration into the blood clot. Mitotic activity begins and fibroblasts proliferate and spread along the fibrin strands into the clot.

Digestion of white cells, bacteria, fibrin, necrotic tissue and other components of the inflammatory exudate by recently formed cells begins within 48 hours of wounding (resolution of the exudate). At the same time that fibroblast proliferation and migration occur, endothelial cells of blood capillaries become hypertrophied and swollen, commence marked mitotic activity, migrate along the fibrin network and become canalized to permit blood flow. Continuity of blood flow is re-established within variable times after wounding, depending upon the nature of the wound, and may be as little as 2 to 3 days for simple skin incisions.

Accumulation of collagen in wounds is not observed until after fibroblast proliferation and, thus, is one of the later stages in the healing sequence. About seven days following a simple cutaneous incision, a highly cellular wound with tensile strength near that of normal tissue has formed. Epithelial cells of the epidermis begin proliferation about 3 days after wounding and, for a simple incision, after approximately seven days have grown across the cut. The time required for scar formation and epithelial closure of the wound is extended for more severe wounds. Continuing scar collagen production produces mechanical pressure which compresses the newly formed delicate walls of the wound capillaries resulting in a reduced vascular system. Thus, after a period of weeks or months, a permanent relatively pale, acellular, collagenous scar covered by epithelium is formed.

TENSILE STRENGTH—COLLAGEN CONTENT

It is essential that the healing of a wound results in repair with adequate strength to withstand normal stresses placed on the tissue. A technique devised for determination of the physical strength of a tissue or wound is burst strength or tensile strength. Burst strength measures the physical force required to disrupt a wound. Tensile strength is a more accurate measure of wound

strenth since this measurement gives the force in load per unit area
required to rupture the wound. Many values given in the literature
as tensile strength values are in fact burst strength measurements.
The well known sigmoid curve of increasing tensile strength during
the course of healing of incised wounds was recorded in 1929 by
Howes, Sooy and Harvey. A lag phase of 3 to 5 days during which
time the wound has little strength is followed by a period of fibro-
plasia and rapidly increasing tensile lasting about 2 weeks. Subsequently tensile strength increases only slowly. Numerous additional
studies have confirmed the above course of events during early wound
healing (Dunphy and Udupa, 1955; Dunphy et al., 1956; Dunphy, 1960;
Sandberg and Zederfeldt, 1963; Adamsons, 1964.

Identification of the substance or substances producing increased
tensile strength is critical to understanding wound healing. Collagen is the major organic component of the dermis and thus it was
logical to determine its contribution to the tensile strength of
normal tissues and wounds. It has been recognized for many years
that soft tissues with the greatest tensile strength, namely tendon
and aponeurosis, have high collagen contents, and that organs such
as the liver and kidneys which have very little collagen have low
tensile strength.

The imino amino acid hydroxyproline is found in high concentration only in collagen. Therefore the quantity of this amino acid
in a tissue may be used as an indicator of the quantity of tissue
collagen. Following development of techniques for quantitative
analysis of hydroxyproline, several investigators determined changes
of collagen content during wound healing and related these changes to
wound tensile strength. Experimental conditions, such as age and
species of the experimental animals and severity of the wounds have
varied greatly among investigators and, thus, the quantitative relationships of collagen content to the stage of wound healing have
also varied among studies. The first studies which showed that accumulation of collagen in the early stages of wound healing closely
parallels increased tensile strength were in 1955 by Dunphy and
Udupa. The collagen content of 3 to 6 day old wounds was less than
15% of that of normal skin but by 12 days wound collagen had increased to 40% that of control values. Viljanto (1964) found a close
relationship between hydroxyproline content of young wound tissue
formed by implantation of cellulose sponges and tensile strength of
the implant wounds. Earlier chemical and histochemical studies
(Edwards et al., 1957) indicated that granular tissue formed by open
wounds and sponge implants represents a similar biologic response
to injury. Further information on collagen accumulation in healing
open wounds was obtained by removal of 2.0 cm circles of skin extending to but not including underlying fascia from the right dorsum
of rats (Reynolds et al., 1963). Median closure time of the wounds
occurred in 15 days. At this time collagen content of the wound
had returned to that of control values.

Incisional wounds may be more readily formed and maintained in a uniform manner than open wounds or granulation tissue thus giving a better controlled experimental design. Consequently, most recent studies of collagen properties and metabolism during wound healing have utilized incisional wounds. Data from different laboratories have shown a correlation between collagen content and increasing tensile strength during the early stages of healing of incised and sutured wounds (Sandberg and Zederfeldt, 1963; Adamsons et al., 1964). During healing of incised wounds, wound collagen content may exceed that of normal tissue (Adamsons et al., 1964, 1966).

The impression commonly is held that incisional wounds gain little additional strength more than 2 to 3 weeks after wounding. Adequate data is now available to refute this misconception. Almost 25 years ago Howes et al. (1939) showed that breaking strength of cutaneous wounds in several species continued to increase for periods in excess of 6 months after injury. Aponeurotic incisions in guinea pigs increased in strength for a year (Douglas, 1952). Moreover, tensile strength of incisional wounds increased for periods up to 13 weeks, a time long after the collagen content had become stable (Levenson et al., 1965; Peacock, 1966). The lack of correlation between collagen content and tensile strength beyond the early stages of wound healing is particularly significant since it discloses that additional factors besides net collagen accumulation contribute to wound tensile strength.

COLLAGEN DYNAMICS DURING SCAR REMODELING

Data showing that collagen content in scar tissue correlates with gain in tensile during the first few weeks of wound healing, but that tensile strength continues to increase for prolonged periods after the collagen content has become stable were summarized above. These total collagen data indicate only the equilibrium state of collagen synthesis and degradation and do not give information concerning kinetics of collagen metabolism. The kinetics of wound collagen metabolism must be known to understand the changes which occur in tensile strength during periods in which collagen content is static. This information would clarify whether the increased scar tensile strength with time results from remodeling of pre-existing collagen fibers or from dynamic remodeling of the scar.

Prolonged remodeling of scar collagen was observed by Levenson et al. (1965) who reported progressive microscopic changes for up to one year in number, caliber and density of rat skin wound collagen fibers. Williams (1970) found that after 84 days the collagen fibrils of skin wounds in guinea pigs were still variable in width and rarely had attained the dimensions of control mature dermal fibers.

Recent scanning electron microscopic studies (Forrester et al., 1970a, 1970b) demonstrated that the structure of healing longitudinal rat skin wounds "evolves from random fibril patterns to large collagen masses." However, the scar collagen architecture was not restored to that of unwounded skin within the 150 day experimental period.

Madden and Peacock (1968) measured the net rate of collagen synthesis and deposition in incised and sutured rat skin wounds by determination of the specific activity of nondialyzable hydroxyproline following injection of radio-active proline. Net rate of collagen synthesis and deposition was greatest at about the fourteenth day and remained significantly higher than skin controls through 70 days post-operative. It is essential to recognize that the radiotracer procedure devised for these experiments measured the combined rate of collagen synthesis and deposition since labeled hydroxyproline found in the wound must be derived from injected proline converted to hydroxyproline in collagen which must in turn remain in the wound. Tensile strength, unlike the total collagen content of primary wounds, was correlated through the first ten weeks of healing with the rate of new collagen synthesis and deposition.

The previous study (Madden and Peacock, 1968) was extended by utilization of radiotracer techniques which measured both net collagen accumulation and net rate of new collagen deposition (Madden and Peacock, 1971). One group of animals received daily injections of tritium labeled proline to give a measure of net collagen accumulation. Rate of new collagen deposition was determined at various times after wounding in a second group of rats by giving a single injection of labeled proline 20 hours prior to sacrifice. Specific activity of collagen hydroxyproline was measured in both groups. Wounds made through the carnosus muscle of the neck were of the same type as in the earlier study (Madden and Peacock, 1968).

It was found during the first 3 weeks of healing that wound collagen accumulated rapidly, but, that after this time there was no additional scar collagen accumulation. However, wound collagen deposition remained significantly elevated throughout the 6 week experimental period. These experiments thus show that beginning at 3 weeks, or perhaps before, both rapid collagen synthesis and removal occurs in dermal scar tissue. No correlation was found between scar collagen content and wound strength subsequent to 3 weeks after wounding. A correlation existed, however, throughout the experimental period between wound breaking strength and the rate of collagen deposition and turnover. These data indicate a biological mechanism for scar remodeling and an explantion for increased wound strength without increased scar collagen content.

SECONDARY WOUNDS–COLLAGEN METABOLISM

Clinical and experimental observations have established that healing, dehisced (split) and resutured wounds gain strength more rapidly than primary wounds during the first few days of healing. Several investigators have searched for an explanation of this phenomenon. Adequate data have been presented to demonstrate that local and non-systemic processes are responsible for the more rapid gain in strength of the dehisced wounds (Taffel et al., 1951; Sandblom and Muren, 1954; Savlov and Dumphy, 1954a). Various proposals have been made as to the nature of the local factor. Suggestions include a local hormone (Savlov and Dunphy, 1954b) alterations in collagen cross-linking or changes in cohesive forces of collagen microstructures (Peacock, 1962; Bryant and Weeks, 1967; Weeks, 1968) and an added increment of collagen (Ogilvie and Douglas, 1964). Gain in tensile strength throughout healing of primary incisional wounds had previously been found to be correlated with the rate at which the wound synthesizes and deposits collagen rather than on total collagen content (Madden and Peacock, 1968). Therefore, Madden and Smith (1970) measured the rate of new collagen deposition in primary wounds and in wounds dehisced after seven days of healing by determining the rate of conversion of radioactivity labeled proline to labeled hydroxyproline. Rate of collagen synthesis of primary wounds was not elevated until 48 hours post-operative. Synthesis was maximal between 7 and 21 days and remained significantly elevated through the 8 weeks of study. The established rate of collagen synthesis and deposition was not changed by dehiscence and resuture of the 7 day wounds. It was concluded that the physical properties of both primary and dehisced wounds may be explained on the basis of their rate of collagen synthesis and deposition.

Healing of secondary wounds was further investigated by Leonard, Madden and Peacock (1971) by use of lathyrism in rats resulting from administration of β-aminopropionitrile. The primary biochemical effect on collagen metabolism in lathyrism is prevention of intermolecular and intramolecular crosslinking of newly synthesized collagen. Other factors involved in collagen metabolism apparently are not altered by administration of β-aminopropionitrile, for example, no effect is seen in previously crosslinked collagen. The effect of this lathyrogen on newly synthesized collagen is permanent, but the effect of this agent on additional newly synthesized collagen disappears within 6 hours after treatment is terminated. Lathyrogenic collagen can be identified mechanically by a large loss in burst strength and can, therefore, be used as a physical label for collagen synthesized during specific periods of wound healing. It had previously been demonstrated that administration of β-aminopropionitrile reduced the tensile strength of fibrils, fibers and new scar tissue (Peacock and Madden, 1966).

Lathyrism was induced at critical times during the primary and secondary healing process in wounds made through the skin and panniculus of the mid-dorsum cervical region of rats (Leonard et al., 1971). Burst strength measurements strongly supported the conclusion that all collagen which imparted strength to secondarily healed wounds was synthesized after dehiscence and resuture of primary wounds. Collagen synthesis during primary healing made no contribution to the burst strength of the resutured wounds. These data are particularly significant in that they indicate that the rapid gain in tensile strength during secondary healing of wounds is not from cross-linking or reutilization of collagen synthesized during the primary healing process.

ZONE OF COLLAGEN CHANGES AFTER WOUNDING

To effectively investigate biochemical changes occurring during wound healing it is necessary to know the zone around the wound in which metabolic changes occur. The area of the reactive zone about an incision is also critical to optimal placement of sutures. Savlov and Dunphy (1954b) reported that excision of 5 mm of tissue surrounding a skin incision did not completely eliminate accelerated healing of secondary wounds. However, excision of 7 mm of tissue from the edges of skin incisions did completely eliminate accelerated healing of secondary wounds (Peacock, 1962).

The reactive zone about an abdominal musculofascial incision in guinea pigs has recently been more thoroughly evaluated by determination of hydroxyproline and hexosamine concentrations at various distances from the wound incision through the fiftieth postoperative day (Adamsons et al., 1966). Ten, 1.85 mm strips of skin on each side of the wound were collected from 3 to 50 days after wounding by means of a special cutting device. Changes in collagen and hexosamine content were limited to a zone 7.5 mm on each side of the incision. However, an area of more intense biochemical activity occurred in a narrower zone, 5.5 mm on each side of the incision. The width of the chemically active zone remained constant throughout the experiment. Biochemical changes in many wound studies have likely been minimized by inclusion of tissue beyond the zone of local changes. This study shows that with the type of wound collected, accurate evaluation of wound biochemical changes are obtained only by analyses of tissue within 6 mm of the scalpel track.

Bevin and Madden (1969) localized the area of collagen synthesis of incised and sutured skin wounds made on the dorsum of 250 gram rats by determining the rate of conversion of labeled proline to nondialyzable hydroxyproline in the wound and at various control sites. The area of increased collagen synthesis and deposition was sharply localized to an area that did not extend beyond 1.5 mm from the wound edge. In contrast to the localized effects on collagen synthesis from an incisional wound, Houck and Jacob (1961) reported decreased dermal collagen distant to a local necrotic lesion induced by croton oil injection. This later observation, however, could not be confirmed by Smith (1963).

Data such as those above indicate that the zone of altered biochemical activity surrounding a wound varies with the type and severity of the injury. It, thus, is probably necessary for each investigator to determine the area of altered activity with his particular experimental conditions.

SKIN GRAFTS—COLLAGEN METABOLISM

Collagen metabolism during healing of skin grafts is in at least some aspects similar to remodeling of scar collagen, and thus may contribute information useful to understanding of wound healing. Histologic evaluations during grafting resulted in opposite conclusions as to graft collagen turnover. Medawar (1945) proposed that collagen was stable during healing of skin pinch autografts and somewhat later Ragnell (1953) observed an apparent increased density of collagen fibers in grafted skin. Other workers (Hinshaw and Miller, 1965; Cramer and Hinshaw, 1965), however, observed virtually complete replacement of mature collagen fibers after grafting of rat and pig dermis. Chemical analyses for collagen in graft tissues have not resolved the differing conclusions from histologic observations. A decrease in collagen content of grafts was found in chemical analyses by Marckmann (1965), but Hilgert (1963) found increased graft collagen content.

Klein and Rudolph (1972) have recently conducted experiments which have lead to significant clarification of graft collagen metabolism. Rate skin isografts were prelabeled with ^{3}H or ^{14}C. By measurement of total collagen mass and radioactivity in the grafts, an assessment was made of the quantity of collagen lost from the grafts and of the amount of newly synthesized collagen present in the grafts. Three types of grafts, split thickness, thick full thickness and thin full thickness, were studied. Four and 20 weeks after grafting each of the 3 types of grafts had lost 46-60% and 82-88%, respectively, of the original collagen.

The loss of old collagen was more rapid than the gain of new collagen for the first two weeks of healing for each of the three types of grafts. Subsequent gain of graft collagen content depended upon the type of graft. Collagen began to accumulate rapidly in thin full thickness grafts at 3 to 4 weeks and increased in amount during the remainder of the 20 week experimental period. Increase in collagen content of split thickness grafts began somewhat later and increased at a slower rate than thin full thickness grafts. Full thickness grafts tended to undergo considerable contraction and scarring and to have a smaller gain of collagen than either of the other two kinds of grafts.

Grafted collagen was lost from each of the 3 types of grafts at the same rate. Therefore, the differences observed in collagen mass were from differences in quantity of newly synthesized collagen. The data demonstrated that large quantities of new collagen are required for normal healing of skin grafts. Furthermore the data showed, contrary to widely held opinions, that synthesis of large quantities of collagen during graft healing, does not result in excessive scarring.

PROTOCOLLAGEN PROLINE HYDROXYLASE IN WOUND HEALING

Since it has been established now that collagen biosynthesis is a critical factor in wound healing, attention is being directed towards the individual processes in the complex collagen biosynthetic pathway. A unique biosynthetic process crucial to the formation of collagen is hydroxylation of proline and lysine to give hydroxyproline and hydroxylysine. The activity of the enzyme which catalyzes the hydroxylation of proline to hydroxyproline, protocollagen proline hydroxylase (PPH), has been found to increase with increased rate of collagen synthesis in several conditions and tissues (Mussini et al., 1967; Lagner and Fuller, 1969; Takeuchi and Prockop, 1969; Uitto et al., 1969; Halme et al., 1970).

Stein and Keiser (1971) have measured the activity of PPH and also the quantity of soluble collagen in granulating wounds of the rat dermis. Full thickness wounds were produced in the backs of rats using a 3 mm dermal punch. The granulating wounds and a surrounding rim of skin were removed at various times by use of 6 mm punch. PPH in rat skin reached maximum activity 5 days after wounding. Enzyme activity increased and attained its greatest activity prior to increased wound neutral salt-soluble collagen. These data with granulating wounds are in complete agreement with earlier studies by Mussini et al. (1967) who found maximum PPH-activity in incised and sutured wounds of rats on day 5 after wounding.

Only a limited number of experimental wound healing studies have been performed on humans. An advantage of the Stein and Keiser technique is that it may also be applied to human experimentation. PPH was assayed in 10 normal humans 5 days after wounding. An increase in PPH activity similar to that observed in the animal experiments occurred in the human subjects. Scleroderma patients had higher baseline PPH-activity than control subjects but showed a normal increase in PPH-activity upon wounding.

COLLAGENASE ACTIVITY IN WOUND HEALING

It is obvious from many investigations on wound healing that
considerable collagen is resorbed during certain phases of the healing process. Collagen resorption occurs not only when necrotic tissue is removed but also during remodeling. For many years, a specific enzyme, collagenase, which degrades collagen was searched for in
mammalian tissues. Such an enzyme, however, was not identified until
introduction of any assay system involving culture of living tissue
on reconstituted collagen gel (Gross and Lapiere, 1962). This technique was used to demonstrate that epithelium is the principal skin
source of the prototype of animal collagenases, anuran tadpole skin
collagenase (Eisen and Gross, 1965).

Early data on activity of collagenase from healing wounds in
mammals was given by Grillo and Gross (1967), who showed intense
collagenase activity in the edge of 15 day old guinea pig dermis
wounds containing both epithelium and new mesenchyme. However,
lysis of collagen gels was observed from only approximately one-
quarter of granulation tissue samples (no epithelium) obtained from
the center of wounds. Marginal wound tissue samples were dissociated
into epithelium and mesenchymal tissue: 80% of the epithelium but
only 54% of the mesenchymal samples lysed collagen gels. Collagenase
activity of normal guinea pig skin was present only in the epithelium
(dermis completely inactive) and appeared only after longer culture
and in lesser quantities than in proliferating marginal wound epithelium. These data demonstrated: (1) The presence of collagenase
in unwounded guinea pig skin (2) That wounding stimulates the epithelium to produce increased enzyme activity; and (3) That collagenase is not found in resting dermal connective tissue of the guinea
pig but is present in newly formed mesenchymal elements of the wound.

Collagenase was detected in human skin wounds from 14 days to
30 years after injury (Riley and Peacock, 1967). Localization and
distribution of human skin collagenase has, however, been found to
be much different from that of guinea pig and tadpole skin (Eisen,
1969). Separate cultures of isolated epidermis, upper dermis and
lower dermis of normal human skin demonstrated that the upper or
papillary layer of the dermis is the major site of collagenase production in normal human skin. No activity was found in the lower
dermis and only minimal activity was observed in the epidermis.
However, as in the guinea pig, the healing wound edge of human epidermis is able to actively synthesize collagenase. Grillo et al.
(1968) had previously shown collagenase activity was much more
frequent in granulation tissue of human wounds than in those of
guinea pigs. In addition granulation tissue from dissociated human
wound edge tissue lyzed collagen gels in 93% of the cultures compared
to only 28% for epithelium from the same source.

CONCLUDING REMARKS

Important advances have been made the past few years towards clarification of the role of collagen metabolism in wound healing. It is now recognized that wounds may continue to gain tensile strength for periods of at least one year. However, wound collagen content and tensile strength do not show a correlation beyond early stages of healing. Prolonged remodeling of wounds occurs including continuing synthesis, deposition and resorption of scar collagen. A correlation exists between wound breaking strength and the rate of collagen turnover and deposition for extended periods after net accumulation of wound collagen has ceased. The prolonged synthesis and deposition of wound collagen provides a biological mechanism for increased wound strength without increased scar collagen content.

An explanation has been sought for clinical and experimental observations that healing dehisced and resutured wounds gain strength more rapidly than primary wounds during the first few days of healing. The established rate of collagen synthesis and deposition apparently is not changed by dehiscence and resuture of wounds. Collagen synthesized during primary healing makes no contribution to the burst strength of resutured wounds and, thus, the rapid gain in tensile strength during secondary wound healing is not from cross-linking or reutilization of collagen synthesized during the primary healing process.

Additional questions related to wound healing have also been explored. The zone of altered biochemical activity surrounding a wound appears to be narrow but also probably varies with the type and severity of the injury. Large quantities of new collagen are synthesized and deposited during normal healing of skin grafts without excessive scarring. Attention is also now being directed towards the individual processes in biosynthesis and degradation of wound collagen. The enzyme which hydroxylates proline to hydroxyproline, protocollagen proline hydroxylase, increases in wounds of both man and experimental animals. Collagenase, an enzyme which may be essential for degradation of wound collagen, likewise increases in wound edges of human and experimental subjects.

The study of wound healing has been of continuing interest to the clinician and to the experimental biologist. It has always been the aim of such studies to devise ways to accelerate wound healing. Little success, however, has been evident in fulfilling this aim. With the increasing pool of detailed knowledge available on the role of collagen in wound healing perhaps the day is close when either a way will be found to accelerate wound healing or it will be recognized that nature already accomplishes the process with maximum efficiency.

REFERENCES

1. Adamsons, R. J., Musco, F., and Enquist, I. F.: The relationship
        of collagen content to wound strength in normal and scor-
        butic animals. Surg. Gyn. Obst. 119:323, 1964.

2. Adamsons, R. J., Musco, F., and Enquist, I. F.: The chemical di-
        mensions of a healing incision. Surg. Gyn. Obst. 123:515,
        1966.

3. Bevin, A. G., and Madden, J. W.: Localization of collagen syn-
        thesis in healing wounds. Surg. Gyn. Obst. 123:515, 1966.

4. Bryant, W. M., and Weeks, P. M.: Secondary wound tensile
        strength gain: A function of collagen and mucopolysac-
        charide interaction. Plast. Reconstr. Surg. 39:84, 1967.

5. Cramer, L. M., and Hinshaw, J. R.: Autograft rejection induced
        by homografting. Plast. Reconstr. Surg. 35:572, 1965.

6. Douglas, D. M.: The healing of aponeurotic incisions. A pheno-
        men intermediate between homograft rejection and autoim-
        munity. Brit. J. Surg. 40:79, 1952.

7. Dunphy, J. E., and Udupa, K. N.: Chemical and histochemical se-
        quences in the normal healing of wounds. New Engl. J. Med.
        253:847, 1955.

8. Dunphy, J. E., Udupa, K. N., and Edwards, L. S.: Wound healing.
        A new perspective with particular reference to ascorbic
        acid deficiency. Ann. Surg. 144:304, 1956.

9. Dunphy, J. E.: On the nature and care of wounds. Ann. Roy.
        Coll. Surg. England. 26:69, 1960.

10. Edwards, L. C., Pernokas, L. N., and Dunphy, J. E.: The use of
        a plastic sponge to sample regenerating tissue in healing
        wounds. Surg. Gyn. Obst. 105:303, 1957.

11. Eisen, A. Z., and Gross, J.: The role of epithelium and mesen-
        chyme in the production of a collagenolytic enzyme and a
        hyaluronidase in the Anuran tadpole. Deve. Biol. 12:408,
        1965.

12. Eisen, A. Z.: Human skin collagenase: Localization and distri-
        bution in human skin. J. Invest. Derm. 52:442, 1969.

13. Forrester, J. C., Zederfeldt, B. H., Hayes, T. L., and Hunt, T.
        K.: Wolff's law in relation to the healing skin wound.
        J. Trauma 10:770, 1970b.

14. Forrester, J. C., Zederfeldt, B. H., Hayes, T. L., and Hunt, T.
     K.: Tape-closed and sutured wounds: A comparison by
     tensiometry and scanning electron microscopy. Brit. J.
     Surg. 57:729, 1970a.

15. Grillo, H. C., and Gross, J.: Collagenolytic activity during
     mammalian wound repair. Dev. Biol. 15:300, 1967.

16. Grillo, H. C., McLennan, J. E., and Wolfort, F. G.: Activity
     and properties of collagenase from healing wounds in mam-
     mals. Centennial Symposium on Wound Healing. Ed., Dunphy,
     J. E. The Blakiston Division, McGraw Hill, New York,
     1968, p. 185.

17. Gross, J., and Lapiere, C. M.: Collagenolytic activity in am
     phibian tissues: A tissue culture assay. Proc. Natl.
     Acad. Sci. 48:1014, 1962.

18. Halme, J., Uitto, J., and Kahanpaa, K.: Protocollagen proline
     hydroxylase in experimental pulmonary fibrosis in rats.
     J. Lab. Clin. Med. 75:535, 1970.

19. Hilgert, I.: Changes in the hydroxyproline and hexosamine con
     tent of grafts after transplantation. Folia Biol. Prague
     9:136, 1963.

20. Hinshaw, J. R., and Miller, E. R.: Histology of healing split-
     thickness, full thickness autogenous skin grafts and donor
     sites. Arch. Surg. 91:658, 1965.

21. Houck, J. C., and Jacob, R. A.: The chemistry of local dermal
     inflammation. J. Invest. Derm. 36:451, 1961.

22. Howes, E. L., Sooy, J. W., and Harvey, S. C.: The healing of
     wounds as determined by their tensile strength. J. Amer.
     Med. Assoc. 92:42, 1929.

23. Howes, E. L., Harvey, S. C., and Hewitt, C.: Rate of fibro-
     plasia and differentiation in the healing of cutaneous
     wounds in different species of animals. Arch. Surg. 38:
     934, 1939.

24. Klein, L., and Rudolph, R.: ^{3}H-Collagen turnover in skin
     grafts. Surg. Gyn. Obst. 135:49, 1972.

25. Lagner, R. O., and Fuller, G. C.: Elevation of proline hydroxy-
     lase activity in diseased rabbit aorta. Biochem. Biophys.
     Res. Commun. 36:559, 1969.

26. Leonard, J.R., Madden, J.W., and Peacock, E.E.: The use of
     lathyrism to study secondary wound healing. Surg. Gyn.
     Obst. 133:247, 1971.

27. Levenson, S.M., Geever, E.F., Crowley, L.V., Oates, J.F.,
     Berard, C.W., and Rosen, H.: The healing of rat skin wounds.
     Ann. Surg. 161:293, 1965.

28. Madden, J.W., and Peacock, E.E.: Studies on the biology of
     collagen during wound healing. I. Rate of collagen synthe-
     sis and deposition in cutaneous wounds of the rat. Surgery
     64:288, 1968.

29. Madden, J.W., and Smith, H.C.: The rate of collagen synthe-
     sis and deposition in dehisced and resutured wounds. Surg.
     Gyn. Obst. 130:487, 1970.

30. Madden, J.W., and Peacock, E.E.: Studies on the biology of
     collagen during wound healing: III. Dynamic metabolism of
     scar collagen and remodeling of dermal wounds. Ann. Surg.
     174:511, 1971.

31. Marckmann, A.: Autologous skin grafts in the rat. Biochemi-
     cal analysis of mucopolysaccharides and hydroxyproline.
     Proc. Soc. Exp. Biol. Med. 119:794, 1965.

32. Medawar, P.B.: A second study of the behavior and fate of
     skin homografts in rabbits. J. Anat. 79:157, 1945.

33. Mussini, E., Hutton, J.J. and Udenfriend, S.: Collagen pro-
     line hydroxylase in wound healing, granuloma formation,
     scurvy and growth. Science 157:927, 1967.

34. Oglivie, R.R., and Douglas, D.M.: Collagen synthesis and pre-
     liminary wounding. Brit. J. Surg. 51:149, 1964.

35. Peacock, E.E.: Some aspects of fibrogenesis during the heal-
     ing of primary and secondary wounds. Surg. Gyn. Obst. 115:
     408, 1962.

36. Peacock, E.E.: Inter- and intramolecular bonding in collagen
     of healing wounds by insertion of methylene and amide cross-
     links into scar tissue. Ann. Surg. 163:1, 1966.

37. Peacock, E.E., and Madden, J.W.: Some studies on the
     effect of B-amino-propionitrile on collagen in healing
     wounds. Surgery 60:7, 1966.

38. Ragnell, A.: The secondary contracting tendency of free
     skin grafts. Brit. J. Plast. Surg. 5:6, 1953.

39. Reynolds, B. L., Leveque, T. F., and Buxton, R. W.: Metabolic parameters in the healing of open skin wounds in animals. Amer. Surg. 29:325, 1963.

40. Riley, W. B., and Peacock, E. E.: Identification, distribution, and significance of a collagenolytic enzyme in human tissues. Proc. Soc. Exp. Biol. Med. 124:207, 1967.

41. Sandberg, N., and Zederfeldt, B.: The tensile strength of healing wounds and collagen formation in rats and rabbits. Acta Chir. Scand. 126:187, 1963.

42. Sandblom, Ph., and Muren, A.: Differences between the rate of healing of wounds inflicted with short time interval. I. Cutaneous incisions. Ann. Surg. 140:449, 1954.

43. Savlov, E. D., and Dunphy, J. E.: Mechanisms of wound healing: comparison of preliminary local and distant incisions. New Engl. J. Med. 250:1062, 1954a.

44. Savlov, E. D., and Dunphy, J. E.: The healing of the disrupted and resutured wound. Surgery 36:362, 1954b.

45. Smith, Q. T.: The response of dermal collagen to croton oil injury. J. Invest. Derm. 40:213, 1963.

46. Stein, H. D., and Keiser, H. R.: Collagen metabolism in granulating wounds. J. Surg. Res. 11:277, 1971.

47. Taffel, M., Donovan, A. J., and Lapinski, L. S.: The effect of trauma on wound healing: An experimental study. Yale J. Biol. Med. 23:482, 1951.

48. Takeuchi, T., and Prockop, D. J.: Protocollagen proline hydroxylase in normal liver and in hepatic fibrosis. Gastroenterology 56:744, 1969.

49. Uitto, J., Halme, J., and Hannuksela, M.: Protocollagen proline hydroxylase activity in the skin of normal subjects and of patients with scleroderma. Scand. J. Clin. Lab. Invest. 23:241, 1969.

50. Viljanto, J.: Biochemical basis of tensile strength in wound healing. Acta Chir. Scand. suppl. 333: 1964.

51. Weeks, P. M.: Functional organization of collagen subunits. Surg. Forum 19:59, 1968.

52. Williams, G.: The late phases of wound healing: Histological and ultrastructural studies of collagen and elastic-tissue formation. J. Path. 102:61, 1970.

BIO-ENGINEERING ANALYSIS OF HEALING TISSUES (Paper in Abstract)

Walter Zingg

University of Toronto, Canada

Repair and regeneration are basic indispensable properties of living tissues. As a result of various insults the function of a given tissue may be impaired; repair and regeneration then take place restoring function partially or completely. For many organs, tests have been devised to assess the degree of restoration of a given function. The study of the mechanical functions of the musculo-skeletal system and skin can be as fascinating - and as sophisticated - as the study of biochemical function. Our understanding of the physical-mechanical components of physiological function and of the mechanisms of disease, though as important as our understanding of the chemical components, lags far behind.

One of the troubles is that in order to understand mechanical studies, we have to understand mechanical terminology. What do we mean by strength of a healing bone, or strength of a healing tendon?

Strength is the force or stress needed to break a thing. Stiffness is not strength; a piece of dry toast is stiff, but weak. By and large these two terms define the properties of a material. Stiffness is a well-defined entity, expressed as Young's modulus of elasticity. Stress is Force (Load) per unit area, Strain is the amount of stretch (under load) per unit length. The strength of a material, or its ability to withstand an external force, depends on the speed with which the external force is applied. A rapidly applied stress is said to have a "high rate of stress" producing a "high rate of strain." Rate indicates stress or strain in relation to time.

Strength, to have any meaning, has to be further defined: a material is strong enough to withstand which force? Tensile strength is easy to define: it is the force necessary to break a string or tendon or a piece of skin by pulling at both ends in opposite directions. The tensile strength of a material usually is expressed as the breaking stress, the breaking load per unit area, for instance pounds per square inch.

Compressive strength cannot be so easily defined. If a rod - or a bone - is increasingly compressed along its axis, sooner or later it will buckle and break.

This type of analysis provides the following data:

Stress $\left(\dfrac{Force}{Area} \right)$

Elongation   (length)

Strain $\left(\dfrac{Elongation}{Original\ length} \right)$

Stiffness $\left(\dfrac{Stress}{Strain} \right)$ (Young's modulus of elasticity)

Energy is the capacity to do work. The total amount of energy within a system is constant: potential energy, kinetic energy, strain energy (energy stored as internal stress).

Toughness of a material is the resistance to the propagation of cracks. In a composite material, such as bone, interfaces may act as crack-stoppers.

Bio-engineering analysis of healing tissues has shown:

1. Different components of the material properties of healing tissues return to normal at different rates.

2. The result of the engineering test depends on the rate of the stress or strain applied.

3. The healing rate is different in different species. It is different in different tissue and at different sites in the same tissue.

4. To complicate the situation further engineering tests give reliable results only when carried out under strictly controlled conditions.

5. Therefore, extrapolation from data from a piece of excised rat skin suspended in a testing machine, to a healing wound in a patient after trauma or operation is risky.

6. The healing of wounds and injuries is a prolonged process. The conditions which allow this proce-s to proceed at an optimal rate are inadequately defined, nor can we accurately describe the activities which can be allowed -- or indeed recommended -- during the healing period.

BIOCHEMISTRY OF PATHOLOGICAL EFFECTS OF LIGHT TRAUMA   (Paper

in Abstract)

W. Louis Fowlks

University of Minnesota

Light can produce a number of different types of trauma. It can, as the result of photochemical alteration of biochemicals, produce foreign molecules in situ. A number of consequences can result from such photochemical reactions:

1. Protein and lipid and polysaccharide components of membranes can be damaged, thus destroying the integrity of these structures. By reducing the activity of membrane bound enzymes involved in transport, light induced trauma could contribute to alterations in the internal milieu with serious consequences to viability of damaged cells.

2. Photochemical changes in nucleic acids (DNA and/or RNA) can interfere with the encoding and the transcription of the genetic code thus resulting in death or mutations. These photochemical reactions consist of dimer formation between photosensitive pyrimidines of the nucleic acid themselves and also between certain types of small molecules which may be present. Also, photochemical addition to the 5,6 double bond of pyrimidines yields water adducts of cross-linked products with proteins, glutathione, or other cellular constituents which contain re- active, OH, NH, or particularly SH groups in their structures.

3. Photochemical damage to proteins inevitably leads to reduction of their enzymic activity. However, enzymic activity results in altered metabolic pathways within the cell which could result in cell death or, at worst, yield cells which function poorly and possibly produce substances which may be deleterious to nearby undamaged cells or other organs.

4. Photochemical changes in chemical compounds which were ingested
    as drugs or applied to the skin may result in the formation of
    compounds with potential cancerogenic activity. A case in point
    is the transformation of para aminobenzoic acid into the poten-
    tial carcinogen 4,4'-azodibenzoic acid.

5. Photochemically altered proteins, polysaccharides, lipids and
    many photo-sensitive small molecules are all potential antigens
    which can result in the allergic and/or the immune response
    sometimes noted in irradiated tissue, particularly the skin.

6. The beneficial effects to be derived from sunbathing because
    of photochemical transformation of Vitamin D precursors to
    the biologically active form of the vitamin are well documented.
    Less well known are photochemical changes which can also occur
    if steroids are present when tissue is irradiated. Such photo-
    chemically altered steroids should have an effect on the normal
    steroid balance of an animal. In addition, substances such as
    histamine, which may be released from lysosomes by photic des-
    truction of their membranes, may also affect the biochemistry
    of the neuro-endocrine system and hence its delicate balance.
    Catechol amines released from the retina during the visual pro-
    cess may also play a similar role and may account for the
    effects of light environment upon the neuro-endocrine system.
    Effects such as sexual maturation, litter mate size and sex,
    and most importantly upon the normal circadian rythmn which man
    shares with other living things, all of which respond to the
    light environment.

    In addition to traumatic effects on tissues or individual cells
which may result from primary photochemical changes in various biochemical components of the living system, emotional and other psychological effects may occur which depend upon exposure to lights
of specific kinds. A well documented example is the mental black-
out which results from photoic driving by a flashing light with the
appropriate flash frequency. Many pilots of small propeller driven
planes were crash victims of this phenonemen before it was adequately
described in the literature. But even subliminal fluctuations in
light intensity which occurs under banks of fluorescence lights
produce vague but real alterations in feelings and in the efficiency
of work output. These effects occur if the frequency of the intensity fluctuations are between 10 and 150 Hz. In this connection,
recent work has implicated light stimulus to the retina as being
involved in the release of melatonin from the pineal gland. The
result of such release, which appears to be regulated by the quantity of light received per day and by the time of day at which the
light is received, may also have profound effects upon brain function or upon the neuro-endocrine system.

TRAUMA FROM VISIBLE AND INFRARED LASERS

Leon Goldman

University of Cincinnati

The introduction of the laser into biology and medicine has stimulated, as it were, greater interest not only in the studies of photo-chemistry but also in studies of the effect of photons on living matter. There are many new applications where the laser is needed in biology and medicine. Now there are selective single wavelengths available, not investigated previously. Tremendous energy and power densities are now possible. There is now, then, a new facet of trauma from light.

THE LASER

The device called the laser, makes use of stimulated emission, briefly, by raising molecules to a higher energy level through various excitation techniques and lasing develops as there is return to the original energy level. Such pumping may be done by light sources, by explosions, chemical reactions, and the like. Now, there is a series of lasers varying from the ultraviolet to the far infrared, 10,600.0 nm. There has been some concern recently about the possibility of development of x-ray lasers. Initial experiments, as yet according to experts, have not produced an x-ray laser with coherent x-rays. Such a laser, which will be available in the future, would have, as it were, an awesome effect on tissue.

In brief, the specific characteristics of laser light as opposed to coherent light are:

1. coherency
2. monochromacity

3.   specific color absorption of some laser systems
4.   tremendous energy and power densities

The changes in tissues include, briefly:

1.   thermal coagulation necrosis
2.   elastic recoil and pressure mechanisms, especially
     with pulsed systems
3.   "changes in electrical fields," especially with
     pulsed systems

Studies, then, can be done with a very specific wavelength, precisely measured output, on a selected target area which can be less
than a micron. The procedure for the application of a laser is as
follows: First, there is a description of the laser system itself,
its characteristics, the details of its transmission, and finally,
the accurate measurement of its output.   Then a safety program is
planned as regards hazards of the laser exposure primarily for the
eyes, then for the skin. Under certain conditions and with certain
types of targets, such as toxic materials, organisms, etc., environmental contamination may be considered and safety measures adopted
for this. With planned safety programs, then, testing and applications are considered.

TECHNIQUES OF ANALYSIS OF REACTIONS

The study of the effect of lasers on living cells start with an
analysis of the effects on tissue cultures.   Then animal studies
are done with adequate controls. Finally, investigative studies on
man develop. As indicated, with proper safety measures such investigative studies on man can be done with safety for the subject as
well as for the operator.

With a laser attached to a microscope, micro-irradiation of
individual cells and tissue cultures can be done. With the precision possible with the collimated beam of the laser, as indicated,
irradiation areas less than 1 micron can be achieved. This permits
micro-irradiation of such cellular components as the nucleus and
nucleolus, chromosomes, mitochondria, etc. The micro-irradiation
techniques of Rounds and Berns on specific foci in living chromosomes,
emphasize the development of the new laser biology and suggest its
application for that which has been called so aptly, molecular
engineering.

Studies on photochemistry, particle size measurement, scanning
techniques of nuclear changes, immunofluorescence techniques, are
also possible with the use of the laser.

Another phase of laser micro-irradiation is the micro-analysis

Table 1. These are the laser systems now available for biology and medicine

Area	Application	Laser	Wavelength (nM)	Output Level	Dose
Ophthalmology	Retinal Tears	Ruby	694.3	5-600 mJ	700μsec
	Vascular diseases (Diabetic Retinopathy)	Argon	488 and 514.5	70-500 mW	100-500μsec
Dermatology	Portwine	YAG+Nd+SHG*	530	70-500 mW	100-500μsec
	Angiomas	Ruby	694.3	50-100 J/cm^2	1-4μsec
	Tattoos	Argon	488-514.5	5-10 W/cm^2	1-60sec
		Ruby	694.3	50-75 J/cm^2	1-4μsec
		Ruby-Q switch	694.3	5-10 J/cm^2	10-30nsec
	Pigmented Lesions (Seborrheic Keratosis)	Ruby	694.3	300-400 J/cm^2	1-4μsec
	Skin Cancer (Melanoma, Basal Cell)	Ruby	694.3	1500-2000 J/cm^2	1-4μsec
Surgery	Incisions on Skin and Internal Organs	Neodymium	1060	2000 J/cm^2	4μsec
		Carbon Dioxide	10,600	1000-10,000 W/cm^2	indefinite
		Argon	488-514	1000-100,000 W/cm^2	
		Nd-YAG	1060	1000-100,000 W/cm^2	
Cellular Biology	Single cell and	Ruby	694.3	50-5000μJ	500μsec
	Cell Colony Irradiation	Argon	488-514.3	in 1-50μM spot 60 HZ @ 1 W peak 0.7 μM spot	
Diagnostics	Spectroscopy	Ruby-Q switch	694.3	4-10 mW	10-50nsec
		Neodymium-Q-switch	1060	in 25-100μM spot	
	Raman	Helium Neon	632.8	70 mW	15-30 min
		Argon	488	250 mW	3-5 min
	Transillumination	Helium Neon	632.8	70 mW	
		Krypton	647.1 647.5	125-300 mW	Indefinite
		Argon and Krypton	568.2 514.5 488.0	500mW	
		Ruby	694.3	400 mJ	500μsec
		Ruby-Q switch		0.25-0.5 J/cm^2	10-30nsec

*SHG: Second Harmonic Generation

Fig. 1. The new flexible high-output CO_2 laser (Laser Industries Ltd., Tel-Aviv, Israel) used for current investigations in laser surgery in the Operating Pavilion of the Medical Center of the University of Cincinnati.

of cellular structures and cells by laser micro-emission spectroscopy. This permits analysis without total destruction of the sample. The laser microprobe, so-called, can be used to detect cations in biological materials even as small as in the nucleus or cytoplasm.

The new techniques of microholography and acoustical holography (ultrasound + laser) offer challenging techniques for examination in third-dimensional imagery, the growth characteristics of the living cell. Acoustical holography makes it possible without x-ray to observe bones, soft tissue, tissue densities of materials, such as plastics which would be transparent to x-rays. This is of interest in regard to the plastic antipersonnel bomb used in Viet Nam.

The topographical effects of laser impacts on living tissue may be observed by:

1. surface microscopy at moderate magnification
2. replica microscopy
3. scanning electron microscopy, SEM

For detailed studies of the reaction in tissue, there are the usual techniques, histopathology, histochemistry and transmission electron microscopy.

In some of our studies, additional examination of the impact area has been done by studies of x-ray diffraction, x-ray fluorescence and element system analysis.

Studies of disturbed function without any structural change have been done in animal eye laser research with the technique of Evoked Potential. As yet, this has not been possible in man.

To study the thermal effect in tissue distribution and transport in tissue, such systems are thermistor, thermocouple techniques have been done as well as the use of liquid crystals.

In this manner, then, a complete picture of the laser effect on living tissue can be obtained, often with the use of laser application itself.

THE TISSUE REACTIONS

The tissue reactions depend, to some extent, on the laser, its output, transmission and the optical and "reactive" characteristics of the target. The target refers to color dependency on wavelength, and the properties, transmission and loss.

The clinical appearance of a laser impact is a reddened area which becomes rapidly crusted, usually dries very well, scarring develops and as a rule, this reaction is calculated to be minimal. Occasionally, however, hypertrophic scarring may develop. Ordinarily, the scar becomes smooth and red, and after a period of some years, repigmentation begins. These mechanisms of regrowth and even repigmentation relate to the pluripotential of the skin appendages.

Studies of hair regrowth in the laser target area now are under investigation by Wright of our Laboratory.

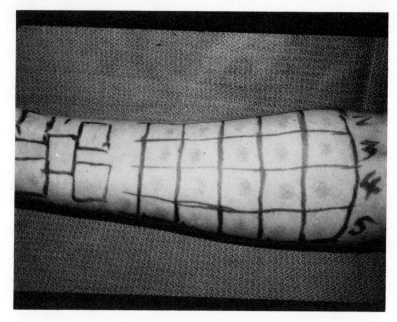

Figure 2

Figure 3

Studies of vaporization of tissues by lasers have been done by Rockwell and by Hall with the CO_2 laser, far infrared. Temperatures of 100°C have been found to be associated with these reactions.

The hypermetabolic studies of thermally burned tissue have not been done as yet in laser investigations. So, it is not known if the laser impacts also destroy, as the "pure" thermal burn does, a hexane soluble lipid in the skin. No studies of increased evaporated water losses from the laser target areas have been done as yet. Most investigations of laser effects relate to small areas of tissue where this evaporative loss would not be significant.

It is evident that much of the cellular (and even vascular?) dynamics of the laser lesion has not been studied. This includes the easily available skin and also the eye. The initial mild tissue reaction shows only minimal changes with essentially a perivascular lymphocytic infiltrate. The epidermal changes after impact of the laser depends on the intensity and distribution of the masses of melanin in this superficial structure. In one instance, the ruby laser produced thrombogenesis of a superficial blood vessel in the mid-dermis of the skin of a test human subject.

As the energy and power densities of the laser increase, one observes a thermocoagulation necrosis whose intensity varies with the parameters of the laser used. The essential tissue reaction element, is a tissue coagulation necrosis, a perivascular lymphocytic infiltrate with occasional polymorphonuclear infiltrate. The skin appendages are more resistant. As the energy and power densities increase, more of the same reactions may be observed with the subsequent development of deep necrotic sloughs. Tissue hemorrhage is uncommon. If there are pigment masses, natural or added, more intense reactions develop here.

With Q-switched impacts, there is more intense reaction in tissues, still non-specific in character. The same characteristics hold in the few studies done on picosecond impacts.

Legend to Fig. 2. Showing technique of determining minimal exposures to lasers on the flexor surface of the forearm: five successive exposures of a single energy density is done for each row.

Legend to Fig. 3. To determine the effect of chronic exposure to lasers. Small doses are given daily to the same spot on the flexor surface of the forearm: picture shows the pruritic erythemato-papular reaction 12 hours after exposure to the impact of a pulsed ruby laser 20 joules/cm^2.

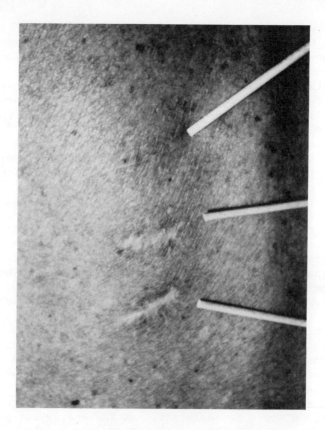

Fig. 4. Studies of wound healing on the back – upper spot scalpel incision, middle spot CO_2 laser incision: inferior spot high-frequency electro-surgical incision.

Free radical formation has been observed in pigmented tissues after laser irradiation. Such reactions have been found also with ultraviolet irradiation of pigmented tissues.

Extensive studies have been done of the effect of the laser in the far infrared in regard to steam development in tissue and the temperatures required for proper excisions. With the ruby laser, a surface coagulation necrosis equivalent to the high-frequency electric current has been noted. With the carbon dioxide laser, by Fidler, the thermocoagulation necrosis was narrower and more effective and less bleeding than high-frequency electrical unit. With the high-output Nd-YAG laser, alternate zones of thermocoagulation necrosis were found within the living tissue. The interesting development with these changes is still under study.

Lesions heal with non-specific scarring. As yet, no carcinomas have been developed in these healing areas or in the area around them. No stigmata of the x-ray response in tissue have been found in the old healed laser scars, endarteritis, telangiectasia or abnormal fibroblasts.

Relatively few attempts have been made to study chronic exposure. This has been done for the eyes with exposures of the cornea and lens and subsequent tissue cultures. Also tissue cultures exposed to laser irradiation have been subcultured to attempt to find abnormal cellular responses. In one series of experiments, Rounds has found abnormal cell development in one series of endothelial cells subcultured repeatedly after exposure to laser micro-irradiation.

For the effects of chronic exposure of the eyes of man, patients have been observed for years, usually after the therapeutic use of the laser, or after accidents. No abnormalities other than non-specific healing have been found in the impact site or in the adjacent areas subject to exposure from reflectance.

Is it possible to have physiological changes without evident tissue damage even after examination with electron microscopy? This is possible with the technique of Evoked Potential. This technique can be controlled in animal eyes but not for man.

Are there only essential differences in the reactions produced by visible and infrared laser systems? As indicated, the visible lasers have greater absorption in color mass. The tissue destruction, therefore, may be more localized and more relative but the essential tissue change is not different from that of infrared lasers. It is not clear that with proper controls, that this property of color absorption would mean more readily free radicle formation. Such reactions have been observed after ultraviolet impacts on melanoma continuing cells as well as after ruby laser impacts on melanoma.

Simply because more infrared lasers are available for biological and medical research more studies have been done on the effect of these lasers on tissues, even tissues of man. Until more detailed studies are available, the laser-induced thermal coagulation by infrared lasers with proper controls is not essentially different from the laser-induced thermocoagulation necrosis by visible lasers. As stressed repeatedly, more studies are needed of the acute and chronic exposure. The need for electron microscopy, both scanning and transmission, is evident as well as studies for disturbances of function and long range studies of cytogenetics.

Do these analyses of functional and morphologic changes suggest

that the pathogenetic mechanisms of the laser reaction is essentially
a thermal effect. As listed above, there are non-thermal effects.
These are often suggested or speculated about, but often not proved.

Laser trauma to tissue with the very short pulses, nanosecond
and picosecond, are suspected of producing non-thermal effects.
The thermal effect, in brief, depends on the duration of laser exposure and to some extent, on the area of exposure. These may be
listed, according to Goldman and Rockwell as:

      1.   absorption and reflection of the tissues
           at the laser wavelength.
      2.   the power density of the laser beam.
      3.   speed of incision or time of exposure.
      4.   the extent of the local vascular flow.
      5.   the degree of tension of the tissue in the
           area of incision.

These refer especially to surgery with the continuous wave (CW)
lasers.

It is assumed that the thermal effect of laser trauma is the
same as that from any thermal destruction of tissue, protein de-
naturation and any changes in cellular metabolic activity as the
primary reaction. The secondary effects relate to vascular disturbances, scarring, etc. Again, more than 10 years of research
in laser biology and medicine has not established the various
laser systems are carcinogenic. It is admitted that at present
there is too meager experience with the ultraviolet lasers in
biology and medicine.

What are some of the non-thermal effects as listed previously?
Elastic recoil and pressure waves were mentioned. From thermally
induced stress, compressional stresses of many thousands of atmospheres can be produced on the cell exposed to the laser. In experimental studies on ruby laser impacts of the femur on rabbits, fat
embolization was produced. This may be a combination of thermally
induced compression stress plus chemical changes in the lipids
in the marrow cavity.

The cavtation phenomenon of sound waves may produce changes in
tissue. It is assumed that there may be conversion here of mechanical energy to chemical energy. It is assumed also that such changes
may be produced by "thermal ionization effects such as those produced in the recoil plume generated at the impact site." Some non-
linear generation effects of light and sound waves include second
harmonic generation, stimulated Raman and Brillouin scattering,
field stripping effects (free radicles, electrons, ionized atoms
and molecules), inverse Brenstrah lung. "This is a process wherein

loosely bound electrons are accelerated by the strong electrical field associated with the laser pulse. Collisions with neighboring atoms and molecules can result in local thermal effects. This localized heating can cause dissociation and ionization which can lend to more local absorption by the plasma produced."

Other non-linear effects include double photon absorption, "self focussing or light beam trapping effect." What is surely lacking in all of these studies of the non-thermal effects of light trauma on tissue is adequate quantitation for living tissue.

At least as far as thermal coagulation necrosis is concerned, there does not appear to be any qualitative reactions in regard to tissue damage produced by infrared light as compared to tissue damage from infrared lasers. As a matter of fact, as detailed above, there is much less data currently available of laser burns than of thermal burns. Perhaps future studies may show even non-linear effects in the mechanisms of thermal burns. Infrared laser burns can be and are more severe, and of course, more precise in their impact. The use of the infrared laser as a surgical tool at burn centes to remove the eschar and provide for immediate graft replacement will be of mutual help in the burn study program. Such a surgical technique is possible because of the relatively bloodless nature of the laser scalpel. Research continues for the development of a precise laser dermatome.

In the setting of this Conference, it is proper to discuss the immunobiology of light trauma. Again, there is much interest in this phase but very few actual investigations, even for cancer immunobiology. In recent experiments, a laser research worker, after deliberate and programmed exposures to the pulsed ruby laser, 694.3 nm, with specified energy densities and target areas, was found to be more reactive than other individuals with a similar skin type. Animal experiments have not been done; primarily for the development of increased reactivity to the various laser systems. These experiments are needed.

In the laser treatment of patients with melanoma, some clinical experiments have been done with laser treatment for immunology. These include:

1. laser irradiation of tissue cultures of melanoma cells and use of "extracts" as tumor antigens;
2. transplants of laser treated melanoma tumors;

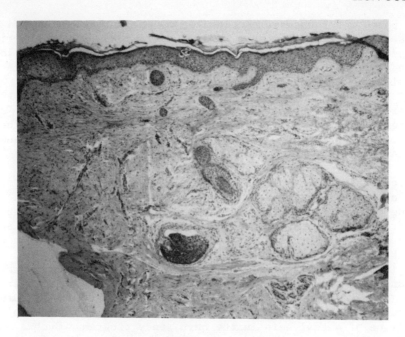

Fig. 5. 110 days after laser treatment of basal cell carcinoma of the wrist showing non-specific fibrosis with no tumor cords. Hematoxylin-eosin X 40.

3. laser treatments as adjuncts to BCG
   vaccination;
4. laser treatment as adjunct to x-ray therapy
   and cancer chemotherapy.

These experiments were done, of course, on metastatic melanoma, without significant changes on the serious prognosis of this stage. Again, as with many phases of laser biology and medicine, such diffuse pilot studies show only that such programs can be done safely and with adequate controls.

Legend to Fig. 6A. (Illustration overpage) Showing the technique of investigation of laser impacts on metastatic melanomas of the back: patient subsequently had BCG vaccination: C - control tumor. Pulsed ruby laser energy densities were 2,000, 4,000 and 6,000 joules/cm^2.

Legend to Fig. 6B. (Illustration overpage) 7 days later: treated lesions healing, patient died subsequently of a cerebral metastases.

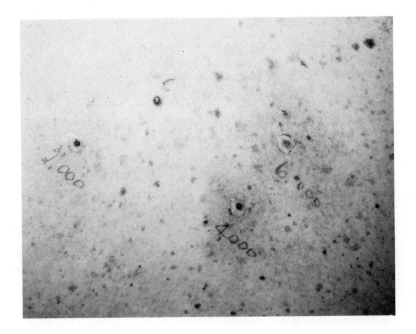

Figure 6A.

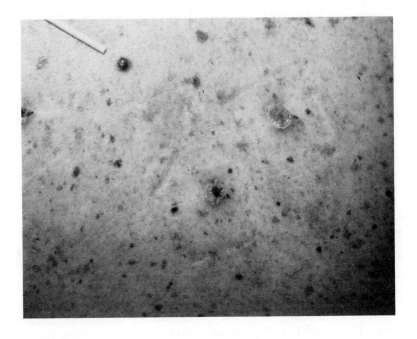

Figure 6B.

SOME MECHANISMS OF TRAUMA FROM LASER

In the course of the development of this technology it is
evident that there can be accidental exposure as well as deliberate
exposure to lasers. In all phases of laser technology, there can
be accidental exposures. These can occur in the research laboratory with pilot and hazardous new equipment, in every field of
applications of the laser and in consumer use and abuse. These
deliberate exposures may be found in the following situations:

1. attempts to determine permissible dose
   exposures for the laser safety program
2. for the actual use of the laser in biology
   and medicine
3. in the use of the laser as weaponry

In the laser safety program it is necessary to know the maximum
permissible exposures (M. P. E.). This is done initially in laboratory experiments by the exposure of cells and tissue cultures to
micro-irradiation with laser systems. Then, follows animal experiments. Here, there is the usual difficulty of the extrapolation of
animal data for man both as regards acute and especially chronic
exposures.

The important and vital areas in the development of the laser
safety program, as indicated, are in the eyes, skin and the respiratory system. Finally, after collection and analysis of controlled
animal data it is necessary to develop standards for man. Here, it
has been found necessary at times to expose the eyes of volunteers,
about the macula and especially the skin of volunteers. The respiratory factors relate to environmental pollution from material processing by lasers, laser-induced thermonuclear fusion, etc. Such
experiments are done in exposure of animals with the usual techniques of such animal toxicity experiments.

Of these three organ systems for exposure, the skin is obviously
the most easily available and the most practical for the deliberate
exposure experiments. This has been shown repeatedly to be of value
not only for acute exposures but also for the important area of
chronic exposure. At present, various organizations are attempting
to develop a series of standard exposure levels which will be
accepted universally. This, obviously, is difficult according to
the goals of the individual group, whether it be for use in industry,
in the military or for consumer protection.

Exposure is done deliberately in the field of experimental biology to study as described micro-irradiation of cells, in chromosomes,
mitochondria, lysosomes, etc., tissue cultures and actual tissues of
animals. By far the vast amount of work in biology has been done

in regard to exposure of the eyes, especially of the rabbits and
monkeys to determine the ocular mechanisms of laser trauma and
the development of safety measures. Animal toxicity investigations,
as well as those of man relate to acute and chronic exposure to
metal drilling, welding, processings of plastics, rubber, textiles,
diamonds, etc.

In medicine, the laser uses deliberate exposure for diagnosis
and also for treatment. Laser diagnosis includes the use of the
laser for micro-irradiation, the laser micro-emission spectroscopy
or the laser microprobe for micro-analysis of tissue including
living tissue. Laser diagnosis is used also in diagnosis for
transillumination of tissue for:

1. transillumination of tissue as incoherent
   light transillumination, mammography or
   xerography
2. holography
   A. interferometric studies
      (1) movements in the chest wall
          in respiration
      (2) movements of the precordium
      (3) for acoustical holography for
          the diagnosis of tumors, espec-
          ially, breast cancer
3. For laser Doppler techniques for measurements
   of circulation
4. For information handling with lasers coupled
   with computers, etc.

For the fields of computer information and rapid retrieval the
lasers are used in closed systems where there is no evident exposures.

In laser treatment of laser ophthalmology, the laser had been
used in thousands of patients for retinal detachments, for the
arrest of progressive diabetic retinopathy, for histoplasmosis of
the eye, central serous, macular edema and other conditions. The
laser has been used in surgery for precision surgery and relatively
bloodless surgery. This is especially true with the recent development of the high-output laser systems, the carbon dioxide laser
and the neogymium-YAG. With the new flexible transmission systems
there is great interest in this in the field of burn surgery, with
the excision of the escar and immediate graft replacement through
bloodless surgery. The laser has been used in man for the arrest of
bleeding in stress ulcers, for liver trauma, carcinoma of the
larynx and papilloma of the larynx, benign and malignant tumors
of the skin, melanomas have been mentioned before, incurable birth-
marks and tattoo marks with the increasing development of laser
surgery.

Laser weaponry is always a possibility for military use of the application of this. The high-output invisible infrared lasers, the increasing interest of the development of coherent x-ray lasers and laser-induced thermonuclear fusion for laser bombs, all indicate the concern about the use of laser weaponry.

Accidental exposure to laser radiation may occur with any open use of the laser systems. It may be inadvertent with the increasing development of atmospheric propagation of laser systems. Such atmospheric propagation is used in ranging, tracking experiments, communication and pollution detection studies by means of laser systems. The concern here is the chronic exposure of tne general public. At present, it is assumed, without adequate information, that such exposures are minimal and not of practical significance. It has been emphasized that too little is known about chronic exposure of the eye especially and more studies are certainly needed.

The apparent neglect of adequate financing for studies of the basic mechanism of the laser action on living tissue is felt most keenly in the field of the study of the long-term effects of laser trauma to tissues. With the increasing developments of laser technology it is no longer possible to neglect this particular field.

A plea has been made repeatedly for the establishment of a multidiscipline Laser Institute of Biology and Medicine. This will assure the rapid and economical progress of the accum lations of basic knowledge of the reactions in tissue of lasers to help develop new and more effective lasers, new safety programs to assure the increase of applications of laser and much needed help for man in biology and medicine.

ASSOCIATED TRAUMA WITH THE USE OF LASER SYSTEMS

For the development of a sophisticated instrument, as the laser, there may be other types of trauma associated with its use. Most important is the trauma from high-power electrical systems necessary for laser instrumentation. Deaths have actually occurred from electrical shock especially in the research laser laboratory. To approve the efficiency of the laser, cryogenics, as liquid nitrogen purges, are used and the hazards of this must be considered.

One of the types of lasers used is the so-called chemical laser where the chemical reaction provides the means for population inversion. At times, these may be highly toxic chemicals such as clenium oxychloride. This is of concern, then, in the accidents which may occur with the breakage of chemical laser.

Fig. 7. Linear scar of the back following Nd-YAG laser treatment 7 months previously: biopsies showed non-specific fibrosis with no tumor.

It is emphasized again that there is increasing development of studies on laser-induced thermonuclear fusion. In these laser systems trauma to tissues may occur from the laser itself, from the neutrons produced and from burns for the hot lithium envelope.

So, trauma with lasers may be associated with exposures, accidents, weaponry and with other types of trauma.

CARE OF LASER INJURIES OF TISSUE

There are no specific types of treatment or care of the laser injury of tissue. There is no evidence at present, that changes similar to those produced by x-ray do not develop from the laser in the tissues of animals or man. There is even no tendency that even chronic exposure or exposure of the new laser systems with nanosecond or picosecond causes the development of x-ray changes in tissue.

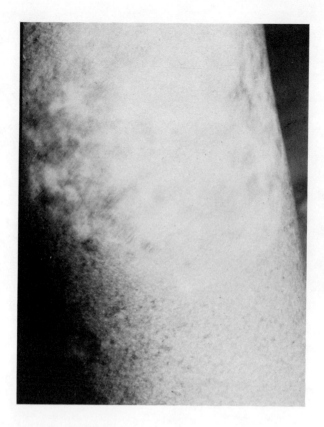

Fig. 8. Progressive gradual repigmentation of scar of the arm
following laser treatment for a tattoo mark.

The burn from lasers is similar to that of an electrical burn
but the laser burn heals more rapidly. For eye injuries, systemic
corticosteroid therapies may be used in all important organs to
attempt to minimize the reaction. Acute skin exposures are mostly
small areas and treated as any type of burn.

CONCLUSIONS

Laser trauma is then another form of man-made trauma which may
become more common and increasingly important for the future. Interest and knowledge of this new form of trauma will do away with
much of the hysteria and anxiety which is often associated with a
new technology.

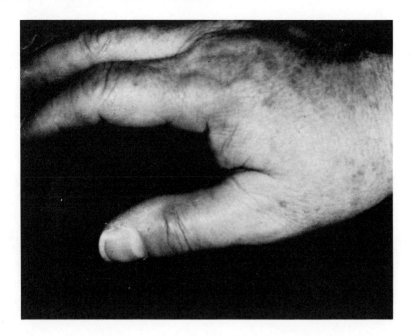

Fig. 9. Edema of the hand following explosion of a bottle of liquid nitrogen, accidental trauma in the Laser Laboratory of the Medical Center of the University of Cincinnati has occurred from electrical shocks and cryogenics, not from lasers.

A significant part of the laser reaction, including lasers in the far infrared with visible light lasers, the reaction is thermal coagulation necrosis usually more severe in the pigmented areas. From the pulsed laser systems other tissue changes, including non-linear effects, are possible. The reactions of the infrared laser does not differ qualitatively from the infrared-induced thermal burn. There is no evidence of carcinogenic activity observed in the past 10 years of repeated exposure to laser impacts. There is still much to be done in this new field of laser biology and medicine. Studies of the pathobiology of laser trauma in tissues should continue, especially with chronic exposure experiments. It may be said with a mixed metaphor technique used often to get attention in class, that from this brief review of new studies photons (and sonons) on tissue, that the field of laser trauma is indeed a virgin field pregnant with many possibilities.

REFERENCES

1. Goldman, Leon: Biomedical Aspects of the Laser, Springer-Verlag,
        New York, 1967.

2. Goldman, Leon and Rockwell, R. J., Jr.: Lasers in Medicine,
        Gordon and Breach Science Publishers, Inc. 1971.

3. Goldman, Leon: Applications of the Laser, Chemical Rubber
        Company (in press).

4. Goldman, Leon, Nath, Gunther, Schindler, Goerge, Fidler, James,
        Rockwell, R. J., Jr.: High-Power Neodymium-YAG Laser
        Surgery, Acta Dermato Venereo. 53:45, 1973.

5. Goldman, Leon: Some New Laser Systems on the Skin (in press).

EFFECTS OF IONIZING RADIATION ON HEMOPOIESIS

Eugene P. Cronkite

Medical Research Center

Brookhaven National Laboratory

INTRODUCTION

Radiation affects most remarkably those tissues of the body that are in a steady state of renewal, such as hemopoietic tissues, gastro-intestinal tract, skin and testes. Radiation effects may be subtle and not expressed for many years, or they may be acute coming on promptly and producing severe symptomatology related to the central nervous system, the gastrointestinal tract or the hematopoietic system. All radiation effects have distinct dose-effect relationships which are reasonably well established. The symptomatology will also vary, depending upon whether the entire body is exposed or whether only a portion of the body has been exposed. Dose rate, the type of radiation, the distribution of absorbed dose within the body, fractionation and oxygen tension of tissues are all factors that influence the intensity and the type of radiation effect that may be observed at various time intervals after the exposure.

DOSIMETRY

Prior to mid 1930's there was no satisfactory unit for the quantitative measurement of radiation to correlate with biological or chemical effects. Accordingly observations prior to late 1930's are qualitative. The basic principle of the scientific toxicology is quantitative measurement of the causative agent and the biological effect produced. In pharmacology, standardization of drugs becomes scientific only when the structure is known and one can measure the drug in appropriate units such as the milligram or some other quantitative reference. With ionizing radiation one is concerned

not with the mass of the agent administered, but with the amount
and distribution of energy that has been absorbed by tissue at the
point of interest, such as the hemopoietic tissue or gonads.

Two dose units are commonly used. The roentgen (R) is a measure of total dose in air of ionizing radiation, and is that quantity
of x-ray or gamma radiation such that the associated corpuscular
emission per 0.001293 g. of air at standard conditions produces in
air ions carrying one electrostatic unit of either sign. For energy
to be deposited, there must be an interaction with matter. Hence,
with x rays passing through a vacuum, no radiation dose is delivered.
In practice one is interested in the energy imparted to various
tissues from a number of different types of radiation, and it is
therefore essential to have a second unit of radiation which over-
comes the limitation of the roentgen. The second unit, the rad,
is the unit of absorbed dose, equal to 100 ergs/g. of absorbed
energy, which applies to any type of radiation in any tissue. For
practical purposes in small pieces of tissue, one R is close to one
rad. However, as irradiated objects become larger and change in
composition, one must consider the diminution in intensity due to
the interaction of radiation with matter (buildup and then exponential
attenuation) and the changing types of interaction (photoelectric
effect, Compton effect, etc.). In tissue this leads to a decreasing
absorbed dose at successive levels after equilibrium is attained.
However, at interfaces such as soft tissue and bone, the absorbed
dose may sharply increase. Thus, in addition to the exposure dose
in roentgens and the absorbed dose in rads, one must be concerned
with the distribution of the absorbed dose in the areas of interest.
For example, if there is sufficient protection of bone marrow by
shielding from one's own tissue to permit marrow regeneration and
survival from what is considered an otherwise fatal exposure dose,
this may lead one to erroneous conclusions about effects or therapy.

THE INFLUENCE OF DENSITY IONIZATION

The density of ionization or the number of ion pairs per unit
track length or electron volts deposited per unit track length varies
depending upon the type of radiation. For example, with alpha particles and high energy neutrons the number of ion pairs per unit
path length per erg absorbed is significantly greater. In general
with more ion pairs per unit path length there is a higher relative
biological effectiveness (RBE) of the radiation per ergs deposited.

THE DOSE RATE

In general, the lower the dose rate, the less will be the
acute somatic effect. Dose rates in excess of approximately

5 rads/min give essentially the same result. However, as the dose
rate falls below 5 rads/min, the effect per unit of radiation
becomes less. As the dose rates fall substantially, biological
repair processes become significant, both for somatic and genetic
effects.[1]

LEVELS OF EXPOSURE FROM NATURAL AND MANMADE SOURCES OF RADIATION

Living things have been exposed to radiation throughout the
entire evolutionary process. Natural radiation sources consist of
cosmic radiation that is partly galactic in origin with the major
part coming from the sun. Cosmic radiation consists primarily of
protons with energies up to 10^7 MeV, alpha particles and heavier
charged atomic nuclei (mainly C,N,O, Mg, Ca and Fe). The preceding
primary cosmic radiations interact with the atmosphere of the air
and produce secondary cosmic radiation. According to Libby[2], cosmic
radiation results in exposure of human beings to about 30-50 mrad/year
at sea level and about 5 times this at about 9000 feet.

Terrestrial radiation comes from natural radioactive materials
in the earth's crust, such as radium, radon, polonium, bismuth,
lead, potassium and thallium. In addition, carbon-14 and tritium
are created in the atmosphere due to cosmic radiation. In addition
to cosmic and terrestrial radiation at sea level, there are certain
building materials which contain higher concentration of radioactive
materials such as bricks and granite so that individuals living in
structures made of these materials have a further increment of
radiation exposure. There are also small amounts of radioactivity
that are in the diet and water, and become incorporated into the
tissues of the body; for example, potassium-40, tritium, and
carbon-14. In some areas of the world radium produces a further
increment in exposure.

To increase the sum total, there is also radioactive fallout
produced by explosion of nuclear bombs.

Natural radiation from all sources delivers an average of
125 mrem per year to the gonads and 122 mrem/year to the bone
marrow. Fallout radioactivity from nuclear bombs adds a few more
mrem/year.

The sum of the exposure from natural radioactivity and fallout
radioactivity, although delivered to essentially the entire world's
population, is small compared to the recommendations of The International Commission on Radiological protection for occupationally
exposed individuals in which the average exposure shall not exceed
5 rem in a year or about 50 times the natural radiation exposure.
In actuality those exposed in radiation industries receive only

about 0.5 rem/year.

At the present time radiation exposure of man is about half from manmade sources and half from natural unavoidable background radiation. Of the manmade sources, about 20% is through the medical use of radiation for diagnosis. The preceding doses of radiation from unavoidable background and manmade sources have not been shown to have any detectible effect. These annual doses are to be contrasted with a single dose of 25 rads to the entire body, the smallest dose that will produce just detectible effects on hemopoiesis. As the dose is further increased, the magnitude of the effect on homopoiesis, gastrointestinal tract, and other tissues increases proportionately. Again however one must contrast the effect of thousands of rads to a finger which will produce only local damage, to such doses to the entire body which would be fatal within a short period of time. The median lethal dose of ionizing radiation measured in the midline of the human body is about 300 rads.[3]

MECHANISM OF ACTION OF IONIZING RADIATION

There is a classic discussion on mechanism of action by Lea.[4] The concepts of direct hits on the target molecules versus the indirect action mediated through products of reaction with the protoplasmic solvent water are all important. Today the most acceptable current view to account for a major part of the biologic effects of radiation may be divided into three interlinked steps. First, photons or particles penetrate the protoplasm, interacting to produce ion pairs. This reaction takes of the order of 10^{-13} seconds. The second step is a primary radiochemical reaction of these ions primarily with water, producing free radicals such as H and OH. These reactions take about 10^{-9} seconds. These free radicals produce a further chain of reactions with themselves, tissue water and oxygen to produce further reactive forms such as H_2O_2 and HO_2 . These products persist for microseconds or in part a few seconds at the most. The last reaction is between these products and critical protoplasmic molecules. The nature of this last reaction is not known with certainty. Since the actual amount of energy imparted to the system is small, it is generally believed that the damage must involve substances of low concentration but major importance to the living system; for example, nucleic acids or possibly some enzymes. The average cell nucleus can be considered to have a mass of 270×10^{-12} gram. The DNA content of the average diploid cell is 6×10^{-12} g. or about 2% or the average nuclear mass. Of the average cellular mass it would be less than 1%. It is of interest that the amount of energy deposited to produce 100% mortality in animals would raise the body temperature by only 0.001^0C. Whatever the mechanism may be, it sets into motion a series of observable histological or chemical lesions that unfold with time.

In addition to the effects observable within days, "bad, invisible
genetic information" may be stored not only in the gonads, but also
in somatic proliferative or nonproliferative cells, presumably
in DNA, that may not be manifested as a disease process for several
years.

BIOLOGICAL EFFECTS AT THE CELL LEVEL[5]

Dose rate effect was mentioned earlier in a general sense.
If dose rate is sufficiently low as with background, it is believed
there is little or no effect. At least one cannot with certainty
prove an effect at levels of background or a few times higher.
However if dose rate is sufficiently high, depositing doses of the
order of several thousands of rads in a short period of time, actual
death of any living cell can be observed promptly in terms of
classical pathologic criteria of cell necrosis. However after
lower doses of radiation (25 to several hundred rads - precise values
vary with the tissue), only disturbances in cell proliferation are
seen. The rate at which cells divide is decreased. DNA synthesis
is impaired in two manners: (1) The rate of DNA synthesis is somewhat
slower generally, (2) cells may continue DNA synthesis and become
polyploid. It is reasonably certain that radiation has effects
other than the outright killing of cells and the interference
with mitosis and DNA synthesis.

The diminution in the production of new cells in tissues that
are undergoing continual renewal, such as the gastrointestinal
mucosa, the blood cell lines, the gonads, etc., results in a progressive hypoplasia to total atrophy, depending upon dose. Some
cells still capable of mitosis that are not killed outright may
be so injured that they will go through 1 or 2 generative cycles,
producing abnormal progeny, such as giant metamyelocytes and hyper-
segmented neutrophils, before dying. The atrophy of these steady
state cell renewal systems and direct injury of other tissues
produce clearly defined clinical syndromes which simplistically can
be divided into three categories.

For teaching purposes these acute radiation syndromes may be
classified generally as cerebral, gastrointestinal, and
hematopoietic.

The cerebral syndrome is produced by extremely high doses of
radiation, order of several thousands of rads, either directly to the
brain or to the entire body. It is always fatal, whether the
radiation is delivered to the brain alone or to the whole body.

The gastrointestinal syndrome occurs when the dose of radiation
is lower in the range of 600-1500 rads. It is characterized by

intractable nausea, vomiting, and diarrhea; these lead to severe
dehydration, diminished plasma volume, vascular collapse and death.
The symptomatology is caused by diffuse necrosis of the proliferating
cells in the crypts of the gastrointestinal tract followed by the
usual orderly migration of the mature cells off of the villus, leaving
denuded villi which is followed by extensive loss of blood and plasma
into the gastrointestinal tract which will result in death unless
there is massive fluid replacement promptly. After doses below
roughly 1300 r, regeneration of the gastrointestinal tract will
commence around the 6th day with complete restoration by 2 weeks.
However, the individuals are yet to experience the sequellae of
bone marrow aplasia.

To understand the effects of radiation on hematopoiesis, it is
necessary to describe the essential characteristics of the hematopoietic system. (1) The bone marrow is distributed in the child
throughout the body and in the adult limited to vertebrae, ribs,
sternum, pelvis, upper parts of the femur, and to a lesser extent,
the calvarium. (2) Continuous cell division in the bone marrow
provides steady supply of mature elements to the peripheral blood
constituting a steady-state cell renewal system. (3) The mature
cells produced by each cytologic subunit of the bone marrow find
their way into the common pool, the blood. (4) The marrow is
subdivided into 4 cellular systems producing granulocytes, erythrocytes, platelets and lymphocytes. The system is regulated by
feedback loops from the periphery that regulate the cell production
within the bone marrow. On an average in the hematopoietic cell
renewal system, the production of cells in the bone marrow is
compensated by the death or disappearance of cells in the peripheral
blood.

A most important radiobiological principle in understanding
radiation damage to the hematopoietic system is the law enunciated
by Bergonie and Tribondeau in 1906. They stated that cells which
divide frequently are radiosensitive and cells that do not divide
are radioresistant. This law holds remarkably well for most cell
systems. There are two notable exceptions - the exquisite radio-
sensitivity of the nondividing oocyte and small lymphocyte.

There are degrees of radiation injury in all proliferating
systems. With dose sufficiently high, cells can be killed outright
and evidence of severe damage can be seen within minutes to hours
following exposure even before the cells commence to divide. This
is described as "direct cell death." In most cell systems the
cell appears to function normally until it attempts to enter division. At this time damage may be seen, consisting of chromosome
damage and nuclear fragmentation. The nucleus may divide, the
absence of cell division leading to polynucleated cells. In some
instances the cells do not divide but continue to grow, producing

giant cells. The radiation sensitivity of the hematopoietic stem
cell has been determined by Till and McCullough to be of the order
of 100 rads. From the nature of the stem cell dose-survival curve
in the mouse, it can be shown that about 2-3 stem cells/thousand
remain capable of proliferation after exposure to an LD_{50}. If human
stem cells have the same dose-survival curve as the mouse, then
about 5 in 100 stem cells survive after the human estimated LD_{50}
of 300 rads.

The bone marrow cell renewal compartment consists of; a) a
stem cell pool, b) a proliferating differentiated pool for each
cell line, and c) a nonproliferating maturing compartment. A as
mentioned earlier is very radiosensitive. B being more mature is
somewhat more radioresistant and C is very radioresistant.
Therefore after exposure to radiation, a dose-dependent fraction
of the stem cells are killed and the surviving fraction remains
to reconstitute hematopoiesis in all cell lines. However there is
a dose-dependent time before the remaining stem cell fraction
commences to replete itself prior to differentiation. In this
interval, the cells in the dividing maturing compartment are very
rapidly reduced to small numbers resulting from their continued
maturation and release into the storage compartment and ultimately
into the blood. As the stem cell compartment repletes itself,
differentiation into each of the cell lines again commences at
various time intervals depending upon the intensity of the stimulus
from the peripheral feedback loops for differentiation of the common
stem cell into the particular cell line.

Another phenomenon that is observed is called the "abortive
rise." Although the abortive rise is not completely understood,
it is believed that the most probable explanation lies in stem
cells that are not grossly damaged, although they are incapable
of sustained proliferation. When they divide they produce a short-
lived progeny of cells that can be seen as the abortive rise in
neutrophils in the peripheral blood.

THE HEMOPOIETIC SYNDROME

With due consideration to the general principles of radio-
biology and characteristics of hematopoiesis, one can consider the
sequence of events in different cell lines. The hematopoietic
syndrome consists of the sequellae of bone marrow suppression.
It is produced by partial or complete aplasia of the hematopoietic
tissues. The aplasia results from two phenomena. First, the more
radiosensitive cells are killed directly. For example, lymphocytes,
erythroblasts and other cells within the marrow and lymphocytic
tissues can be found undergoing pyknosis, karyorrhexis and karyo-
lysis. In addition, those cells that are not killed outright may
suffer severe injury to their chromosomal material resulting in

aberrations of mitosis characterized by tripolar mitoses, chromosomal
bridges, fragmentation, micronuclei and other well-known cytologic
abnormalities. The result is a decreased production of new cells
that upsets the normal steady-state equilibrium between production,
utilization and senescence. When production ceases, as with total
aplasia of the marrow, the diminution in concentration of the
various elements in the peripheral blood is a function of the normal
lifespan of the cells involved. Thus, for example, the red cell
with a lifespan of approximately 120 days shows a diminution in the
red cell mass of about 0.83%/day unless accelerated by hemorrhage.
Platelets with lifespans estimated in the vicinity of 10 days
diminish slowly as long as megakaryocytes still remain in the marrow.
Upon disappearance of the last megakaryocyte, the diminution in the
platelet count is more abrupt and approaches zero within about 10
days after the disappearance of the last megakaryocyte. Granulocytes
have a half-life in the bloodstream of about 6-12 hours with a
random disappearance from the bloodstream. Thus after high doses
of radiation, which eradicate the progenitor cells for granulocytes,
there is a very sharp diminution in the granulocyte count in the
peripheral blood as the preformed pool of cells in the marrow are
extruded into the bloodstream and utilized. Granulocytes attain
minimum values in about 5 days under these conditions. After lower
doses of radiation that do not totally eradicate granulocytopoiesis,
there is a slower diminution in the granulocyte count as production
is impaired and the peripheral level decreases and fluctuates with
"abortive rises." It may take as long as 30-40 days for the minimum
granulocyte levels to be reached. During this period of time,
abnormal granulocytes characterized by hypersegmentation and, at
times, nuclear satellites are observed in the blood and marrow.

The destruction of lymphocytes is dose-dependent up to about
300 rad. Further increases in dose decrease the peripheral level
of lymphocytes no further. The diminution in peripheral concentration of lymphocytes develops rapidly and is maximal by about
48-72 hours after exposure with the bulk of the diminution occurring
within 24 hours after exposure. Concomitant with the atrophy of the
lymphocytic tissues and the developing lymphopenia, there is severe
diminution in the capability of new antibody production and suppre-
sion of the anamnestic response.

As a result of the progressive granulocytopenia, thrombopenia
and lymphopenia, a typical symtomatology develops. The individual
becomes susceptible to infection as a result of the severe granulocytopenia and the impairment of antibody production. Within about
7 days, infections from commensal or exogenous pathogens may develop.
Ulcerations of the oral mucosa are characteristic and may be scattered
throughout the gastrointestinal tract to the anus, producing a second
bout of gastrointestinal signs and symptoms. When the platelet
count falls below 25,000/mm^3, a tendency to bleed develops. If the
thrombopenia becomes worse with platelets falling below 10,000/mm^3 ,

the purpura will be severe, and fatal hemorrhage into the brain or myocardium may develop. Hemorrhages into the bowel will accentuate GI dysfunction and may act as a focus for intussusception. The bleeding greatly accelerates the onset of anemia which may be fatal unless treated with adequate blood transfusions. The clinical picture is similar to that seen from aplasia of the marrow due to any cause and its treatment is identical.

In the real world the probability of having totally homogeneous whole body radiation which results in a uniform pattern as described above, is very unlikely. Several human radiation casualties have been observed from radiation accidents in industry or in the laboratory. These have all been marked by inhomogeneities of the absorbed dose.[7] In some of the individuals, many thousands of rads were received by the hands with almost total destruction of the epithelium. Superficial skin over the abdomen and the face also received tremendous amounts of radiation with the absorbed dose grading off toward the back and down to the legs which had some shielding. Very severe gastrointestinal injury was observed in two of these individuals. In the fatal cases, there was severe injury to the gastrointestinal tract and marked necrosis of the skin of the hands and the face, with prominent edema and erythema of the side of the body facing the criticality event. Of course in addition there was severe hematopoietic depression, as described above.

In another criticality accident,[8] the dose to the head and the thorax was measured in the thousands of roentgens. Immediately after the exposure, the individual was disoriented, amnesic and shortly thereafter went into collapse and was prevented from immediate death by extensive fluid and electrolyte replacement. He died within a short period of time and at autopsy there was serious injury to the myocardium in addition to the brain.

In another incident, a group of individuals were exposed to x rays being emitted from an unshielded Klystron tube that was not functioning properly. The technicians trying to adjust the radiofrequency output received serious radiation exposure from unfiltered 200 kV x rays. The dose to the face and upper part of the body was severe. The details of this incident are of considerable interest.[9] The dose distribution to the head of one individual showed marked inhomogeneities in the absorbed dose. The initial neurological symptoms were minimal but later severe neurological symptoms developed, demonstrating that neurologic injury per se is not necessarily fatal unless the absorbed dose is uniform throughout the head and greater than the minimum necessary to produce a neurologic syndrome in animals.

Lastly, in the event of nuclear warfare and widescale radioactive fallout, there will be marked inhomogeneities in the absorbed dose as observed in the one fallout accident.[10] In this case fallout material was deposited directly upon the skin in some individuals. In these people, the dose to the first few mm was greatly in excess to the whole-body dose. These individuals developed severe beta burns. On the day of exposure, they experienced severe itching and burning of the skin which subsided within a period of 24-36 hours. Thereafter, for a period of about 10 days, there were no symptoms referable to the skin. After the 10-day latency period, itching and burning of the skin returned with spotty pigmentation that increased in size and coalesced to form large plaques. This was followed by either dry desquamation, comparable to sunburn in some individuals, or to ulceration of the exposed areas in others. It is difficult to assign a meaningful number for the dose to the superficial layers that were contaminated by fallout. However, it is most likely in excess of 2,000 rads at the level of the Malpighian layer of the skin.

The above incidents clearly demonstrate that in accidents involving ionizing radiation, one should expect marked inhomogeneities in the absorbed dose. Accordingly, the symptom complex observed may not necessarily follow the oversimplification presented under the three classic symptom complexes induced in animals by uniform whole-body irradiation exposure. After radiation accidents one should be on the alert for the unusual and should, after the health physicists work out the exposure conditions and distribution to the body, be alert for signs and symptoms to develop in the organ systems most heavily exposed.

THERAPY OF RADIATION INJURY

Therapy of all disease processes is based on proved clinical principles that apply to the signs and symptoms that are present in the course of the disorder. Accordingly, there is no mystery about the management of radiation injury. Standard principles are involved and one treats the nausea, vomiting with standard therapy. Similarly electrolyte balance is also managed in the established way. Granulocytopenia and thrombopenia per se need no therapy. Severe granulopenia will dictate reverse isolation and possible administration of prophylactic wide spectrum antibiotics in addition to nonabsorbable intestinal antibiotics. Hemorrhage and infection are treatable by fresh platelet transfusions and antibiotics respectively.

Very high exposure groups in which the probability of survival is very low require considerations of bone marrow transplantation from HLA compatible individuals. Discussions of this are beyond the purview of this chapter.

Table I - Chance of Serious Injury or Death - Per Year*

Auto accident (disability)	1 chance in	100
Mortality attributable to smoking	1 "	175
Cancer, all types and causes	1 "	700
Auto death	1 "	4,000
Air pollution	1 "	10,000
Fire death	1 "	25,000
The "Pill" death	1 "	25,000
Drowning	1 "	30,000
Firearms	1 "	50,000
Electrocution	1 "	200,000
Airplane round trip, New York City - San Francisco and return. Radiation cancer risk	Less than 1 "	1,000,000
Reactor emanations; site boundary (5 to 10 mrem/yr)	Less than 1 "	1,000,000
Average for population within 50 miles of reactor	Less than 1 "	10,000,000

*From Bond[13]

LATE EFFECTS OF RADIATION

All types of leukemia with the exception of chronic lymphocytic leukemia, cancer of the female breast, lung cancer, thyroid
cancer, cancer of the bowel, and possibly other cancers have been
observed following exposure to high doses and high dose rates.
Admittedly there are few data. However the data are sufficient to
allow reasonable predictions of what would happen in an irradiated
population exposed to high doses at high dose rates. The data on
which the following statements are based is published by The United
Nations,[11] and The National Academy of Science.[12] These data lead
one to the conclusion that there will be one to two excess leukemias
per 10^6 people per year per rem exposure for at least a period of
20-25 years after the exposure. The incidence of other types of
cancer is of the order of 7 or 8 times that of leukemia alone.
Accordingly, one can say that there will be approximately 10 excess
cases of cancer per 10^6 people per year per rem exposure. Whether
the incidence will be less with chronic exposure is not known,
however, it is very likely that it will be.

In concluding a discussion on radiation injury in a conference
on pathobiology of trauma, it is worthwhile contrasting radiation with
other traumatic events in modern life. Bond[13] has collated information on probability of serious injury or death from numerous
factors in modern life. These are listed in Table 1 as probability
per year. The probability of injury in an automobile accident is
1 in 100 per year. Example from table show the probability decreases
1 in 4,000 deaths from autos, 1 in 25,000 deaths from the "Pill,"
1 in 50,000 from firearms, and 1 in 200,000 from electrocution.
Radiation from existing possible sources is the least hazardous.
For example, if each person in the United States took one round
trip transcontinental jet flight per year, the additional exposure
of 1 millirem per year assuming linearity and no threshold at
equilibrium for the 200,000,000 population would increase the
probability of cancer by less than 1 in 1,000,000. This is not to
say that radiation can be ignored. The hazard potential is present.
The study, caution, surveillance and regulation that has characterized development of nuclear industries and power production
are commendable. They are models for application for control of
other hazardous factors of modern life and in new industrial
processes.

REFERENCES

1.   Russell, W.L., The effect of radiation dose rate and frac-
       tionation on mutation in mice. In: Repair from Genetic
       Radiation Damage. Ed. by F. Sobels, Pergamon Press
       (Oxford) 1963, pp. 205-217, 231-235.

2.   Libby, W.F., Proc. National Acad. Sci. 45, 959-976, 1959.

3.   Cronkite, E.P., Radiation injury in man. Chap. V in The
     Biological Basis of Radiation Therapy. Ed. by E.E.
     Schwartz, J.B. Lippincott Co., Philadelphia and Toronto,
     1966.

4.   Lea, D.E., Action of Radiation on Living Cells. MacMillan Co.,
     New York, 1947.

5.   Bond, V.P., T.M. Fliedner, and J.O. Archambeau, Mammalian
     Radiation Lethality. Academic Press, New York and
     London, 1965.

6.   Till, J.E. and E.A. McCullough, A direct measurement of the
     radiation sensitivity of normal mouse bone marrow cells.
     Rad. Res. 14, 213-222, 1961.

7.   Hempelmann, L.H., H. Lisco, and J.C. Hoffman, The acute
     radiation syndrome: a study of nine cases and a review of
     the problem. Ann. Int. Med. 36, 279-510, 1952.

8.   Shipman, T.L., A radiation fatality resulting from massive over-
     exposure to neutrons and gamma rays in Diagnosis and
     Treatment of Radiation Injury. pp. 113-133, WHO,
     Geneva, 1961.

9.   Howland, J.W., M. Ingram and H. Mermagen, The Lockport
     Incident: Accidental partial body exposure of humans to
     large doses of x irradiation in Diagnosis and Treatment
     of Radiation Injury. pp.11-26, WHO, Geneva, 1961.

10. Cronkite, E.P., V.P.Bond and C.L. Dunham, Some Effects of
     Ionizing Radiation on Human Beings. U.S. Atomic Energy
     Com . TID 5358, Supt. of Documents, U.S. Gov't Printing
     Office, Wash., D. C., 1956.

11. Ionizing Radiation: Levels and Effects. A report of the
     U.N. Scientific Committee on the Effects of Atomic
     Radiation to the General Assembly. Vol II, 1972,
     New York and Geneva.

12. The Effects on Populations of Exposure to Low Levels of Ionizing
     Radiation. Report of the Advisory Committee on the
     Biological Effects of Ionizing Radiations. (BEIR report),
     Division of Medical Sciences, National Acad. of Sci., NRC,
     Wash., D. C. , 1972.

13.   Bond, V.P.   Radiation effects data and their interpretation.
        Presented: American Bar Association, National Institute/
        Environmental Litigation, Dallas, Texas, 12-14 April 1973.

THE PROBLEM OF ACUTE SEVERE TRAUMA AND SHOCK

Robert M. Hardaway III

William Beaumont Army Medical Center

In the twelve years since January 1961, the United States has suffered the loss of 45,937 killed in Southeast Asia plus another 303,622 wounded. This makes the war in Vietnam the fourth most bloody war in United States history, placing behind World War I, World War II, and the Civil War.

However, wars mercifully and hopefully have an end. Other causes of bodily injury unfortunately seem to be on the steady increase. For instance, in the single year of 1970 there were 55,200 fatal automobile accidents or nearly 10,000 more fatalities in the USA due to auto accidents alone than were killed in the entire twelve years of the Vietnam War. In 1970 alone there were 5,100,000 injuries due to accidents alone -- almost 17 times as many as were wounded in twelve years in Vietnam. One out of every eight hospital beds in the USA is occupied by an injury case. There were ten and one-half million disabling accidents in 1965 with over 400,000 permanent disabilities. Accidents are the leading cause of death in the USA between the ages of 1 and 37. These young people are the future of our country. Experts have estimated that 25 per cent of the 800,000 persons who die each year in the USA of medical and surgical emergencies would not die if even present knowledge were generally available. [6] In the USA in 1969 alone there were 116,000 fatal accidents, almost three times the entire twelve year toll for the Vietnam War. In the USA in 1969 alone there were over 50 million injuries (of which over 2,000,000 were hospitalized), more than 177 times the entire total for the twelve year Vietnam War. About 50% of both automobile injuries and deaths and other traumatic injuries and deaths involve the use of alcohol. An increasing number are associated with the use of other drugs. Motorcycles are rapidly increasing in number and have an accident rate over four times that

of automobiles. All vehicles have a death rate of 4.7 deaths per
100 million miles. For motorcycles the figure is 20 deaths per
100 million miles. Since the Vietnam War started there have been
1,128,408 accidental deaths in the USA. This is 25 times the number
of Americans killed in Southeast Asia. Four hundred five thousand
eight hundred fifty were killed by automobiles alone. This is over
11 times the number of Americans killed in Southeast Asia. In the
12 years of the Vietnam War there were over 600 million injuries
in the USA. This is 2,000 times as many as Americans who were
wounded in Southeast Asia during the same time. Compared with the
civilian toll of killed and wounded, the U. S. Armed Forces toll in
Vietnam is a drop in the bucket.

Of course, the best treatment for this tremendous problem is
prevention. But I will leave this important and difficult task for
others and talk about what we can do after the injury has taken
place. The Army Medical Department has had a great deal of experience in the treatment of trauma and has devoted a great deal of
effort to the study of wounds and their treatment. In this it has
been eminently successful. The U. S. Army Medical Service in the
Korean War was the best in history up until that time. There was
tremendous improvement over the excellent medical care provided in
World War II. Umlimited use of whole blood was universally available. The mortality of hospitalized wounded was reduced from 4.5%
to 2.5%. Highly qualified surgeons, modern equipment, advanced
surgical techniques, and many other factors seemed to make any
future improvement difficult, and unlikely to produce any very
dramatic change. However, this was not the case. The improvement
of results in Vietnam over Korea was a dramatic break-through.

It is estimated that over 22,000 lives have been saved in Vietnam, in addition to those who would have survived if the medical
care had been only as good as the highly efficient medical care
given the wounded during the Korean Conflict. This estimation is
as follows: The ratio of hospitalized wounded to number of men
killed in action in Korea was 2.8 to one. In Vietnam it is 4.15
to one. This is in spite of Air Force and other aircraft casualties
which have had many more killed in Vietnam. If the Korean War ratio
had prevailed during the present conflict, the number of battle
deaths would be about 48 per cent higher. [1] There have been 45,937
killed in Vietnam. There would, therefore, have been 48 per cent,
or 22,050 more killed in Vietnam if the medical care had been the
same as that available in Korea which was the best in history up to
that time.

This dramatic achievement was largely brought about by the more
prompt and efficient treatment of hemorrhagic and traumatic shock.
Certainly the most important factor was the speed with which the
patient could be treated.   In World War I Santy observed several

hundred wounded on whom he had documented the time which elapsed between the reception of the wound and definitive treatment in a hospital. The mortality was as follows:

Hours Intervening	Mortality %
1	10
2	11
3	12
4	33
5	36
6	41
8	75
10	75

With the availability of prompt helicopter evacuation in Vietnam, the time between wounding and effective treatment was reduced to one to two hours. (In Korea it was four to six hours.) Actually, most individuals who are going to die of wounds will do so within four to six hours, if they are untreated. Wounded, who in any other war would have died on the battlefield, arrived at medical facilities alive, although barely so. However, with prompt and effective treatment they lived. With the exception of massive wounds of the head and those killed outright, most all of the wounded now survive.

Speeding ambulances are not the answer to the speed-in-treatment problem as they usually further endanger the patient and others.

The availability and financial resources to supply helicopter service to civilian hospitals are severely restricted. However, a serious attempt to accomplish this with military assistance is being made at the present time in San Antonio, Texas. Another approach to the problem is bringing treatment facilities to the patient by means of specially equipped and staffed truck ambulances modified into a special treatment facility. This has been done very effectively in Heidelberg, Germany. It is possible to give effective treatment for shock from a regular ambulance if a qualified individual is available in the ambulance and I. V. fluids, oxygen, and other equipment is available in the ambulance.

The influence of time on shock is dramatically illustrated by case reports of two soldiers wounded in Vietnam. A 29 year old white male was injured by a hand grenade at 0910 on 4 March 1967 near the 3d Surgical Hospital in Vietnam. He suffered a traumatic amputation of both legs below the knee. Within five minutes, at 0915, he was admitted to the 3d Surgical Hospital. His measured blood loss was 2600 ml. It was estimated that he lost an additional 2000 ml of blood during cleaning and examination of wounds, at x-ray, and during surgery. At 1010, his blood pressure was 120/80 mm Hg and central venous pressure was +7 cm H_2O. He had been given 4000 ml blood and 1000 ml of Ringer's lactate. Thus, his total measured blood loss was 4600 ml. His replacement was about 4000 ml of blood plus 1000 ml of lactate. He thus received only 400 ml more fluid than he lost, and not quite as much blood, but he still responded well to treatment, and withstood anesthesia and surgery well.

Contrast this with 30 year old white male who was wounded at 1020 on 12 February 1967, north of Bien Hoa, Vietnam. As a result of mortar fragments, he suffered intra-abdominal injury to the left kidney, spleen, vena cava, colon, and small bowel. He was admitted to the 3d Surgical Hospital at 1120, one hour later. On admission, his blood pressure by cuff was zero, pulse 160, central venous pressure zero, hematocrit 32. His measured blood loss was 3000 ml. He was immediately given 2000 ml of Ringer's lactate and 4500 ml of blood. He lost an additional 2000 ml during inspection of wounds, transportation to surgery, and during surgery. During surgery, which started at 1230, he was given an additional 3000 ml of blood and 1500 ml of Ringer's lactate. This makes a measured total loss of 5000 ml of blood, with replacement of 7500 ml of blood plus 3500 ml or Ringer's lactate, or a total replacement of 11,000 ml. This is 6000 ml more intravenous fluid replacement than the measured loss, all within five hours of injury. At the completion of this, his blood pressure was 100/60, pulse 112, and central venous pressure +8. Delay in treatment will necessitate the administration of much more blood and fluid than was lost.[2]

A new type of surgical patient of great importance appeared in Vietnam. This is the critically wounded patient suffering rapid blood loss from vascular or organ injury who, under other circumstances, would have been "killed in action." Now he is delivered to a hospital shortly after injury. He may arrive with unobtainable blood pressure, lactate levels of 100 mg per cent or more, and severe metabolic disturbances. He may be in severe alkalosis or acidosis. His serum pH may be anywhere from 6.9 to 7.7. One will not know what it is if it cannot be tested. Many patients are in marked respiratory alkalosis. The massive addition of sodium bicarbonate to these patients only makes this worse and overloads the patient with sodium. His pO_2 may be 40, but there may be no clinical evidence for this and no respiratory injury. He may suffer a cardiac arrest from this factor alone, if it is recognized. The patient is on the verge of death and, if effective treatment is not immediately

forthcoming, will die. This type of patient has seldom before
appeared in any military or civilian hospital. The resuscitation
of these patients requires much more sophisticated laboratory and
treatment facilities that have ever been used in a combat zone, or
even in a stateside teaching hospital.

The prompt and scientific treatment of shock is thus the key to
survival after severe injury. This has been the object of extensive
research work both in the military services and in civilian insti-
tutions. The following is based on intensive study and treatment
of shock in a special shock study unit at Walter Reed Army Medical
Center, Washington, D. C. and by a research team from the center
working in Vietnam.[2]

Classification of Shock

Shock is defined as inadequate capillary perfusion. It may be
classified into reversible, refractory, and irreversible shock.
(Fig. 1).

Reversible Shock. This is further subdivided into early reversible
shock (stage of vasoconstriction); late reversible shock (stage of
capillary and venule dilation); and refractory shock (stage of disseminated intravascular coagulation). The first two respond relatively well to volume administration.

Early reversible shock: (vasoconstriction). This stage is
characterized by elevated catecholamine levels and evidence of
vasoconstriction. Blood pressure may be elevated, normal, or low.
Blood pH is often high. This shock is easily treated by a modest
amount of blood or other fluid.

Late reversible shock: (capillary and venule dilation produces
expansion of vascular space). This is characterized by a delay in
treatment or more severe hemorrhage or injury. If 1500 ml of blood
is lost and the resulting shock left untreated or inadequately
treated for a period of time, it may require much more than 1500 ml
of blood and fluid to bring conditions back to normal. This is due,
in part, to expansion of the vascular space (Fig. 1). Blood pressure is low. There is continued vasoconstriction and evidence of
decompensation of blood pressure maintaining mechanism. It is
relatively easily treated by adequate volume administration, which
often amounts to more than the volume lost. When adequate amounts
of fluid volume are given, the vascular space is filled, and the
catecholamine level decreases.[4] (Fig. 2) With the fall in catecholamine level, a normal physiological vasodilation takes place,
opening up arterioles and producing an adequate capillary flow. If
this does not take place, as may occur in refractory shock, vasodilation may be indicated.

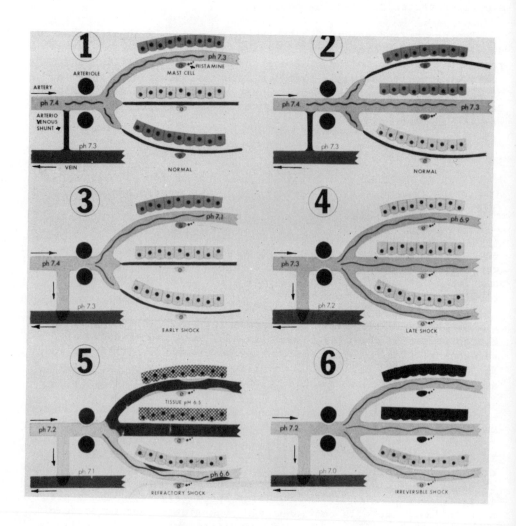

Figure 1

FIGURE 1. Stages in the development of Irreversible Shock. Diagram
represents a small artery ending in an arteriole with sphincter-like
action. Arteriole feeds three capillaries, each nourishing a group
of cells.

   a. Upper left. Normal condition. Arteriole is fairly widely
opened. Only one capillar is being perfused while others rest.
Capillaries open in rotation on demand of cells adjacent to them.
Histamine secreted by mast cells along the capillaries cause capillary sphincters to open. Constant perfusion is not necessary. Blood
flow through the capillaries is rapid; pH drop along the capillaries
is minimal. Cells of middle capillary are becoming slightly anoxic.
Arterio-venous shunt is closed.

   b. Upper right. Normal condition. Center cells have become
slightly anoxic and mast cells have secreted histamine causing central capillary to open. Upper capillary is now closed as mast cells
have stopped secreting histamine. Arteriole is fairly widely opened.

   c. Second row left. Shock, Phase I (Reversible). Elevated
catecholamines have caused vasoconstriction. The arteriovenous
shunt has opened as a part of the catecholamine action. There may
be a loss of arterial blood pressure due to poor venous return to
the heart. All of these cause a slow capillary flow in the upper
capillary.

   d. Second row right. Phase II Shock (Reversible). Due to the
long time required for adequate perfusion by the cells because of
the slow capillary flow, all capillaries are now open. Capillary
flow is extremely slow and pH fall along the capillary is marked
due to anaerobic metabolism with lactic acid production.

   e. Lower left. Phase III Shock (Refractory). Stage of disseminated intravascular coagulation (DIC). Stagnant acid blood in
presence of sepsis or tissue injury has begun to coagulate, causing
sticking of red cells -- occluding upper two capillaries. This stops
perfusion in these capillaries completely. Cells nourished by these
two capillaries are dying. Blood through the lower capillary is
sluggish. Circulating blood is incoagulable. Lyses of clotted
blood in upper capillary has already started due to endogenous
fibrinolysin.

   f. Lower right. Phase IV, Irreversible Shock. Cells nourished
by upper two capillaries have died producing focal tissue necrosis.
Capillary clots have lysed and circulation has been restored. However, large areas of necrosis are producing multiple organ failure
which result in death of the organism.

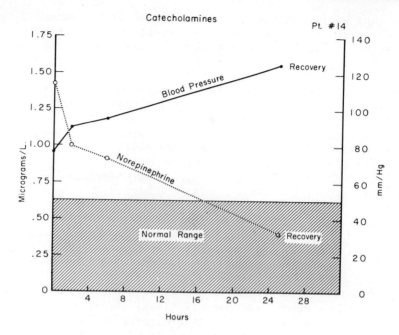

Figure 2

Refractory shock: This shock is difficult to treat and responds poorly to routine treatment by blood, plasma expander, electrolytes, or other substances in the usual amounts. It is shock usually complicated by marked trauma, infection, neglect, heart failure, kidney failure, liver failure, pulmonary failure, etc. It is almost always associated with disseminated intravascular coagulation [2] (DIC). Its onset is marked by the sudden occurrence of a clotting defect. It is difficult to treat and requires large volumes of fluid given under central venous pressure control and frequently vasodilators, in addition. Vasodilators must be given only in the presence of a high central venous pressure.

Irreversible or fatal shock. The line of demarkation between refractory shock and irreversible shock is not sharp. Clinically it should never be assumed to have taken place until death supervenes. It is associated with progressive cellular death and necrosis in various vital organs such as the kidney, liver, heart and lungs, which advances to organ failure. Because of large natural safety factors, considerable necrosis can be tolerated, without organ failure or death. Perhaps three-quarters of normal kidney or liver parenchyma may be destroyed without producing organ failure. If natural safety margins are decreased by renal disease, liver disease, heart disease, etc., a much smaller amount of cellular death will be sufficient to cause organ failure. Cellular death starts in the refractory shock phase and, only when relatively advanced, does organ failure result. Even with considerable organ failure, effective treatment (renal dialysis, etc.) may enable the organ to regenerate, if tissue necrosis can be halted by restoring capillary perfusion. However, even if one organ can be tided over (kidney by dialysis), death may result from failure of other organs (liver, heart, lungs, etc.)

The following procedures are indicated in the treatment of all types of severe shock including hemmorrhagic, traumatic, and septic, or combination of these:

1. Placement of a central venous catheter for pressure measurements and fluid administration. Supraclavicular insertion into the subclavian vein is a very satisfactory method (Fig. 3). The tip of the catheter should be in the right atrium, or, at least, in the chest. This is used to (a) obtain a venous blood sample, (b) administer intravenous fluids, (c) measure central venous pressure. Administer intravenous fluid volume until arterial blood pressure rises to normal, and until central venous pressure rises to about five cm H_2O. This is discussed below.

2. Maintain an adequate air way. Nasal O_2 may be needed for a short time. Respiratory assistance if arterial pO_2 is below 70 mm Hg. Tracheal intubation or tracheotomy should be carried out and connected to a respirator. Careful respiratory management is essential. A volume cycled respirator may be required. "Shock lung" is also a "stiff lung" and considerable pressure may be necessary to insure adequate ventilation. O_2 concentration should be kept as low as possible to maintain pO_2 of 70 mm Hg. A respirator with room air is often adequate. The most common cause of death in adequately treated shock is acute pulmonary failure, due to "shock lung" (congestion, hemorrhage, edema, fibrin desposition).[4]

3. Monitor ECG.

4. Insert indwelling urethral catheter to monitor urinary output and as an index of organ perfusion.

5. Maintenance of normal blood pH. Although $NaHCO_3$ may be used if pH is very low, the best way of treating metabolic acidosis is to restore capillary perfusion by adequate IV fluids. In early hemorrhagic or traumatic shock, however, pH is usually high due to excess blowing off of CO_2 by hyperpnoea. The pH should be frequently measured.

6. Administration of one mg/Kg of Dibenzyline in 100 to 200 ml of fluid, if central venous pressure reaches 15 cm H_2O when arterial pressure is still inadequate. Isuprel may be given as a vasodilator, but it should not be given if the pulse is over 120 as it is an inotropic agent and cardiac arrhythmias may be produced. Regitine, a short term vasodilator, may be used in lieu of Dibenzyline. A vasodilator should never be used in the absence of a high normal central venous pressure (10–15 cm H_2O).

7. Generally accepted medical and surgical procedures are essential. These include the prompt control of hemorrhage by ligature of bleeding vessels (by laparotomy if necessary), the surgical drainage of abcesses, appropriate antibiotic therapy and specific treatment for complications such as cardiac arrest, acute renal failure, bleeding or perforated stress ulcer and hemorrhagic diathesis. Patients should be kept at relatively normal temperature, with the aid of automatic cooling and warming blankets. Vasoconstrictor drugs have little or no place in the treatment of shock. [2] Steroids are yet unproven as to value. Endogenous levels of both of these types of substances are already very high in shock, if adrenals are normal.

Intravenous fluid administration is the single most important factor in the treatment of any type of shock. There are two vital questions in proper volume administration: (1) how much fluid? and (2) what kind of fluid?

The amount of fluid required is a much more complicated problem than is at first apparent. It cannot be dismissed as merely "replacing lost volume." Not only is it usually impossible to estimate or measure lost volume, but the total blood volume required in shock is often more than a "normal" blood volume.

The most important method of determining the volume requirements in shock and burns is the accurate determination of central venous pressure. This is the pressure in the right atrium or great veins of the chest. It is not a measurement of blood volume. It is not even a measurement of volume requirements. It merely indicates whether the heart is capable of pumping additional blood volume, and if adequate blood volume is being presented to the heart for its pumping action. There are a number of methods of inserting a central venous catheter. It may be inserted into almost any peripheral vein. The important thing is that the tip of the catheter must be placed within the chest cavity (but not in the right ventricle), preferably in or near the right atrium. Leg veins are less desirable because of the length of the catheter needed and the increased danger of infection. Arm veins are better, but are usually collapsed in shock and may be difficult to fill. Jugular veins are satisfactory. Pro-

bably the best vein is the subclavian. It is large, always present
in a constant position, and easily dilated by a slight Trendelenberg
position. This position is also important to prevent air embolism.
The vein may be approached either from below or above the clavicle.
Probably the supraclavicular approach is the easiest and is productive of the fewest complications. [2,5] It has proven very satisfactory for routine use in Vietnam in over 800 procedures, with only
two small and inconsequential pneumothoraces and an occasional minor
ecchymosis. The "intracath" is recommended for convenience. We
prefer to use the 14 guage two inch needle with the 18 gauge radio-
opaque tubing. The 12 inch variety most frequently reaches the
right atrium.

TECHNIQUE OF CATHERIZATION

The patient's neck is first prepared with antiseptic solution
and a point one cm posterior to the junction (Fig. 3) of the lateral border of the sternocleidomastoid muscle and the clavicle is
infiltrated with two cc of one-half per cent Xylocaine. The operator, using sterile gloves, then pulls the 14 gauge needle from the
intracath set and attaches it to a sterile 10 cc syringe containing
two cc of saline. The skin puncture is made, and one cc of the
saline in the syringe is injected, to clear the needle of any tissue
plugs. With gentle aspiration, the needle is advanced at a 45 degree
angle between the clavicle and the sternocleidomastoid muscle, in
the horizontal plane. The approach is from above the clavicle with
the patient's head tilted slightly away from the side of the procedure, and the feet of the patient elevated six to eight inches to
prevent air embolus and to distend the subclavian veins. When the
vein is punctured, blood is freely aspirated and the plastic tubing,
still within its sheath, is inserted and threaded down, leaving about
three inches of catheter on the skin. The sheath which is provided
with the catheter should be slipped in place over the end of the
needle to prevent shearing of the catheter. The needle hub and the
catheter, using tincture benzoin compound, should be taped in place
and may be sutured there.

It is important in preventing catheter emboli that the needle
barrel and initial polyethylene tubing be fastened with one piece of
tape. The catheter is connected to a three-way stopcock. One arm
goes to a water manometer and the other to a bottle of intravenous
fluid. Fluid is administered until it is desired to measure intravenous pressure. Then the stopcock is turned to divert fluid into
the manometer until it is high. The stopcock is then turned, to
connect the catheter with the manometer. The level in the manometer
then falls to the central venous pressure. The tip of the catheter
must be in the chest, preferably in or near the right atrium. If
the catheter tip is in an arm vein, inferior vena cava, right ventricle, or other extrathoracic sites, readings will be meaningless or,

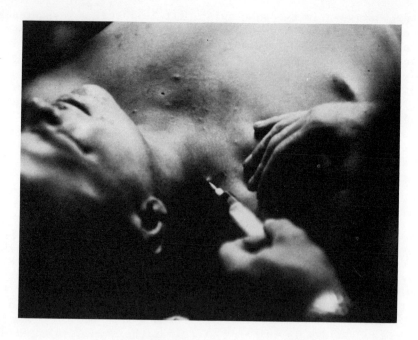

Figure 3

even worse, will give false information.

Taking "0" level (level of 3-way stopcock) at the patient's mid-axillary line and reading, usually in centimeters of water, we interpret zero to five cm H2O as low, five to ten cm as normal, 10-15 cm as high normal, and above 15 cm as evidence that pulmonary edema may be imminent. The catheter is used to measure central venous pressure, administer intravenous fluids, obtain venous blood samples, and for phlebotomy, if necessary. The catheter should be cleared by withdrawing blood before taking a sample. When administering intravenous fluids, note that it takes time for the system to equilibrate, so that not only are serial readings more useful than single ones, but when giving large volumes of fluid rapidly they should be given in 300-500 cc increments, with 15 to 20 minutes between volumes, to see if the venous pressure changes. Of course, pressure measurements are taken intermittently, when the intravenous fluid is interrupted. If no intravenous fluids are being given, the catheter must be kept open with a slow drip of any solution, to which 15 mg of heparin and 10 mg of one per cent Xylocaine per 1000 ml are added.

Several points in the proper evaluation of central venous pressure need emphasizing. Vasopressors artifically elevate the central venous pressure. Accurate central venous pressure measurements must be taken after vasopressors have been discontinued. Stopping vasopressors lowers central pressure. Additional intravenous fluid may then be given, if indicated. Blood pressure can be maintained as high or higher than with the use of vasoconstrictors by following this procedure (Fig. 4).

Care should be taken that the tip of the central venous pressure catheter is in the chest. If it is in a peripheral vein or the inferior vena cava, the reading is of no value. Care should be taken that the tip of the central venous pressure catheter is not in the right ventricle. If it is, it will read too high.

Central venous pressure is usually taken in centimeters of water. In general, readings of less than five cm H_2O indicate inadequate venous return to the heart and inadequate blood volume. Twenty-five per cent of the patient's normal body volume may be deficient and the body still maintain a normal system blood pressure. Therefore, a normal blood pressure is maintained by catecholamines and vasoconstriction. If the patient is subject to spinal or other types of anesthesia, the pressue may drop precipitously (Fig. 5). He should be given fluid until the central venous pressure is adequate. Readings of five cm H_2O to 10 cm H_2O indicate adequate venous return to the heart. If systemic blood pressure is then normal, no more volume is required. However, if systemic blood pressure is inadequate, it is relatively safe to give more fluid slowly and intermittently up to a level of 15 cm H_2O. There are exceptions to this, however. When adequate blood volume is obtained and the right atrium is receiving adequate return, the normal right heart pumps this volume into the pulmonary artery. If vasopressors and capillary thromboses are still present and have not been relieved by physiologic vasodilation, the pulmonary pressure may rise precipitously and pulmonary edema may result, even in the presence of a normal central venous pressure of eight cm H_2O.

It may be only with right heart failure that the central venous pressure rises. This points up the importance of slow and intermittent volume administration, when the central venous pressure is above six cm H_2O. Readings of 15 cm of H_2O and above are too high and pulmonary edema may be produced. Fluid volume administration should be halted. If a vasopressor is being given, it should be stopped. Attempts may be made to improve cardiac action by adequate oxygenation, digitalis or inotropic agents. A vasodilator such as Dibenzyline will lower the central venous and pulmonary artery pressure. Isoproterinol may be helpful. It may be necessary to remove blood, but usually this is less desirable and unnecessary. The most accurate but difficult index of volume administration is pulmonary artery

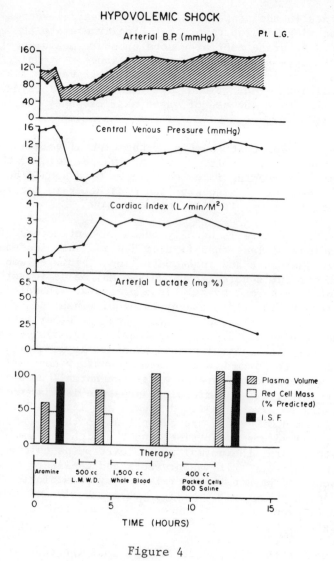

Figure 4

pressure, but this is possible only in a special shock study unit.

In addition to the central venous pressure, other criteria of volume administration may be used. Pulmonary artery pressure is the most accurate method of determining how much fluid may be safely given intravenously. Of course, this requires cardiac catheterization but many intensive care units now can do this without x-ray

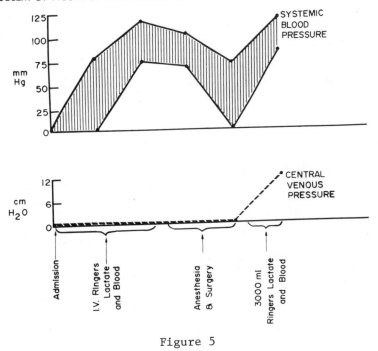

Figure 5

control, using the new flexible catheter and careful pressure and other controls. Urine output, systemic blood pressure, and other clinical determinations are helpful. Lactate levels are a good index of adequate tissue perfusion. Signs of peripheral vasoconstriction, such as pale, sweaty skin, thready pulse, etc., indicate the need for intravenous fluids. The difference between rectal and skin temperatures, if high (3°C.), indicates vasoconstriction. If the difference is low (less than 1°C.), fluids are not indicated. The indications of over-hydration may come on very suddenly. Of course, frequent auscultation of the lungs should be done, to detect early pulmonary edema. Chest x-ray should be done when indicated.

Blood volume measurements, either by tagged red cells or radioactive albumin, are essentially useless in determining volume requirements in acute shock. However, the determination of red cell mass is the best index of red cell requirements, and the detection of occult blood losses.

Lack of urine secretion or rapid weight gain should not deter the fluid administration, if the central venous pressure and other indications are followed. The old policy of using urinary output and

overt and insensible fluid and blood loss as a measure of fluid volume requirements has no place in the treatment of shock.

Regarding the kind of fluid, initially this is of less importance than the amount required. Perfusion of the tissues is all important, even if the type of perfusate is inadequate. Dogs' brains
can do without oxygen for twice as long as they can do without perfusion. Therefore, fluid, any fluid, should be started immediately,
to buy time for obtaining the proper kind of fluid, e. g., crossmatched blood, electrolyte, etc. However, for severe or prolonged
shock, the type of fluid given becomes increasingly important. In
general, no one fluid is the magic fluid for the treatment of shock.
It is a matter of tailoring the types of fluid to the needs of the
patients to produce and maintain normal blood constituents.

Blood: Except in unusual circumstances it is better to wait for
properly crossmatched blood rather than use uncrossmatched blood.
This is because (1) any fluid is satisfactory for a while. (2) Most
uncrossmatched blood produces a mild amount of hemalysis which in a
normal individual is harmless, but in a low flow shock state may
produce disseminated intravascular coagulation, renal failure, and
other undesirable complications.[2] Much more blood may be required
than was lost.[2] It may be desirable for a while in the acute shock
state to accept a slightly low hematocrit of 35 or so to promote
faster flow in the microcirculation. It is possible that more red
cells per minute will pass through a capillary if this is the case.

Crystalloids: There has been much recent debate on the role of
buffered saline and other solutions in the treatment of shock. Ringer's lactate has been widely and successfully used in Vietnam. Experimentally, it is possible to replace a very large proportion of
a dog's blood with Ringer's lactate and save the animal from hemorrhagic shock. Certainly any intravenous fluid is life saving. Ringer's lactate will raise blood pressure, lower lactate levels and
benefit acidosis, by its volume and flow promoting effect. However,
all crystalloids are quickly lost from the circulation, even though
they may be retained longer than usual in hypovolemia. There is
nothing magical about them in shock treatment. They do, however,
have an essential place in the treatment of shock. They are needed
to replenish depleted water and electrolytes, both in the blood and
in the extracellular fluid. Blood pH should be monitored. It is
commonly thought that all shocked patients are acidotic. However,
this is not the case. Many, if not most cases in early shock, may
be alkalotic, with pH's of up to 7.6 or over. The reason for this
is not entirely clear, but may often be due to hyperpnea with blowing off of CO_2, which may almost disappear from the blood. This
high pH is in the face of lactate levels which may become very high.
The giving of Na bicarbonate empirically to these patients is certainly contraindicated, and may result in an Na overload. Later in

shock, the pH falls and may reach very low levels. There may be
some benefit in giving bicarbonate. However, the best way of
treating this metabolic acidosis is to improve capillary flow by
adequate volume administration and vasodilators, if necessary. This
rapidly lowers the lactate level.

The administration of fluid volume in shock should have as its
goal: (1) an adequate blood volume based on central venous pressure measurements; and (2) a normal red cell mass, adequate protein
and colloid, the replacement of deficiencies in extracellular fluid,
the correction of electrolyte and pH abnormalities in the blood and
supply of caloric requirements. No one fluid is adequate, but a
combination of different fluids, tailormade to keep a normal blood
composition, is needed. Protein and other colloids are essential
to maintain water in the blood and prevent edema. Whole blood, colloid, and electrolytes are all needed in proper amount and proportion.

The use of vasodilators in shock is an extensive subject and can
not be discussed in detail here.[2] Their use is limited to refractory shock in the presence of a high central venous pressure. Refractory shock is defined as that shock which does not respond with
an adequate systemic blood pressure and urinary output to an adequate
blood volume, oxygen and other supportive therapy. They are most
useful in septic shock or shock complicated by infection, massive
tissue injury or hemolysis. They routinely produce a rise in systemic blood pressure (never a fall if given with a high CVP), a
great increase in cardiac output, a marked diuresis, a fall in peripheral resistance, a fall in mean transit time, and rise in stroke
volume (Fig. 6 and 7). They will lower both pulmonary artery pressure and central venous pressure. In reversible shock vasodilation
in both lungs and systemic arterioles takes place automatically as
catecholamines fall with proper fluid replacement. However, in refractory shock when disseminated intravascular clotting is present,
or when extremely high levels of lactic acid are present, the arterioles may not automatically open with adequate blood volume and a
vasodilator may be required. The best vasodilator is Dibenzyline
in a dose of one mg/Kg in 200 ml of saline, given over a one or two
hour period. It is an alpha-sympathetic blocker. It causes complete
paralysis of the sympathetic endings for a period of 24 hours or
more and is easily given in one or two hundred ml of any IV fluid
over a half hour period. If hypovolemia is present and the blood
pressure falls, it should be halted and more IV fluids given.

Although digitalis has been recommended to lower CVP, the chances
are that the high CVP is due to a high pulmonary pressure secondary
to a pulmonary microcirculation, partially obstructed by vasoconstriction and capillary thrombi. This obstruction is opened up by
vasodilators.

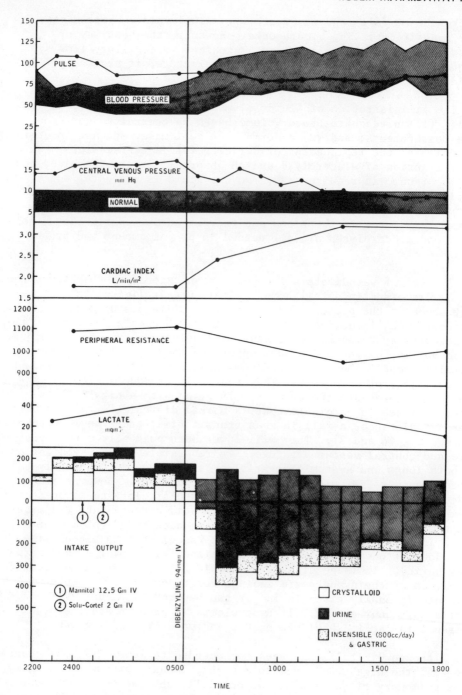

Figure 6

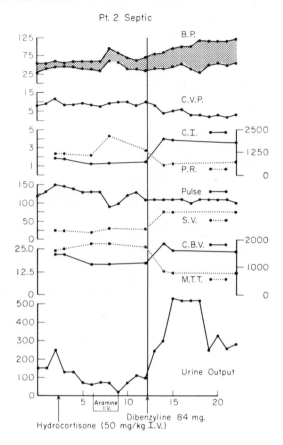

Figure 7

Isoproterenol is useful as a vasodilator, but has an inotrophic action on the heart. It is a beta sympathetic stimulator and tends to speed the heart, making it less effective if the pulse rises to 140 or so (Fig. 8). It should not be used if the pulse is over 100 or so.

Regitine is a reasonable substitute for Dibenzyline as a vasodilator, if the latter is not available. It is also an alpha blocker and has no direct effect on the heart. However, it is harder to administer as it requires a continuous intravenous drip.

The problem of acute respiratory failure, or shock lung, is perhaps the most difficult one to combat of any aspect of shock. It is probably related, at least in part, to disseminated intravascular

TACHYCARDIA DUE TO ISOPROTERENOL

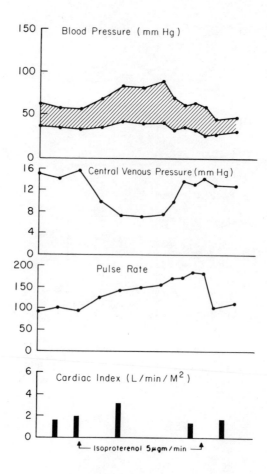

Figure 8

coagulation. Other factors which may play a part are fluid overload, 100 per cent O_2 administration over a long period, fat and tissue embolism, loss of surfactant, arteriovenous shunting, and others. It is characterized by low pO_2 levels in the presence of low pCO_2 levels. There is a deficient ability to absorb oxygen, but no difficulty in excreting CO_2. The pO_2 levels may be extremely low, down to 30 mm Hg, without clinical evidence of the fact. Cardiac arrest may occur due to anoxemia, and be totally unexpected, unless determinations of pO_2 are made. In fact, cardiac arrest and "wet lung" or "shock lung," are the two most common causes of early death among hospitalized wounded in Vietnam.

SUMMARY

The most important treatment for all types of shock is the prompt and adequate administration of intravenous fluid in amounts governed by informed interpretation of central venous pressure. This amount may greatly exceed normal blood volume or fluid losses. Any fluid is adequate for emergency short term use. When available, blood, electrolyte, or other fluid should be used to maintain normal constituents of the blood. If vasopressors are being given, they should be carefully discontinued. Central venous pressure will fall and more fluid can be given.

Low blood oxygen tension may be present without clinical evidence. This must be adequately treated. If blood pressure and urinary output do not respond to the above treatment, as may occur in sepsis, vasodilators often produce beneficial results. They should be given only in the presence ofcentral venous pressure near the upper limits of normal. Under these circumstances they never cause a fall in systemic pressure. There is essentially no indication for vasopressors in the treatment of shock. The usefulness of steroids is unproven.

REFERENCES

1. Hamilton, E. L.: Director of Medical Statistics Agency, Office of The Surgeon General, U. S. Army, Washington.

2. Hardaway, R. M.: Clinical Management of Shock. Charles Thomas, Springfield, Ill., 1968.

3. Matsumoto, T., Hardaway, R. M., Heisterkamp, C. A., Pani, K. C., Leonard, F., and Margetis, P. M.: Individually Disposable Spray Unit (H-Bomb) for Tissue Adhesives. Amer. Surg., 33:671, 1967.

4. Hardaway, R. M.: Intensive Study and Treatment of Shock in Man. J. A. M. A. 199:779-790, 1967.

5. James, P. M., Bredenberg, C. E., Lentsky, S., Anderson, R. W., Collins, J., and Hardaway, R. M.: Central Venous Pressure: Its Use and Abuse. Amer. Med. Assn., Clinical Congress, New York, 1970.

6. Safar, Peter. Editorial, Critical Care Medicine. 1:48, 1973.

ENDOCRINE ASPECTS OF TRAUMA

Maynard E. Jacobson

Department of Medicine

The University of Minnesota

INTRODUCTION

Trauma has a generalized efffect on the whole organism or man in addition to well recognized direct effects on specific tissues. The nervous and endocrine systems coordinate and integrate a whole organism response to these generalized effects of trauma. These are in part efforts to facilitate and coordinate for the organism protection, stabilization and adaptation to environmental change. The adaptation response of the organism to bring about adjustments to meet environmental change was a concept first developed and organized as a thesis by Hans Selye of this conference. It remains a cornerstone of our concepts. Effecting such adaptations is the major purpose of the nervous and endocrine system in response to trauma or any stress.

These two systems have many similarities but the primary difference is that the nervous system transfers its signal by a network of neurons with synaptic linkage. The endorgan must be intact with this neuron linkage for a response. The endocrine systems elaborate a hormone that is transported by the systemic blood circulatory system to distant target sites. The response requires usually stored hormone release into the circulation, synthesis and release to blood concentrations commensurate with an appropriate degree of response, an intact circulatory system, and eventually diffusion to hormone receptor sites. These hormone receptor sites may be common to all cells of the organism or only specific cells with specific receptor sites for a specific hormone. The nervous system response is immediate and the hormone system response is slower and more sustained.

As we consider the hormonal system as it facilitates adaptation
to trauma it is not possible to ignore the integral role that the
central nervous system (CNS) plays. It is often the only sensor
system that detects and initiates a signal to some endocrine systems
in response to trauma. The control of that response may be integrated
and modulated by many complex nervous functions. The same nervous
system can also integrate the response with other effectors such
as smooth muscle, skeletal muscle, and exocrine glands.

The nervous system interacts with the endocrine system in
essentially three types of control of endocrine function:

1. secretion of hormones by neurons into the circulatory
    system as in the case of antidiuretic hormone (ADH);
2. secretion of "neuroendocrine" hormones into a selective
    circulatory network as in the case of the secretion of
    hypothalamic releasing or inhibiting factors into the
    portal-hypophyseal vessels in the control of the anterior
    pituitary hormones;
3. the direct innervation of gland cells in the case of the
    adrenal medulla.

As we concern ourselves with the management of trauma in man,
a given hormone response may be impaired by diseases of or associated
trauma to the operative mechanisms of the sensory, integrating and
mediating nervous system.

The endocrine responses to trauma are significantly variable
in pattern, kind, degree and even what hormonal system that responds
to a select type of trauma. Obviously the immediate patterns of
response to a broken leg, hemorrhage, whole body burn, or stressful
visual stimuli are vastly different. Furthermore, the initial
response may effect changes that subsequently may induce secondary
and tertiary responses, habituation that modulate subsequent responses,
and development of a new "steady state" that in itself may subsequently change. It becomes too complex to comprehend unless we
attempt to categorize these by hormonal systems and areas of functional response.

The specific aspects in the role that the endocrine system plays
in trauma will be divided into major primary responses, secondary
and interrelated aspects and miscellaneous aspects.

MAJOR PRIMARY ENDOCRINE RESPONSES TO TRAUMA

We shall consider as primary endocrine responses to trauma
the following areas:

A.   The Hypothalamic-Pituitary System

B.   1. CRF   ACTH   cortisol

     2. vasopressin

B.   Sympathetic-Adrenal Medulla System

C.   Renin-angiotensin-aldosterone system

                The Hypothalamic-Pituitary System

     The response of the hypothalamic-pituitary system is one of
the most complicated and extensively investigated areas in endo-
crinology. It is of major importance in trauma. Within the hypothalamus are numerous neuron centers with defined mass of neurons
with specific endocrine related function and are referred to as
hypothalamic nuclei. Many ill-defined centers and pathways are
also present. As it relates to direct relationships to the pituitary gland these nuclei are found chiefly in the ventral hypothalamus.[1] At the hypothalamic level it operates by neuro-endocrine
mechanisms integrated by numerous neural pathways with the limbic
system, reticular formation, brain stem, and subcortex. These
pathways may provide the sensory input for stimulation, inhibition
or modulating control of pituitary hormone release. Some stimuli
such as endotoxins may have a direct effect.

     The hypothalamus as one of the most primitive parts of the
nervous system is now recognized to effect its endocrine action
by neuro-secretion of multiple short chain polypeptides known as
releasing and inhibiting factors that control the six established
hormones of the anterior pituitary.[2] Separately oxytocin and vasopressin are secreted by neurons whose axons end primarily in the
posterior pituitary. The six releasing and inhibiting factors
now established (of which two have been synthesized) are:

     1.   corticotropin releasing factor (CRF)
     2.   thyrotropin releasing factor (TRF)
     3.   growth hormone releasing factor
     4.   follicle-stimulating hormone releasing factor (FSHRF)
     5.   luteinizing hormone releasing factor (LRF)
     6.   prolactin inhibiting factor (PIF)

These factors are released into a special vascular system called
the portal-hypophyseal system with drainage into the anterior
pituitary gland. As these factors reach the anterior pituitary
gland they stimulate the release into the circulation and synthesis
of a corresponding hormone of the anterior pituitary. The function
of release is closely related to synthesis. The exception is

prolactin inhibiting factor (PIF) which acts to inhibit the release
of prolactin from the pituitary. Thus in man decreased PIF facilitates human prolactin (hPr) release and secretion in the pituitary.
Thus some forms of injury and disease of the hypothalamus with
an intact pituitary gland can manifest itself by galactorrhea.[3]
There continues to exist some controversy as to the discrete existence
of other inhibiting factors or whether such inhibiting influence
is modulated with the hypothalamus to reduce releasing factor
secretion.

The stimuli that elaborate these factors are varied and many.
As trauma is a stimulus, the pattern of factors secreted by these
hypothalamic "transducer" neurons vary with the stimulus. In turn
the hormones of the anterior pituitary released into the circulation
by these factors vary. There are numerous examples in biology
of various stimuli inducing a variety of responses. In man hypoglycemia results primarily in the release of adrenocorticotropic
hormone (ACTH) and growth hormone (hGH).[2,4,5] Hypovolemic shock
results in release of ACTH and vasopressin (ADH).[4] Pain, surgery,
and strenuous exercise releases ACTH, hGH, prolactin (hPr) and
ADH.[6,7] Isolated hyperosmolaremia releases only ADH.[4] Emotional
stimuli may release only ACTH.[5] Some stimuli may inhibit release.
Sleeplessness inhibits hGH release. Many forms of stress may
inhibit release of gonadotropins when stimulating the release of
other factors.

As conceived by Hans Selye, stress is in concept nonspecific.
In turn many stimuli creating a response may be varied and
nonspecifically induce a response. Nevertheless, the hormones of
primary importance in trauma are adrenocorticotrophic hormone
(ACTH) and vasopressin (ADH). We shall consider these individually.

Adrenocorticotrophic hormone (ACTH). ACTH is released by
three primary means. These are: (a) circadian modulation as
controlled by CNS influence above the ventral hypothalamus, (b)
negative feedback servo mechanisms mediated by glucocorticosteroids
of the adrenal gland, (c) stress. Circadian influence is the
primary operative mechanism in the nonstressed basal state of man
with a regular diurnal wake-sleep activity pattern.[8] The highest
levels are reached before awaking and lowest levels with onset of
sleep.

Isolated negative feedback may be a selective stimulus. This
is illustrated by ACTH release to decreases in circulating glucocorticosteroids as induced with metyrapone (an 11β hydroxylase
inhibitor or cortisol synthesis) or adrenalectomy.[9] Decreased
levels of glucocorticosteroids acting primarily on the hypothalamic
level induces corticotropin releasing factor secretion and subsequent ACTH release. This rise in circulating ACTH in turn increases

adrenal production of glucocorticoids. This control develops a long
loop servo mechanism for control as these increased glucocorticoids
feed back to the hypothalamic-pituitary system in accordance with
some set point control level. The converse response can be seen
with high levels of glucocorticosteroids or the administration of
biologically active glucocorticoids such as dexamethasone. The
circadian modulation of this response is a time-phase suppression
sensitivity on the release of ACTH. Dexamethasone in man in modest
dosages, as 1.0 mgm orally at the awaking phase (8:oo a.m.) has a
trivial impact on ACTH release. This is at the height of circadian
ACTH release. In contrast, dexamethasone has a pronounced effect
when administered in the same dose at the nadir of ACTH secretion.
The subsequent expected rise in ACTH release may be completely
suppressed.[10] Clinically, any glucocorticosteroid therapy program
which in doses and dosage administration pattern exerts this suppre-
sive effect over a prolonged period of time does produce ACTH
suppression and adrenal atrophy. This may be of profound clinical
significance in trauma if adrenocorticoid response is necessary in
the adaptation response.

Trauma induces ACTH release by the mechanism of stress.[5] In
man, this mechanism can override the other controlling mechanisms.
This can be demonstrated by the observation that circadian modulation
and short duration dexamethasone suppression are overridden as
disease and trauma is increasingly severe and acute. The subsequent
elevated circulating corticosteroid levels may show no detectable
circadian or suppression influence.

The central nervous system above the level of the ventral
hypothalamus can augment, inhibit or block a response to some forms
of trauma.[4] Extremity denervation, section of the lateral spinothalamic tract of the spinal cord and higher CNS lesions have been
demonstrated in animals to block cortisol response to trauma isolated
to that extremity.[11] Quadraplegia or spinal anesthesia in man may
inhibit the cortisol rise in response to abdominal surgery.[12]
Endotoxin stress acts directly on CRF release and does not require
higher CNS mediation. At higher CNS levels the limbic system
appears to have modulating effect on stress response with both
inhibition and augmentation effects.[4,13] In some animals the
hippocampus has an inhibiting influence and the amygdaloid nuclear
centers an augmenting influence. As a stressful stimulus becomes
repetitive or acute trauma persists, gradually a modulating effect
appears reducing the continued high output response or repetitive
response.

The primary site of action of ACTH is on the adrenal cortex
initiating the synthesis and release of glucocorticosteroids.[14] In
man this hormone is primarily cortisol. In addition ACTH can be
demonstrated at levels above maximal cortisol response to augment

the output of a second hormone of the adrenal cortex called aldosterone.[4] There is a valid widespread view that ACTH has no significant effect on aldosterone secretion. This may be true in the
usual physiological operative level of ACTH. The exception is acute
severe trauma producing very high levels of ACTH. At such levels
ACTH can induce a degree of hyperaldosteronism.

One extra-adrenal effect of ACTH and shared by growth hormones,
which is of some interest in the adaptive responses to stress, is
the mobilization of free fatty acids. This effect is often referred
to as a lipolytic effect. This may be helpful in utilization of
fat and conservation of glucose and proteins as energy sources for
adaptation. ACTH has also been demonstrated to impair liver hydro-
lating enzymes that inactivate cortisol.[15] This may be the major
explanation for the very prolonged half-life of cortisol that has
been observed in dying patients.

As noted earlier, the well delineated and extensively investigated ACTH response in man is on cortisol synthesis in the adrenal
cortex. Cortisol is one of the primary hormonal messengers to
facilitate adaptation to stress. There have been extensive reviews
of the adrenal cortex and action of steroid hormones.[14,16,17] As
trauma is a major insult to man cortisol secretion rates may increase
from basal levels of 20-30 mg/24 hours to 300-400mg/24 hours. As a
given dose of ACTH is repetitive or continuous its graded response
in cortisol synthesis increases and adrenocortical hyperplasia
develops. This is noted as successive days of ACTH stimulation
in man is associated with successive increments in cortisol secretion.[14] This is why adrenocortical hyperplasia is a frequent autopsy
finding after prolonged stressful illnesses. This phenomenon
adds to the responsiveness of the adrenal cortex to stress. With
ACTH, cortisol is secreted at a trickle rate. Cortisol is a hormone
in man essential for survival. Hypophysectomized or chronically
ACTH suppressed animals or man may not increase cortisol in response
to trauma.[4] The importance of this cortisol effect is well recognized in biology by the lethality of trauma in animals and man without
cortisol.[15] Survival with low minimal levels of cortisol secretion
may be possible in the basal unstressed state. However, stress in
terms of trauma, disease, or deprivation may become quickly lethal.[18]
The specificity of cortisol is recognized by the protection it affords
from such potentially lethal stimuli.

Trauma may place special needs on cellular matabolism of a
nature that requires continued or increased cellular needs.
Deprivation or increased energy needs may depend upon these needs
substantially from protein and fat. Glucose and glycogen are a
limited source. Thus gluconeogenesis and fatty acid mobilization
become important for what may be markedly increased cellular needs
with trauma. While cortisol shares this function with several

hormones such as growth hormone, epinephrine, glucagon, it does promote gluconeogenesis. In the absence of cortisol, gluconeogenesis in spite of other operative mechanisms does not provide enough glucose to avoid hypoglycemia. The other intracellular needs during trauma are less well understood. Cortisol binds with specific steroid proteins complexes which are found in the cell cytoplasm and nucleus. Our current concept speculates that they thus influence ribonucleic acid formation(mRNA) as a mechanism to ultimately accomplish specific cellular activity. <u>On this basis increased cortisol secretion in major trauma is an essential adaptation response.</u> The clinical administration of cortisol in augmented doses in man is of specific importance with stress of trauma or disease when man cannot be expected to appropriately respond to that stress with an increase in ACTH or cortisol secretion.[18,19,20] An intact hypothalamic-pituitary-adrenal system in general is effective in secreting cortisol at high enough rates to effect protective and adaptative advantages. There may be some exceptions which has led to therapeutic controversy in selected conditions. Usually even in endotoxin shock or the dying patient high circulating levels of cortisol are observed.[21] Whether even higher levels could effect greater benefit is uncertain. As an exception low circulating levels of cortisol have been observed in end stage exhaustion in man after acute high altitude exposure. Other controversial aspects include the use of massive doses of glucocorticoids considerably above any physiological capacity to secrete cortisol as a therapeutic measure in potentially lethal shock syndromes.[22]

There have been widespread and varied tissue effects of glucocorticoids that can be described in advantageous terms in the adaptative process to acute trauma.[17] The tissues which have been described to have <u>cortisol receptors include fibroblasts</u>, lymphoid cells, liver, <u>brain, heart, intestines, muscle, skin, and fat.</u> In some tissues metabolic pathways which utilize glucose and others that synthesize protein and fat are inhibited. The liver specifically increases gluconeogenesis. These are effects geared to conserve and provide a continuous availability of glucose. Many of these functions parallel or potentiate cyclic AMP in some cells. These characteristics may be associated with the observed synergistic effect corticosteroids have with catecholamines.

Cortisol has a major effect on suppression of the inflammatory reactions and suppression of selected immunological responses.[17] Cortisol decreases heat, fever, tissue swelling, inflammatory edema, vascular extravasation, cellular migration and fibroblastic proliferation. Cortisol has been described as stabilizing the lysosomal membranes which could modify tissue destruction. Though cortisol does not inhibit antibody production at ordinarily attained physiological levels, cortisol is associated with lymphoid involution. This includes observations with stress-induced levels of

cortisol. The younger lymphoid so-called B cells and thymus-
derivative T cells may be cortisol sensitive and completely destroyed
by a shut-off of energy generating metabolic pathways utilizing
glucose. Older mature lymphoid cells are resistant. By this
means cellular immunity of the delayed hypersensitivity type and
development of a new antibody clone of cells can be modified.
Acutely this may be useful in hypersensitivity and inflammatory
disorders.

Some of these effects on selective target receptors can be
considered disadvantageous. As a generalization, the disadvantages
of excessive cortisol become increasingly apparent as trauma or
disease becomes increasingly chronic and high levels of cortisol
persist continuously. On a long term basis some effects of cortisol
may enhance the invasiveness of infection and inhibit such tissue
repair as wound healing. On a long term basis excessive cortisol
produces hypertension, impaired collagen formation in skin and
connective tissues, osteoporosis, muscle wasting, and many other
findings associated with Cushing's syndrome. These are not usually
seen with acute or intermittent stress.

In addition cortisol can inhibit sleep, has a permissive effect
on catecholamine induced bronchodilatation, inhibit action of
parathormone on bone and vitamin D on the gut, and enhance the
pressor effect of catecholamines. It may also enhance hepatic
synthesis of angiotensinogen and exert a modest mineralocortocoid
effect. The common clinical observation of multiple gastric
ulcerations and bleeding in the stressed patient may be due to the
adverse effect of cortisol on enhancing cholinergic secretion
of H^+ by gastric mucosa coupled with an effect on gastric mucin and
gastric mucosa. Recently in experimental animals exogenous cortisone
plus immediate abdominal surgery resulted in a high incidence of
gastric ulcers and perforation. When the same cortisone administration was delayed to the 50th post-operative day, infrequent ulceration
was observed.

It is not possible to provide here a more detailed and exhaustive
review of cortisol effects that may play a role in trauma. It does
appear that cortisol plays an important role in maintaining the
circulatory system, modifying the inflammatory response, modifying
cellular destruction, and altering some cellular metabolic pathways
to facilitate a survival. These effects are protective and adaptative in nature and with major stress the increased cortisol is
essential for survival.

Vasopressin or antidiuretic hormone (ADH). The second major
hormone of the hypothalamic-pituitary system is vasopressin or
antidiuretic hormone (ADH). ADH is really a neuroendocrine hormone
synthesized in ventral hypothalamic neurons and stored in the

posterior pituitary (neurohypophysis) for release into the systemic
circulation.[1] Release is controlled by two primary mechanisms:
(a) volume receptors located in the carotid system and atria of the
heart with a vagal afferent pathway to the hypothalamus, (b)
osmo-receptors in the central nervous system. Thus hypovolemia
and hyperosmolality of the blood are the usual stimuli that release
ADH.[4] Water dehydration, hemorrhage, positive and negative pressure
ventilation and orthostasis may all invoke ADH release by these
mechanisms. Selective forms of trauma to the CNS, thoracotomy or
thoracic trauma may also produce a form of inappropriate secretion
of ADH. Other stimuli that may be a recognized part of trauma that
can also release ADH include pain, visceral afferent stimulation,
anoxia, and some drugs.

     The only response of vasopressin of certain physiological
significance is the inhibition of water diuresis by activating
cyclic AMP induced water reabsorption in the collecting tubules
of the kidney. Certain renal-vascular disorders as circulatory
shock or acute renal tubular necrosis can remove the importance or
target response of ADH. Nevertheless, the importance of ADH in the
protective and adaptative response to trauma is obvious. These are
basically to conserve water in body fluids, facilitate circulatory
adequacy and maintain the cellular environment. Such a syndrome can
produce dilutional hypoosmolemia, urinary hyperosmolality, sodium
diuresis chiefly mediated by an undefined "third factor" and
inability to tolerate further water loading. In an advanced stage
water loading may produce mental confusion, coma and death. Such
a state requires water restriction.

                    Sympathetic-Adrenal Medulla System

     Much of the sympathetic system is solely neurosecretory in
nature and not a part of classical hormonal physiology. The parasympathetic and sympathetic system activity is mediated by a neuron
network producing acetylcholine and/or norepinephrine serving
directly as neural "transducers" or "effectors." Sympathetic innervation of the heart and blood vessels provides an important regulatory
system for rapid cardiovascular adjustments to a variety of stressful
situations, such as exercise, shock, temperature regulation, emotional
stress, hypoxia, etc.[25] Its adrenergic activity is mediated by the
release of norepinephrine as a neurotransmitter. It has an activation effect on cardiac rate and myocardial contractility. It has
a vasoconstrictive effect on blood vessels in skin, adipose tissue,
splanchnic bed and kidneys. These effects would facilitate a
greater proportion of cardiac output that would be distributed
to particularly the heart and brain.

     Epinephrine and norepinephrine are stored in chromafin granules
of the adrenal medulla. Sympathetic neurons also enervate that

adrenal medulla and in this way becomes a part of the endocrine
system. These hormones are released by mechanism with activation
of the sympathetic nervous system. These provide a more generalized
and prolonged systemic response to the hormonal activity of epinephrine and norepinephrine. In general they facilitate the responses
attributed to the sympathetic nervous system. Clinically there is
no notable deprivation effect from removal of the adrenal medulla
with total adrenalectomy. This deprivation is compensated for by
the sympathetic nervous system. Norepinephrine does leak into the
circulatory system from its neuron synaptic sites.

A complex concept of the action of these hormones has been
developed by the formulation of an adrenergic amine receptor site
concept. Alquist has proposed that there are two types - alpha and
beta adrenergic receptor sites.[25] According to this concept epinephrine is a beta adrenergic agent to skeletal blood vessels causing
vasodilatation; norepinephrine and epinephrine cause vasoconstriction
in the skin and kidney by stimulating alpha-receptors. Alpha-
adrenergic blockers such as phenoxybenzamine (Dibenzyline) and
phentolamine (Regitine) antagonize only vasoconstrictor effects;
whereas the beta-blockers such as propranolol (Inderal) antagonize
only vasodilator action. The beta blocks are also found to antagonize the positive chronotropic and inotropic effects of catecholamines on the heart. The cardiac responses are referred to as
being mediated through beta adrenergic receptors. The distribution
of alpha- and beta-receptors in the vascular system varies widely.
This provides for selective circulatory control to various tissues
and organs.

Epinephrine also activated adenylcylase increases of cyclic
AMP. This is a common mechanism of action for many hormones. The
specificity is receptor site related. These sites include epinephrine activation of glycogenosis which increases available glucose
and separately a lipolytic effect to facilitate utilization of
fatty acids as an energy source.

Renin-Angiotensin-Aldesterone System

A common protective need in response to trauma is the preservation of the circulatory system. The renin-angiotensin-aldosterone
system in fluid and electrolyte balance is necessary for circulatory
homeostasis and cellular matabolism. Any event that reduces effective circulatory volume increases the activity of this system. The
current concept is that baroreceptors closely related to sodium
or effective perfusion in the kidney releases renin from juxtaglo-
merular cells in the walls of the afferent arterioles in close
proximation to the maculodensa of the kidney tubule.[26] Renin is
an enzyme which acts upon a substrate in the plasma   globulin
fraction called angiotensinogen forming a short chain decapeptide

called angiotensin I. Angiotensin I is cleaved by a converting
enzyme in the blood to a hormonally active octopeptide called
angiotensin II.[27] This substance is the most powerful vasopresssor
agent known. Angiotensin II at recognized physiological concentrations is too small to effect a vasopressor response but could potentiate the sympathetic-adrenergic pressor effect. In part this effect
may be operative in states of pathological sodium balance. For
example, sodium depletion may reduce this effect and the converse
may be true in some increased sodium states as may be the case
in high-renin types of hypertension.

The major recognized effect of angiotensin II is on aldosterone
secretion of the adrenal cortex. The regulation of aldosterone
secretion is controlled by angiotensin II, serum potassium concentration, and ACTH.[28] The role of the central nervous system apart
from ACTH is controversial. ACTH may be significant only at very
high levels as may occur in trauma. A deficiency of potassium
reduces the effect of other stimuli and a high serum potassium
concentration increases aldosterone secretion. This may serve as
a minor servomechanism in aldosterone control. Nevertheless, the
major operative control of aldosterone secretion is angiotensin II.

Aldosterone is a mineralocorticosteroid which effects the
renal reabsorption of the sodium ion in the distal tubule and
tubular water flow dependent potassium diuresis.[29] This can have
an obvious effect on sodium and potassium balance. As sodium is
the prime osmole of the serum and as it facilitates water retention
its importance in the homeostasis of the circulatory system is
apparent. A closed loop regulatory system is developed as this
circulatory effect in turn exerts an effect on secretion of renin.
As blood loss, fluid loss, fluid extravasation, vasodilation, volume
pooling, and any other mechanism that reduces effective perfusion
volume, this mechanism is important in restoring circulation volume.
As pathology permits continued fluid extravasation by tissue and
vascular injury, heart failure, hepatic insufficiency, nephrotic
syndrome or any similar mechanism, significant sodium and water
retention facilitated by hyperaldosteronism may add to the edematous
state. The priority of the adaptation response is to restore an
effective circulatory state.

SECONDARY HORMONAL ASPECTS OF TRAUMA

Fluid and
Fluid and Electrolyte Balance

The secretion of ADH as it results in water antidiuresis and
expansion of fluid volume may reduce aldosterone excretion and
increase sodium excretion. Conversely aldosterone induced sodium
and water retention may reduce ADH to facilitate homeostasis.

Trauma may independently maintain ADH secretion and add to an adaptative process capable of establishing an increased level of body water and sodium. As the traumatic events do not increase the renin-aldosterone sodium retention, and ADH secretion persists it may add to the development of dilutional hyponatremia.

The adequacy of the circulatory-fluid-electrolyte balance is of major importance to kidney function. Cinical cortisol insufficiency is associated with two well known observations. Hypotension as a primary manifestation in early phases provoked by orthostatic change can create eventually a fatal shock syndrome. Secondly, cortisol insufficiency has a well recognized effect that reduces free water clearance of the kidney. Thus cortisol deprivation by itself can alter kidney function and circulatory adequacy.

Carbohydrate Metabolism

Glucocortocoids, ACTH, growth hormone, glucagon and catecholamines all modify carbohydrate metabolism and induce hyperglycemia and/or forms of carbohydrate intolerance. Glucocorticoids and growth hormone in particular induce peripheral cellular insulin "resistance." These may contribute to reduced glucose utilization. Epinephrine and glucagon inducing cyclic AMP activation of glycogen phosphoralase phosphatase also produces increased availability of glucose. These may all with other previously described effects of cortisol produce hyperglycemia and a type of "stress" diabetes mellitus. Clinically severe persistent stress in man particularly with a diabetic predisposition and therapeutic use of glucocorticoids may result in severe hyperglycemia. This hyperglycemia often in the absence of significant ketoacidosis may produce a hyperosmolar state of a severity that occasionally causes coma. Any associated dehydration, sodium retention or uremia in particular may have a significant contribution to the severity of hyperosmolaremia.

Insulin "resistance" in stress may be influenced by other factors of unknown mechanisms including infection, fever, and hypoxia. This coupled with increased fatty acid metabolism may account for ketosis occasionally clinically observed in stress.

Hypoglycemia when observed in trauma is the effect of inadequate or insufficient hormonal response to stress and requires a search for the etiology of the hormonal disorder.

MISCELLANEOUS HORMONAL ASPECTS OF TRAUMA

As trauma and stress may be of a selective nature other aspects can play a role in the response to trauma. A partial list includes the following:

1. The importance of other hypothalamic-pituitary hormones
    in trauma is not well established. Variably thyrotropin
    (TSH) has been described to increase with a cold environ-
    ment.

2. Human growth hormone and prolactin increase to many common
    stress stimuli which parallel ACTH such as hypoglycemia,
    abdominal surgery and strenuous exercise.[6,7] The signi-
    ficance of these hormones in trauma remain to be established.
    They do contribute to alterated carbohydrate metabolism
    as previously noted.

3. Selected stress results in inhibition of gonadotropin
    release from the pituitary that may alter gonadal function
    including resultant amenorrhea in the female.[4] With some
    trauma there is some teleological wisdom in this response.

4. Thyroxine has a synergistic effect on adrenergic activity.[25]

5. Weightlessness and immobilization inactivity enhances
    calcium mobilization from bone reducing parathormone
    secretion, phosphaturia and increasing calcuria. Rarely
    hypercalcemia or hyper-hypercalcuria may be of major impor-
    tance.

6. Weightlessness and the supine position reduce aldosterone
    secretion and increase saluresis. Subsequent return to
    gravity or for the sick person resuming upright activity
    will require the reverse adaptation. Over time such states
    modify the adaptation responsiveness.

7. Prostaglandins: In recent years a great deal of interest
    and investigation has developed in a group of compounds
    synthesized from essential fatty acids. In fact there are
    several series of compounds classified as prostaglandins. A
    wide variety of tissues possess enzymes to synthesize
    these hormones. The various prostaglandins have diverse
    and probably in man ubiquitous activity.[31] This area must
    be considered a frontier in biology. Some groups of pros-
    taglandins activate specific tissue kinin systems and play
    a role in vasoconstriction, vasodilation, vascular permea-
    bility, water and electrolyte transport and may even mediate
    some synaptic transmission in the nervous system. As such
    a new frontier in the biology of trauma has developed.

CONCLUSION

The hormone system is a major effector in responses of protection
and adaptation to trauma. These responses facilitate the organism

to conform to new demands. The nervous system is an integral
part of many of these hormonal responses as a sensor, integrator,
and neuroendocrine mediator particularly in the hypothalamic and
sympathetic system. Vasopressin, cortisol, aldosterone, and
catecholamines are of primary importance in responses geared to
main cardiovascular circulatory adequacy, particularly for vital
organs. These are closely related to conservation of fluid and
electrolytes by antidiuresis, sodium retention, and modification
of the inflammatory reaction. ACTH, glucocorticoids, growth
hormone, glucagon all may play a role in cellular metabolism to
facilitate conservation of energy sources and activation of alternate metabolic pathways for energy. The basic site of action of
hormones is on the cellular level. Other complex effects of
hormones may be of selective benefit in the adaptive responses
to trauma.

REFERENCES

1. Daniel, P.M.: The Anatomy of the Hypothalamus and Pituitary
       Gland, Neuroendocrinology (Martini and Ganong), Academic
       Press, 1966.

2. Frohman, L.A.: Clinical Neuropharmacology of Hypothalamic
       Releasing Factors. NEJM 286: 1391, 29 June 72.

3. Ehni, G., and Eckles, N.E.: Interruption of pituitary stalk
       in patients with mammary cancer. J. Neurosurg. 16:
       628-652, 1959.

4. Martini, L. and Ganong, W.F.: Neuroendocrinology, Academic
       Press, 1966.

5. Nelson, D.H.: Regulation of Glucocorticoid Release. Am J. Med.
       53: 590, 1972.

6. Noel, G.L. , Slih, H.K., Stone, J.G. and Frantz, A.G.: Human
       Prolactin and Growth Hormone Release During Surgery and
       Other Conditions of Stress. J. Clin. Endo. 35:840, 1972.

7. Frantz, A.G. , Kleinberg, D.L. and Noel, G.L.: Recent Progr.
       Horm. Res. 28:527, 1972.

8. Krieger, D.T., Allen, W., Rizzo, F., Krieger, H.P.: Charac-
       terization of the Normal Temporal Pattern of Plasma
       Corticosteroids. J. Clin. Endocr. 32:266, 1971.

9. Liddle, G.W., Estep, H. L., Kendall, J. W., Jr.. William, W. C.,
       Jr. and Townes, A W.: Clinical Application of a New Test of

Pituitary Reserve. J. Clin. Endocr. 19:875, 1959.

10. Little, G. W., Island, D., Meador, C. K.: Normal and Abnormal
    Regulation of Corticotropin Secretion in Man. Recent
    Progr. Horm. Res. 18:125, 1962.

11. Gibbs, F. P.: Central Nervous System Lesions That Block Re-
    lease of ACTH Caused by Traumatic Stress. Am. J. Physiol.
    217:78-83, 1969.

12. Newsome, H. H. and Rose, J. C.: The Response of Human ACTH
    and hGH to Surgical Stress. J. Clin. Endocr. 33:481,
    1971.

13. Kawakami, M., Kimura, F., Ishida, S., and Yamase, M.: Changes
    in the Activity of the Limbic-Hypothalamic Neural Pathways
    Under Repeated Immobilization Stress. Endocr. Jap. 18:
    469, Dec. 1971.

14. Yates, F. E.: Physiological Control of Adrenal Cortical Hor-
    mone Secretion. Adrenal Cortex (A. B. Eisenstein), Little
    Brown, 1967.

15. Bajusz, E.: Physiology and Pathology of Adaptive Mechanisms.
    Pergamom Press, 1969.

16. Weber, Georg: Action of Glucocorticoid Hormone on the Molecu-
    lar Level. Functions of the Adrenal Cortex II (K. W.
    McKerns), Appleton-Century Crofts, 1968.

17. Baxter, J. D. and Forsham, P. H.: Tissue Effects of Gluco-
    corticoids. Am. J. Med. 53:573, Nov. 1972.

18. Graber, A. L., Ney, R. L., Nicholson, W. E., Island, D. P. and
    Liddle, G. W.: Natural History of Pituitary-Adrenal Re-
    covery Following Long-Term Su-pression with Corticoster-
    oids. J. Clin. Endocr. 25:11-16, 1965.

19. Salassa, R. M., Bennett, W. A., Keating, F. R. and Sprague,
    R. G.: Post-Operative Adrenal Cortical Insufficiency;
    Occurrene in Patients Previously Treated with Cortisone.
    JAMA 152: 1509, 1953.

20. Bayliss, R. I. S.: Surgical Collapse During and After Corti-
    costeroid Therapy. Brit. Med. J. 2:935, 1958.

21. Melby, J. . and Spink, W. W.: Comparative Studies on Adrenal
    Cortical Function and Cortisol Metabolism in Healthy
    Patients and Patients in Shock Due to Infection. J. Clin
    Invest. 37:1791-98, 1958.

22. Lillehie, R. C., Longerbeam, J. K., Bloch, J. H. and Manax, W.
    G.: The Nature of Irreversible Shock. Annals Surg. 160:
    682, 1964.

23. Lucas, C. E., Sugawa, C., Riddle, J., Rector, F., Rosenberg, B.
    and Walt, A. J.: Natural History and Surgical Dilemma
    of "Stress" Gastric Bleeding. Arch. Surg. 102:266, 1971.

24. Okabe, S., Saziki, R., and Takagi, K.: Cortisone Acetate and
    Stress on the Healing Process of Chronic Gastric Ulcer in
    Rats. J. Appl. Physiol. 30:793, 1971.

25. Abboud, F. M.: The Role of Catecholamines in Circulatory Dis-
    eases. Advances in Int. Med. 15:17-49, 1969.

26. Tobian, L.: Interrelationships of Electrolytes, Juxtaglome-
    rular Cells and Hypertension. Physiol. Rev. 40:280, 1960.

27. Laragh, J. H., Baer, L., Brunner, H. R., Buhler, F. R., Sealey,
    J. E. and Vaughan, E. D.: Renin, Angiotensin, and Aldo-
    sterone System. Am. J. Med. 52:633, 1972.

28. Davis, J. O.: The Regulation of Aldosterone. The Adrenal
    Cortex (A. B. Eisenstein) Little Brown, 1967.

29. Mulrow, P. J., Forman, B. H.: Effects of Tissue Mineralo-
    corticoids. Am. J. Med. 53:561, 1972.

30. Fisher, D. A. and Odell, W. D.: Effect of Cold on TSH Secre-
    tion in Man. JCEM 33:859, 1971.

31. Hinman, J. W.: Prostaglandins. Ann. Rev. of Biochem. 41:161,
    1972.

MICROBIOLOGY OF TRAUMA

John A. Washington II

Mayo Clinic, Rochester

GENERAL CONSIDERATIONS

To the degree that trauma imposes a derangement in host defense
mechanisms and thereby provides opportunity for infection to begin,
the microbiology of trauma represents a survey of opportunistic
microorganisms. The term "opportunistic infection," however, generally is used only for those infections due to organisms not usually
considered to be pathogenic, a qualification which is partially applicable to trauma.[1]

Normal anatomic barriers, such as skin and mucosal surfaces,
serve as obstacles to invasion by microorganisms. Disruption of
them by trauma provides a new portal of entry for microorganisms;
the foreign body concomitantly introduced may serve as a nidus for
infection. Additionally, alteration of the normal or autochthonous
microbiota of the skin or mucous membranes as a consequence of
trauma may in itself provide opportunity for infection by other
organisms. Open, fresh, traumatically acquired wounds are generally
considered to be contaminated wounds.[2]

Despite the fact that surgical procedures disrupt normal anatomic barriers and occasionally introduce foreign bodies (prostheses), they are carried out under aseptic conditions. Although the
amount and type of microbial contamination of the surgical wound
under these circumstances appear to be related to the risk of development of wound sepsis,[2-5] other factors peculiar to the hospital
and operating room environments[2] impose a sufficiently different
set of problems and considerations from those directly relevant to
the microbiology of trauma that they will not be considered in this
review.

123

It is known that infection is 5 times as frequent in contaminated wounds as in refined clean wounds (elective, not drained, and
primarily closed wounds[2]). Whether or not infection develops in a
contaminated wound depends on the number and pathogenicity of the
contaminating organisms; the amount of devitalized tissue present
in the wound; the presence of foreign material; the nature, location,
and duration of the wound; the status of the host's defense mechanisms; the type and thoroughness of the treatment administered; and
the general condition and nutrition of the patient.[6]

The size of the inoculum and the degree of pathogenicity of the
organism introduced into a wound are known determinants of whether
or not sepsis results. Unquestionably, however, host defense mechanisms also affect the outcome of wound contamination. A converse
consideration is that the mere presence, in a wound, of an organism
usually considered to be pathogenic is not synonymous with infection.
The presence of necrotic tissue is of no use to the healing process
because it supports the growth of microorganisms. Foreign bodies
may not only introduce organisms into tissues but may also serve as
a nidus for infection. An extensive wound with a large amount of
devitalized tissue provides an excellent site in which infection
may begin, particularly if it is contiguous or in direct communication with an epithelial or mucosal surface harboring autochthonous
microbiota. The severity of the wound may influence the development
of sepsis, especially if there is hemorrhage, shock, or some other
serious complication of the trauma.

The basic underlying condition of the host, including defense
mechanisms, nutrition, and physical and physiologic states, intro-
duces other important variables in both management and outcome of
therapy. Primary or secondary deficiency of immunocytes in the T
(thymus-dependent) or B (thymus-independent or plasma cell) cell
systems may result in enhanced susceptibility to infection with
fungi or with bacteria (including mycobacteria), inasmuch as the T
system has significant activity in relation to viral, fungal, and
mycobacterial infections and the B system, which is responsible for
antibody production, fulfills a number of functions related to infections and especially those due to bacteria.[7]

The role of nutrition in natural resistance to infection is a
complex problem because, experimentally, malnutrition may favor or
may protect against infection; furthermore, the effects of malnutrition not only vary from species to species but also are not uniform
within a given species.[8] Nutritional factors, particularly vitamins,
are essential to the integrity of epithelial and mucosal surfaces,
so that deficiencies in these factors may lead to pathologic changes
in these surfaces and secondary infection. Circulating antibodies

have been found to be diminished in a variety of nutritional deficiency states.[9,10] Finally, glycolysis, stimulation of the hexose
monophosphate shunt, and bactericidal activities of leukocytes isolated from children suffering from protein-calorie malnutrition have
been reported to be significantly lower than in normal children.[11]

The inflammatory response of the host may be altered by diabetes
mellitus, renal failure, diseases of the hematopoietic or reticuloendothelial system, and steroid antimetabolite, irradiation, or
other therapy.[1] Acidosis, whether uremic or diabetic in origin, has
been reported by some investigators to affect the early phases of
the inflammatory response;[12-15] however, other workers have been unable to relate abnormalities in motility of and phagocytosis by leukocytes to changes in serum pH.[16-19] In uremia there is abnormal
antibody production, impaired delayed hypersensitivity, and suppression of antibody response,[20] while in diabetes mellitus there
appears to be a delay in healing of an acute inflammatory lesion,[20a]
possibly because of a deficiency in chemotaxis of polymorphonuclear
leukocytes.[19]

With respect to establishment of anaerobic infection, additional
factors must be considered, the most important of which is that the
oxidation-reduction potential ($\underline{E}_h$) in an area of the wound must be
decreased. As summarized by MacLennan[21] in his excellent review of
histotoxic clostridial infections in man, this decrease in $\underline{E}_h$ may
commonly be accomplished by the following: (1) failure of the blood
supply to the area, as a result of trauma to the vessels, pressure
of tourniquets, casts, or dressings, or effects of cold, shock, or
edema; (2) presence of foreign bodies in the wound; (3) occurrence
of tissue necrosis and hemorrhage in the wound, of infection, or of
necrotizing agents in soil (for example, calcium chloride) or medications; or (4) presence and multiplication of other bacteria in the
wound. The mechanism by which a decreased $\underline{E}_h$ appears to favor
anaerobic infection is related to the incomplete oxidation of pyruvate in muscle and its subsequent reduction to lactate. The decrease in $\underline{E}_h$ and pH enables clostridia to germinate and enhances
the activity of the organisms' proteolytic enzymes, resulting in
release of amino acids into the lesion and producing a favorable
environment for bacterial multiplication.[21]

Contaminated wounds may be classified as being primary or secondary, depending on whether the contamination occurred at the time of
or shortly after injury (primary) or 24 hours or longer after injury
(secondary). Infection of traumatically acquired wounds may also be
classified as primary or secondary inasmuch as a primary infection
resulting from organisms introduced at the time of injury may become
superinfected at a later date with other organisms.[6] In the hospital
environment, such superinfecting organisms not infrequently are resistant to multiple antimicrobial agents.

Microbial contamination from trauma may originate endogenously from the autochthonous microbiota of the host or exogenously from the environment in which the injury occurred.[22,23] Egress of microbes from the gastrointestinal or respiratory tract or entry of microbes from the skin as a result of laceration, perforation, or rupture may result in infection in other areas contaminated in this process. Dissemination of microorganisms by the bloodstream or the lymphatics may secondarily infect a large hematoma, hemothorax, or any area with devitalized tissue in which phagocytosis is impaired and opportunity for activity of humoral antibody is limited. Burn wounds are frequently initially colonized with the autochthonous skin microflora and subsequently, for reasons which are less clear, with Pseudomonas aeruginosa.

MICROBIOLOGIC CONSIDERATIONS

Inasmuch as most traumatically acquired infections, with the exception of rabies, are due to bacteria (aerobes, facultative anaerobes, and mycobacteria) or fungi (yeastlike and filamentous), the remainder of this discussion will be confined to these two groups of microorganisms.

Specimen Collection and Transport. Proper specimen selection is critical for microbiologic examination. The following recommendations for culture of tissues were made by Weed[24] in 1958 and are worth review. In some instances, viable organisms may be few in number and unevenly distributed in the lesion, especially in a chronic one; therefore, multiple specimens should be obtained from large lesions or when several lesions are present. When there is an abscess, a portion of the wall should be obtained along with the pus; if granules are present in the pus, they should be washed free of surface contamination and examined separately for evidence of Actinomyces. Material from an ulcer should include a generous amount of the base and the margin with surrounding normal tissue. Sinus tracts should be curetted after careful cleansing of their orifices with alcohol to decrease surface contamination.

Tissue should be minced with sterile scissors and then ground in a sterile mortar with an abrasive. A 10 to 20% suspension of the homogenate in broth is prepared for inoculation of cultures or animals. A portion of the tissue also should be examined histologically to establish whether or not the lesion is inflammatory and, if it is, whether or not the process is suppurative or granulomatous. Special stains may be helpful in providing clues as to the etiology of the process.

Since anaerobic bacteria are important in infections resulting from trauma (Table 1), certain special precautions should be taken

Table 1.--Types of Infections in Which Anaerobes Are Predominant
        or Common Pathogens*

Brain abscess	Appendicitis
Otogenic meningitis, extradural	Subphrenic abscess
or subdural empyema	Other intra-abdominal abscess
Chronic otitis media	Wound infections after bowel
Dental infections	surgery or trauma
Pneumonia secondary to	Puerperal sepsis
obstructive process	Postabortal sepsis
Aspiration pneumonia	Endometritis
Lung abscess	Tubo-ovarian abscess
Bronchiectasis	Other gynecologic infections
Thoracic empyema	Perirectal abscess
Breast abscess	Gas-forming cellulitis
Liver abscess	Gas gangrene
Pyelophlebitis	
Peritonitis	

*From Sutter VL, Attebery HR, Rosenblatt JE, et al: Anaerobic
Bacteriology Manual. Anaerobic Bacteriology Laboratory, Wadsworth
Hospital Center, Veterans Administration, Los Angeles, The Regents
of the University of California, 1972. By permission.

to ensure their recovery. Because anaerobes constitute a major part
of the autochthonous microbiota in man (Table 2), material for culture should be collected carefully so as to prevent contamination
with normal flora. Furthermore, material from sources with autochthonous anaerobic microflora (for example, throat, sputum, feces,
voided urine, vaginal material, and gastric and small bowel contents) should not be routinely cultured anaerobically. Ideally,
material for anaerobic culture should be aspirated through a needle
into a sterile syringe, taking care to expel any air bubbles in the
syringe. The material should then be transferred to a butyl-rubber-
stoppered tube containing an oxygen-free gas ("anaerobe tube") such
as carbon dioxide or nitrogen[25] A less ideal method is to drop a
swab specimen into the anaerobe tube and then quickly stopper it
again. Tissue specimens also may be dropped rapidly into an anaerobe
tube. Sutter et al.[26] recommended inserting the swab into prereduced
and anaerobically sterilized (PRAS) semisolid Carey-Blair medium
covered with a "head" of oxygen-free carbon dioxide; however, Holdeman and Moore[27] advised against using any transport medium because
overgrowth of facultative bacteria, with suppression of anaerobes,
may result (PRAS media are not as reduced as many infection sites).

In any event, cultures of a traumatically acquired wound may be
helpful for establishing the extent of contamination. If surgical
debridement has been delayed for 6 to 8 hours, Gram-stained smears
of the wound may be helpful in assessing the degree and type of
contamination.[28]

Selection of Media. The media selected should provide optimal
conditions for isolation of potentially pathogenic bacteria and fungi. For recovery of aerobic and facultatively anaerobic bacteria,
the following should be inoculated: a nutrient agar medium with
added sheep blood; differential agar media, such as eosin-methylene
blue and phenylethyl alcohol with blood, for isolation of gram-negative and gram-positive bacteria, respectively; and a nutrient broth
medium, such as thioglycollate, supplemented with ascitic fluid or
rabbit serum. For recovery of anaerobic bacteria, anaerobic jar
techniques are satisfactory,[26,29] provided that the specimen has
been properly collected and transported to the laboratory and that
the laboratory takes suitable precautions to prevent prolonged exposure of inoculated media to the air.[30]

For recovery of mycobacteria, an appropriate decontamination procedure should be performed when necessary. The duration of this procedure is determined by the extent of nonmycobacterial contamination
of the specimen; however, it should be remembered that this process
is also lethal to mycobacteria and should not be prolonged. Rarely,
it may be necessary to inoculate guinea pigs in an attempt to recover
mycobacteria in the presence of other bacteria resistant to decontamination. In general, Lowenstein-Jensen and Middlebrook 7H10 agar
media are satisfactory for recovery of mycobacteria.

Table 2.—Anaerobes as Normal Flora in Humans

Site	Clostridia	Nonsporulating bacilli								Cocci	
		Gram-positive					Gram-negative				
		Actino-myces	Bifido-bacteria	Eubac-teria	Lacto-bacilli‡	Propioni-bacteria	Bacte-roides	Fuso-bacteria	Vibrios	Gram-positive	Gram-negative
Skin	0	0	0	U	0	2	0	0	0	1	0
Upper respiratory tract§	0	1	0	±	0	1	1	1	1	1	1
Mouth	±	1	1	1	1	±	2	2	1	2	2
Intestine	2	±	2	2	1	±	2	1	±	2	1
External genitalia	0	0	0	U	0	U	1	1	0	1	0
Urethra	±	0	0	U	±	0	1	1	±	±	U
Vagina	±	0	2	U	2	0	1	±	1	1	1

*From Sutter VL, Attebery HR, Rosenblatt JE, et al: Anaerobic Bacteriology Manual. Anaerobic Bacteriology Laboratory, Wadsworth Hospital Center, Veterans Administration, Los Angeles, The Regents of the University of California, 1972. By permission.

†Symbols: U = unknown; 0 = not found or rare; ± = irregular; 1 = usually present; 2 = usually present in large numbers.

‡Includes anaerobic, microaerophilic, and facultative strains.

§Includes nasal passages, nasopharynx, oropharynx, and tonsils.

Fungi may be readily recovered by inoculating brain heart infusion blood agar without and with antibiotics. A combination of chloramphenicol and gentamicin has been found useful in suppressing bacterial contaminants. Furthermore, placing the agar into petri dishes rather than tubes or bottles provides a larger surface on which to detect and isolate colonies of fungi.

CLINICAL CONSIDERATIONS

The distinction between wound contamination and wound infection by a particular microorganism may not be clear-cut. In the first place, many microorganisms are widely distributed in nature. For example, Clostridium perfringens has been encountered in soil, sand, dust, air, on the clothing and skin of man, and in the alimentary tract of man and animals.[21,31-34] Other microorganisms represent autochthonous microbiota (Table 2). Therefore, it is entirely possible to isolate an organism, such as C. perfringens, from a healthy wound. MacLennan[31] pointed out that all but the most trivial war wounds in Western Europe contained clostridia, in contrast to those in the Middle East where these organisms were found in only about 30% of war wounds; the difference was attributed to the shortened survival of anaerobic microflora in the desert. Williams and Miles[35] proposes that the term "contamination" be used to describe the presence of pathogenic bacteria in wounds less than 8 hours old and that any wound more than 8 hours old yielding pathogenic bacteria be called "infected," regardless of the number of microbes isolated or the clinical state of the wound. In their view, infected wounds could be frankly septic, could exhibit no signs of infection except for delayed healing, or could exhibit no clinical evidence whatsoever of infection.

Another problem in ascribing an etiologic role to a particular microorganism is the frequency with which mixed wound infections, particularly those involving anaerobic bacteria, are encountered.[29,36-39] The extent to which mixtures of bacteria interact synergistically in infection is not clearly known; however, bacterial synergism between a few species was shown in experimental infections by Meleney[40] and, subsequently, by Hite et al.[41] MacLennan[31] found that the likelihood of survival in cases of gas gangrene was diminished when other histotoxic clostridia were present with C. perfringens, compared to when C. perfringens was present alone.

Although wound sepsis, gangrene, erysipelas, and tetanus have long been recognized as the "scourges of surgery," it was not until the latter half of the 19th century that the work of Pasteur and Koch established the microbial etiology of infection. Not unexpectedly, major wars have stimulated considerable interest in the bacteriology of trauma,[43] particularly in regard to anaerobes. The concomitant

evolution of contaminated wound management since the 18th century
has been reviewed by Wangensteen and Wangensteen.[44]

It was recognized during World War I that the bacteria principally involved during the first week after injury were the sporula-
ting anaerobes and nonsporulating aerobes of fecal origin; during
the second week, the stage of secondary infection, pyogenic cocci,
P. aeruginosa, and various coliforms became predominant.[43] Although
pseudomonads still are prominent secondary invaders in both war and
civilian wounds, the prominence of anaerobic bacteria and of gas
gangrene began to wane during and after the Korean conflict.

Because of the frequency with which mixed infections are encountered, it is not always possible to segregate clearly the various
types of responsible organisms; however, insofar as it is possible,
the remaining discussion will be divided according to the type of
organism causing infection -- anaerobic bacteria, aerobic and facultatively anaerobic bacteria, mycobacteria, and fungi.

Anaerobic Bacteria. On the basis of his studies of anaerobic infections of war wounds in the Middle East, MacLennan[31] proposed the
following classification of traumatic histotoxic infections of man:
(1) simple contamination, (2) anaerobic cellulitis, and (3) anaerobic
myositis of either clostridial or streptococcal origin. Because the
muscle lesion is more necrotic than inflammatory, Robb-Smith[45] proposed that the term "anaerobic myonecrosis" replace the term "anaerobic myositis." In simple contamination, clostridia multiply in
necrotic tissue, in cellular debris, or in a hematoma, frequently
without observable clinical change, occasionally delaying wound healing, and often producing damage by proteolysis.[21] Most isolates of
C. perfringens in civilian practice represent simple contamination.[21,
37,46,47] In anaerobic cellulitis there is multiplication and spread
in already necrotic tissue with gas production but without progressive involvement and destruction of muscle; living muscle remains intact.[21,33] Anaerobic myonecrosis represents an acute and rapidly
progressing invasion of healthy, undamaged tissue.[21] Although gas
is not ordinarily apparent early in the course of the infection, it
usually becomes apparent as the pathologic process and toxemia advance. It is critical that anaerobic cellulitis and anaerobic myonecrosis be distinguished from one another so that unnecessarily
aggressive surgery can be avoided.

Equally important is the necessity of recognizing other nonclostridial myonecrosis crepitant infections which may mimic gas gangrene.[21,37] As pointed out by Altemeier and Fullen,[48] there are many
other bacterial and nonbacterial lesions which may simulate clostridial myonecrosis. Among these are the "aerobic aerogenic infections,"
due to Enterobacteriaceae such as Escherichia coli or Klebsiella
pneumoniae, which are usually polymicrobic, are much less severe than

Table 3.--Clostridia Isolated From Gas Gangrenous Wounds: Percentages of Cases in Various Reports

Organism	MacLennan[31]	Smith and George[66]	Dhayagude and Purandare[67]	DeHaven and Evarts[68]	Altemeier and Fuller[48]
C. perfringens	56	39	52	80	95
C. novyi	37	32	0	50	8
C. septicum	19	0	36	3	4
C. histolyticum	6	0	16	0	...
C. sporogenes	...	54	...	63	...
C. bifermentans*	4	54	...	23	...

*Distinction between C. bifermentans and C. sordellii not clear-cut.[69]

clostridial myonecrosis, and usually involve the subcutaneous or epifascial compartments.[48] Streptococcal myonecrosis resembles that due to clostridia in many respects.[48-53] Indeed, anaerobic streptococci have been implicated in the etiology of anaerobic cellulitis.[54] Characteristically, in this type of myonecrosis there is marked cutaneous erythema, there is no muscle death except in neglected cases, and Gram-stained smears of muscle show many gram-positive cocci.[48] B-Hemolytic streptococci may produce an epifascial, spreading, subcutaneous gangrene with thrombosis of vessels and subsequent sloughing of skin.[31,48,50,52,53] Bacteroidaceae have produced lesions simulating myonecrosis.[48,55-57] Although staphylococcal myositis or myonecrosis is rare in the United States,[58] it is encountered not infrequently in tropical countries.[59,60]

Gaseous infiltration of tissue has been seen in injuries resulting from application of compressed air hoses to body orifices, from irrigation of wounds with hydrogen peroxide, and from accidental injection of gas-producing chemicals such as benzene.[48] Again, differentiation of these infections and lesions from clostridial myonecrosis is imperative because conservative management of the former types of infections is usually adequate. A Gram-stained smear of wound exudate or of muscle is a simple and rapid means of distinguishing these from clostridial and streptococcal myonecrosis.

Comparable in severity to gas gangrene are uterine infections due to C. perfringens,[21,61] especially when accompanied by intense hemolysis.[62] These infections usually occur after attempted mechanically induced abortion or, less commonly, after spontaneous abortion, prolonged or instrumental delivery, or delivery in unhygienic conditions. In such cases the clinical course appears to represent a direct result of α-toxin or lecithinase activity.[61] That clostridial infection may occur after relatively minor injury is attested to by the number of cases of myonecrosis complicating parenteral injections, frequently of epinephrin and caffeine.[63]

The microbiology of the clostridia has been completely described elsewhere.[26,27] The clostridia elaborate a variety of biologically active compounds, many of which are toxins and enzymes, but their activities are not clearly understood.[64]

The most important species of clostridia associated with myonecrosis are C. perfringens, C. novyi, and C. septicum; of less importance are C. histolyticum, C. sporogenes, C. sordelli, C. bifermentans, and C. fallax.[65] However, different investigators have reported somewhat different distributions of clostridial species isolated from gas gangrenous wounds (Table 3).

That gas gangrene is not limited to war or other mass casualty situations was discussed by MacLennan,[21] and that it continues to be a problem in civilian life, including in postoperative infections, has been suggested by DeHaven and Evarts.[68]

Wound infection with C. botulinum type A is rarely recognized, there having been only nine (four fatal) reported cases in the United States by 1972.[70-76] That more such cases have not been recognized is surprising in view of the worldwide distribution of this organism in soil and water; however, this species' principal notoriety is as an etiologic agent of food poisoning due to its production of the most potent neurotoxins known.[77]

Another potent producer of a neurotoxin is C. tetani which is also widely distributed in soil and in the intestinal tracts of man and animals. Tetanus and botulism are probably more correctly termed "intoxications" inasmuch as potent neurotoxins are primarily responsible for their morbidity and mortality. Tetanus was described by Hippocrates; however, knowledge about its etiology is relatively recent.[78] The morbidity and mortality due to tetanus in the United States have followed a gradual downward trend since 1950,[79] undoubtedly due to the widespread use of tetanus toxoid vaccine. Peak incidence of the disease in recent years has occurred among the very young and the elderly, in residents of the lower Mississippi Valley and the Southeast, during the summer months, and in nonwhites.[79] Puncture wounds, especially from nails or slivers of wood, lacerations, and abrasions, usually of the hands and feet, are commonly associated with tetanus.[79-81] The median incubation period varies between 7 and 9 days, and there is an inverse relationship between the length of this period and the case-fatality ratio (this ratio remains high in any event).[79,80] The rate of recovery of C. tetani from the wound has varied between 28 and 32%,[79,80] presumably because it is a fastidious anaerobe, because of interference with the chances of its recovery by overgrowth of contaminants, or because of its suppression by prior antimicrobial therapy.

Without minimizing the importance and gravity of clostridial infections, it has become apparent, in recent years particularly, that infections due to the nonsporulating anaerobic bacteria occur frequently.[57] An increasing awareness of this problem, based in part on the recognition of bacteriologic clues to the possible presence of anaerobic infection (Table 4), and better utilization of proper techniques of specimen collection and transport have led to more complete isolation and accurate identification of the etiologic agents involved as well as to more appropriate and specific therapy.

In 1941, Pulaski et al.,[39] using thioglycollate broth as the primary medium for recovery of anaerobes, isolated anaerobic gram-negative bacilli from only 3 of 200 cases of fresh, accidental wounds. Citing his own unpublished studies, in 1942 Altemeier[43] reported isolating Bacteroidaceae and anaerobic streptococci from 11 and 8%, res-

Table 4.--Bacteriologic Clues to Anaerobic Infection

1. Foul odor of specimen
2. Location of infection in proximity to a mucosal surface
3. Infections secondary to human or animal bite
4. Gas in specimen
5. Previous therapy with aminoglycoside antibiotics (in particular kanamycin, neomycin, or gentamicin)
6. Black discoloration of blood-containing exudates; these exudates may fluoresce red under ultraviolet light (B. melaninogenicus infections)
7. Presence of "sulfur granules" in discharges (actinomycosis)
8. Unique morphology on Gram stain
9. Failure of organisms seen on Gram stain of original exudate to grow aerobically. But, failure to obtain growth in fluid thioglycollate medium is not adequate assurance that anaerobes were not present
10. Growth in anaerobic zone of fluid media or of agar deeps
11. Growth anaerobically on media containing kanamycin, neomycin, or paromomycin at 100 μg/ml (or medium also containing vancomycin at 7.5 μg/ml in the case of gram-negative anaerobic bacilli)
12. Characteristic colonies on agar plates anaerobically
13. Young colonies of B. melaninogenicus may fluoresce red under ultraviolet light (blood agar plate)

*From Sutter VL, Attebery HR, Rosenblatt JE, et al: Anaerobic Bacteriology Manual. Anaerobic Bacteriology Laboratory, Wadsworth Hospital Center, Veterans Administration, Los Angeles, The Regents of the University of California, 1972. By permission.

pectively, of 99 fresh accidental wounds. In his study of 708 wounds, De Waal[82] found clostridia in only 4% of the cultures; the anaerobic methods were not described, and no mention was made of other anaerobic isolates. In 1956 Gunn[83] summarized the world's literature of over 330 reports of Bacteroides infections; there were 148 cases of septicemia in which the primary lesion was recorded, and 24 (16 fatal) of these represented gunshot wounds. Bodner et al.[84] reported three post-traumatic cases of Bacteroides bacteremias; one patient died subsequently of Klebsiella sepsis after recovery from his Bacteroides infection and another died in shock after receiving inappropriate antimicrobial therapy.

In most recent series of septicemias due to organisms of the family Bacteroidaceae, the predominant primary lesion has been in the gastrointestinal tract.[84-86] Since the Bacteroidaceae represent a major portion of the autochthonous microflora of the large intestine (Table 2), it is not surprising that, in traumatic perforating abdominal trauma, infection with anaerobic bacteria and particularly with gram-negative non-spore-forming bacilli is common. In a prospective study of infections in cases of penetrating abdominal trauma, Thadepalli et al.[36] found that anaerobic bacteria were present in approximately 50% of the wounds by the end of 7 days. Bacteroides and Clostridium were the most frequent isolates from initial wound cultures in their study. In his review of actinomycosis, Guidry[87] cited reports of abdominal infection due to this agent after traumatic perforation of the bowel. In a review of 501 cases of intraabdominal abscesses, Altemeier and co-workers[88] found that 17 (3%) were related to trauma; of these 17 cases, Bacteroides were isolated from 24% and anaerobic streptococci from 6%.

Post-traumatic musculoskeletal infections, including osteomyelitis, due to nonsporulating anaerobes have been reported by Beerens and Tahon-Castel,[89] Nettles et al.,[90] Ziment et al.,[91] Pearson and Harvey,[92] and Leake.[93] Isolation of one or more species of Bacteroides with anaerobic streptococci were common findings in these studies. Anaerobic infections after human bites may result in severe tissue destruction, osteomyelitis, and deformity.[49,92] Reports of actinomycosis after human bite wounds have been cited by Guidry.[87]

Rotheram and Schick[94] found bacteremia in 61% of patients whose blood was cultured after septic abortion. Most of the cultures contained anaerobic or microaerophilic bacteria, principally streptococci and Bacteroidaceae. In a high percentage of instances, the same bacteria were recovered from a paired cervical culture. The frequency of anaerobic bacteremia after septic abortion was also noted by Smith et al.[95] and by Gelb and Seligman.[96]

The prominence of anaerobic bacteria, and especially of the clostridia, in war wounds persisted in published studies[28,31,38,43,97,98]

until and including the time of the Korean conflict, at which time
several major differences were noted.[99] First, most clostridia isolated were proteolytic rather than saccharolytic and, as a result,
there was a relatively low incidence of gas gangrene. Second, most
hemolytic streptococci were of the fecal variety. Third, pathogenic
staphylococci were isolated from nearly a quarter of the cases, col-
iform bacilli from a third of the cases, and P. aeruginosa from many.
The decreased incidence of gas gangrene probably was not due to any
decrease in wound contamination but rather to earlier debridement,
arterial repair, and antimicrobial therapy.[98]

Studies of the bacterial flora of wounds incurred during the
Vietnamese conflict have demonstrated the predominance of staphylococci, P. aeruginosa, and Enterobacteriaceae, principally Escherichia
coli, Klebsiella-Enterobacter-Serratia group, and Proteus.[100-103]
Anaerobes were rarely isolated in any of these studies. Although he
did not give specific bacteriologic data, Feltis[104] reported that
sepsis represented the second most common cause of death in Vietnam,
accounting for 20% of the deaths overall (in 6,927 casualties) and
for 5.9% of the deaths within the first 24 hours after injury. Twenty of 24 casualties died from sepsis after laparotomy, and 16 of
these had colon injuries. In 30 casualties with severe wounds of
extremities, recently reported by Tong,[103] 40% developed bacteremia,
principally due to gram-negative bacilli with a predominance of
Enterobacter and Mimeae. In view of the frequency of abdominal war
injuries,[102] it is surprising that infection with nonsporulating
anaerobic gram-negative bacilli has not been recognized more frequently.

Although anaerobes were not isolated from contaminated or infected wounds in Williams and Miles'[35] bacteriologic study of infection and sepsis in industrial wounds of the hand or in Nicholls'
[105] study of 91 pyogenic hand infections, unpublished observations,
[106] between 1967 and 1971 at the Mayo Clinic, of wound isolates from
mutilating hand injuries demonstrated the presence of such bacteria
in 24% of 38 injuries with positive cultures. Among the clostridia
isolated were four C. perfringens, two C. sporogenes, and one each
of C. fallax, C. botulinum, C. sphenoides, and C. hastiforme. In
addition, there were two isolates of Bacteroides fragilis and one
and one each of Fusobacterium fusiforme, Peptococcus sp., and Pepto-
streptococcus sp.

Certainly one of the most unusual reports of anaerobic infection
was that by Stanley et al.,[107] of a man who was admitted to a Western
Australian hospital with the complaint of a "numb bum" following a
bite by a brolga (Grus rubicunda), a 160-cm bird with a predilection
for bottlecaps. Cultures of the wound at the time of surgical debridement grew C. perfringens predominantly with C. tetani and a new
species tentatively named C. brolgaseptica by the authors.

Aerobic and Facultatively Anaerobic Bacteria. The variety and number of species of such bacteria which may be recovered in pure or mixed cultures from post-traumatically contaminated or infected wounds are almost infinite. Their wide endogenous and exogenous distributions provide ample opportunity for entry into traumatically acquired wounds. Their frequent presence in mixed cultures makes the assessment of the clinical significance of a single species ex-temely difficult. Nevertheless, the frequency of wound infection with gram-negative bacilli has increased in recent years.[98,108] According to Altemeier,[108] between 1942 and 1956, two-thirds of invasive wound infections seen in surgical practice were due to gram-positive bacteria; however, between 1956 and 1970, there was a 14-fold increase in the number of gram-negative bacilli, so that now two-thirds of the infections are of this type. The most commonly encountered organisms in gram-negative sepsis are E. coli, Klebsi-elleae, Proteus, and P. aeruginosa. Although trauma may have been the primary event, many of these infections are nosocomial in origin, and, in published reports it can be difficult to separate those ac-quired nosocomially from those of community origin. Nonetheless, infection with many of these organisms may be directly related to trauma because they are widely distributed in nature as well as in the autochthonous microflora of man and animals.

Bacterial ulcers of the cornea may be traumatic in origin. The most commonly isolated organisms are Streptococcus pneumoniae, P. aeruginosa, Moraxella liquefaciens, S. aureus, S. pyogenes, and Pro-teus;[109,110] however, Neisseria, Corynebacterium, Haemophilus, Kleb-siella pneumoniae, Serratia marcescens, E. coli, and Mycobacterium fortuitum are among the other less common bacteria which have been isolated from such lesions. Similar agents have been found to cause intraocular infections secondary to penetrating trauma of the eye.[111]

Among the aerobic gram-positive bacilli, Corynebacterium diph-theriae is known to cause cutaneous infections, commonly in tropical countries[112,113] and only occasionally in North America.[114-116] Re-cently, Cockcroft et al.[117] reported 44 cases of cutaneous diphtheria (in Vancouver, British Columbia) among indigent patients, 11 of whom had infected lacerations and abrasions; in most cases, S. aureus and streptococci were isolated concurrently. Skin trauma and abrasions were noted by Liebow et al.[112] to be contributory factors in the de-velopment of cutaneous diphtheria, as were crowding and poor perso-nal hygiene.

In Table 5 are listed the percentage occurences of various bac-teria in initial wound cultures in several studies between 1940 and 1971. Williams and Miles' report[35] represented a bacteriologic study of industrial wounds of the hand. The reports by Miles et al.,[97] Pulaski et al.,[39] and De Waal[82] represented studies of a variety of traumatic wounds in both ambulatory and hospitalized patients; those

Table 5.—Occurrence (%) of Various Bacteria in Initial Wound Cultures

Organism	Miles et al.[97]	Pulaski et al.[39]	De Waal[82]	Williams and Miles[35]	Lindberg et al.[98]†	Matsumoto et al.[102]	Adler et al.[118]
Staphylococcus aureus	53	52	19	18	17; 46	25	31
S. epidermidis	25	81	28	...	12; 33	3	...
Streptococcus							
β-Hemolytic	25	13	13	1	0; 61		12
Other aerobic	14	29	19	...	57; 76		15
Anaerobic‡	...	11	7	...			...
Gram-negative bacilli	35	23	...	3	...	...	...
Escherichia coli	...	7	10	...	21; 1	15	9
Klebsiella-Enterobacter	...	...	...	...	19; 1	6	5
Proteus	6	7	2	...	17; 6	6	5
Pseudomonas aeruginosa	7	1	1	...	0; 2	16	11
Anaerobic nonsporulating	...	2	...	...			
Sporulating gram-positive bacilli							
Aerobic	15	42	...		10; 62		
Anaerobic	32	23	8		84; 36		
Nonsporulating gram-positive bacilli	11	4	...		5; 18		

*isolation rates from total initial wound

*Percentage calculated on basis of /cultures. Absence of entry means no isolates stated.

†First number represents occurrence in summer; second number represents occurrence in winter.

‡Includes microaerophilic.

Table 6.--Occurrence (%) of Various Bacteria Isolated from
War Wounds, Vietnam, 1968-1972

Organism	Kovaric et al.[100]	Heggers et al.[101]	Tong[103]
Staphylococcus aureus	24	20	2
S. epidermidis	4	...	30
Streptococcus			
β-Hemolytic	...	5	...
Enterococcus	5	...	3
Other	...	...	1
Escherichia coli	9	3	9
Klebsiella-Enterobacter	30	12	20
Serratia marcescens	<1	...	5
Proteus	11	13	1
Pseudomonas	11	47	2
Other gram-negative bacilli	<1		15†
Sporulating bacilli			
Aerobic	...		17
Anaerobic	<1		<1
Bacteroides			<1

*Percentage calculated on basis of total bacteria isolated. Absence of entry
means no isolates stated.

†All isolates represented Mimeae.

of Lindberg et al.[98] and Matsumoto et al.[102] represented studies of
Korean and Vietnamese war wounds, respectively; and that of Adler et
al.[118] represented studies of community acquired infections of skin,
subcutaneous, and deep tissues seen at the Boston City Hospital in
January 1970. In general, staphylococci and streptococci were commonly isolated; however, gram-negative bacilli frequently occurred in
all of the studies except the one by Williams and Miles.[35] Seasonal
variations in occurrences of coliforms were reported by Lindberg et
al.[98] and by Matsumoto et al.[102] Whereas only 11% of bacterial isolates from war wounds in Italy reported by Rustigian and Cipriani[38]
in 1947 represented gram-negative bacilli, 50 to 60% of bacterial
isolates from war wounds in Vietnam[100,101,103] have been gram-negative bacilli (Table 6). Waldvogel and associates[119] reviewed 155
cases of osteomyelitis secondary to a contiguous septic focus. Se-
venty-nine of these cases represented infection after open reduction
of fractures. Most of these types of osteomyelitis in the mandible
and pelvis and many of them in small bones were due to gram-negative
bacilli, principally P. aeruginosa, Proteus, E. coli, and Klebsiella-
Enterobacter. S. aureus was the predominant species isolated from
this type of osteomyelitis.

In mutilating hand injuries seen at the Mayo Clinic between 1967
and 1971 there were one or more positive cultures in 38 cases and 28
contained Enterobacteriaceae.[106] In the 38 positive cultures there
were 127 bacterial isolates, of which 40% represented Enterobacteriaceae, 5% represented pseudomonads or other nonfermenting gram-
negative bacilli, 44% represented gram-positive aerobes or facultative anaerobes, and 12% represented anaerobes (Table 7). Of the 38
wounds with positive cultures, 83% of those due to farm implements
had mixed cultures, in contrast to 40% of those due to other mechanical equipment. When pure cultures were encountered, all of those
wounds due to farm implements contained gram-negative bacilli, whereas virtually all of those associated with other mechanical equipment
contained gram-positive cocci. These data are in marked contrast to
those of Williams and Miles,[35] in which there was a 3% occurrence of
coliforms in industrial hand wounds, and to those of Nicholls,[105] in
which 11% of pyogenic hand infections harbored gram-negative bacilli.

Of interest in Table 7 is the frequency of isolation of members
of the tribe Klebsielleae, especially from wounds resulting from farm
implements. Forty-two of the 50 isolates of Enterobacteriaceae be-
longed to this tribe. Of relevance to this problem is a recent report by Duncan and Razzell[120] that 71% of isolates of Enterobacteriaceae from forest-related samples and fresh produce were Klebsiella
with Enterobacter representing 19%, Citrobacter representing 8%, and
E. coli representing 2%. Rustigian and Cipriani[38] made note of the
isolation of "Aerobacter" or "Paracolon aerobacter" in their studies
of 36 war wounds. Reports of isolation of members of the Klebsiella-
Enterobacter (Aerobacter)-Serratia group are cited in Tables 5 and 6.
Differentiation between isolation rates of K. pneumoniae and Entero-

Table 7.--Bacteria Isolated from Mutilating Hand Injuries, Mayo
        Clinic, 1967-1971

		Gram-positive aerobes and facultative	
Enterobacteriaceae:		anaerobes:	
Escherichia coli	6(3)*	Staphylococcus epidermidis	21(11)
Citrobacter freundii	2(1)	S. aureus	12(7)
Klebsiella pneumoniae	10(1)	Streptococcus	
Enterobacter cloacae	6	Group D	14(5)
E. aerogenes	13(2)	Viridans	2
E. hafniae	1	Corynebacterium	2(1)
E. agglomerans	3	Bacillus	6(2)
Pectobacterium carotovorum	1(1)		
Serratia marcescens	8(1)		
	50(40%)		55(44%)
Other gram-negative bacilli:		Anaerobes:	
Pseudomonas aeruginosa	2(1)	Clostridium perfringens	4(2)
P. maltophilia	1	C. sporogenes	2(1)
P. fluorescens	1	C. fallax	1
Acinetobacter calcoaceticus	1(1)	C. botulinum	1
Mima polymorpha	1(1)	C. sphenoides	1
Alcaligenes faecalis	1	C. hastiforme	1
	7(5%)	Bacteroides fragilis	2
		Fusobacterium necrophorum	1
		Peptococcus	1
		Peptostreptococcus	1
Total = 127 isolates			15(12%)

*Numbers in parentheses represent number of injuries due to mechanical equipment
other than farm implements.

bacter (Aerobacter) aerogenes was made by Matsumoto et al.[102] who
also encountered Enterobacter hafniae ("Hafnia group") and Citrobacter in initial cultures of wounds incurred in Vietnam.   Enterobacter species were frequently isolated in studies of wounds incurred in Vietnam reported by Kovaric et al.[100] and by Tong.[103] Klebsielleae were isolated in 18% of 17 cases of intra-abdominal abscess
related to trauma reported by Altemeier et al.[88] E. liquefaciens
was isolated from a post-traumatic heel abscess by Washington et
al.,[121] but no post-traumatic isolates of E. hafniae were encountered in that study.   Isolation of E. agglomerans, formerly described
as the Herbicola-Lathyri group in the genus Erwinia, from wounds,
particularly those related to outdoor accidents, has been reported
by Gilardi et al.,[122] von Graevenitz[123] and Pien et al.[124] In the
study by Pien et al.,[124] farm injuries and outdoor accidents were
the predominant sources of these isolates, usually in mixed cultures,
strongly suggesting a relationship to the known presence of E. agglomerans in plants.[125,126]

   Serratia marcescens is a widely distributed saprophytic organism
with a long history[127] and with recent notoriety because of its role
in causing serious, frequently fatal, nosocomial infection.   Its isolation from post-traumatic wounds has received less attention recently but was briefly noted by Wilkowske et al.[128] and Crowder et
al.[129]   Bacteremia due to S. marcescens after trauma has been reported by Henjyoji et al.;[130] however, in most cases the bacteremia was
nosocomially acquired.   S. marcescens was frequently isolated from
combat wounds in Vietnam by Tong.[103] In the study[106] of mutilating
hand injuries at the Mayo Clinic, S. marcescens represented 18% of
the Enterobacteriaceae isolated and was commonly associated with
injuries caused by farm implements.   In other studies[131] at the Mayo
Clinic of 41 patients with musculoskeletal infections associated with
S. marcescens, 10 patients had the infection after severe trauma
incurred in their occupations as farmers.   Endophthalmitis due to
S. marcescens, after removal of an intraocular foreign body, has
been reported by Bigger et al.[132]

   Citrobacter has been isolated from war wounds by Matsumoto et
al.[102] and from mutilating hand injuries.[106] Of interest in this
regard is the number of isolates of Citrobacter from animals, water
and sewage, and foods and feeds recorded by Ewing and Davis.[133] Although usually associated with diarrheal disease, Edwardsiella tarda
is a rarely recognized species of the Enterobacteriaceae which has
been isolated from a post-traumatic subgaleal abscess[134] and from
two infected lacerations.[135]

   Although infection with P. aeruginosa is usually considered to
be of nosocomial origin, this organism is widely distributed in nature, especially in moist environments, and is part of the autochthonous microflora of man, occurring particularly in feces and in

saliva. Recently, Kominos et al.[136] isolated the organism from to-
matoes and other vegetables and demonstrated that these sources represented an important vehicle for its introduction into the hospital.
Its isolation from traumatically acquired wounds has been cited in a
number of reports (Tables 5, 6, and 7). Because of its wide distribution and physiologic characteristics which enable it to survive in
many diverse environments, the organism may be isolated from a variety of hospital sources. Although early colonization of the surface
of a burn wound is predominantly by gram-positive organisms, gram-
negative species, principally P. aeruginosa, become dominant by the
third day and sepsis is always a threat.[137] Those interested in
further information on this complex topic are referred to the excellent recent review by Moncrief.[137]

P. maltophilia, P. putida, P. putrefaciens, P. cepacia, and P.
stutzeri were isolated in pure cultures by Gilardi[138] from several
post-traumatic wound infections. Taplin et al.[139] isolated P. cepacia from swamp water in Georgia and described a correlation between
colonization of toewebs by this species and a form of foot rot seen
in troops under combat conditions in Vietnam. Ederer and Matsen[140]
isolated P. cepacia from two patients with major trauma treated by
extensive surgery. As emphasized by Gilardi,[138] pseudomonads are
widely distributed in water, soil, and plants, are frequently contaminants, and have a low degree of virulence and limited invasive
capabilities; however, they may act as primary or opportunistic
pathogens and they may produce extensive suppurative lesions. Their
recognition is important not only because much remains to be learned
about their pathogenicity but also because of their frequent resistance to the commonly used antipseudomonadal antibiotics.

In recent years there has been considerable interest in P. pseudomallei, the etiologic agent of melioidosis. This is a relatively
rare glanderslike infection which was originally described in India
by Whitmore and Krishnaswami[141] in 1912 and has since been recognized
in certain areas of East Asia, the Philippines, Central and South
America, the West Indies, and Turkey. Current interest in P. pseudomallei is related to the appearance of the infection, occasionally
in its fatal form, in men returning from Vietnam. The pseudomallei
group of bacteria was the subject of an excellent review by Howe and
co-workers[142] in 1971. In brief, this organism has been isolated
from stagnant water and rice paddies and can survive for long periods
in tap water. Presumably, it gains entry to the body by ingestion or
through broken skin. In its subclinical form, melioidosis is widespread in Southeast Asia. Although the most common form of infection
is pulmonary, rapidly fatal septicemias, localized or multiple suppurative lesions, and chronic infections have been described. Re-
activation of the infection, sometimes after many years of latency,
may occur in anyone who has ever contracted melioidoses. Although
generally susceptible to readily attainable serum concentrations of

tetracycline, chloramphenicol, or novobiocin, most strains have been found to be resistant to the penicillins, cephalosporins, polymyxins, and aminoglycosides.

Another member of the family Pseudomonadaceae, the genus Aeromonas, is widely distributed in nature, particularly in water, marine animals, and foods. Although usually present in mixed cultures, A. hydrophila was reported[143] to have been isolated from post-traumatic wounds of nine patients; in eight of these, the injury had been contaminated by river or lake water or soil.

Although frequently isolated from extremity wounds by Tong,[103] the clinical significance of the Mimeae is just as uncertain as the taxonomy and nomenclature of these organisms. That they represent normal skin flora[144] undoubtedly accounts for the frequency of their isolation from wounds. Gardner et al.[145] found that Mima polymorpha was frequently not hospital-acquired and could seldom be causally related to infection. Their wound isolates were from chronic open wounds and were thought to represent surface contamination. The isolation of Mimeae from slowly healing wounds and burns was also mentioned by Reynolds and Cluff[146] who suggested that these bacteria were of low virulence but opportunistic in nature. Two cases of post-traumatic infections with Herellea vaginicola were described by Daly et al.;[147] however, in one instance the portal of entry of the organism (which produced an abscess above the elbow) was unknown, and in the other instance the organism appeared to be responsible for causing meningitis after surgical evacuation of a subdural hematoma. In contrast to their findings with M. polymorpha, the isolates of H. vaginicola studied by Gardner et al.[145] were generally nosocomial in origin; however, like M. polymorpha, their wound isolates of H. vaginicola seldom appeared to be of any significance.

Much has been written in recent years regarding infection due to pasteurellae in wounds inflicted by animals. Pasteurella multocida generally has been recognized as part of the normal microflora of the upper respiratory tract of both domestic and wild animals. Epizootics due to this organism have occurred in various animal species. Human infections with P. multocida usually follow animal bites or scratches and may remain localized to the site of injury or may become systemic. Most commonly, a localized infection is seen after cat or dog bites and cat scratches.[148] An extensive bibliography on this subject, principally since 1945, was published by Tindall and Harrison.[149] In this review of P. multocida infections, reports of septicemia, osteomyelitis, tenosynovitis, and meningitis were cited. To those cases reported in the literature and cited in their review, these authors added 11 of their own, all but 1 of which occurred after wounding by a cat or dog (the only exception representing an oppossum bite) and none of which included septicemia or osteomyelitis. Septic arthritis with P. multocida after wounding by an

animal has been reported only rarely.[150-152]   Infection with P.
pneumotropica also has been only rarely reported: in a case reported by Olson and Meadows,[153] a soft-tissue infection of the hand
followed a cat bite; in a report[148] of Pasteurella infections in the
United Kingdom, it was mentioned that P. pneumotropica was isolated
from two infected animal bites; in an article by Miller,[154] two
cases were mentioned as personal communications, in one of which
there developed soft-tissue infection after a dog bite and in the
other of which there developed a fatal septicemia after a dog bite.

Infection with Chromobacterium violaceum is rare and in most
cases has been limited to tropical and subtropical climates. The
organism is a saprophyte of soil and water, and in 5 of the 17 reported cases the infection occurred after a cutaneous injury.[155]
Dauphinais and Robben[156] reported a fatal septicemia in a patient
in Florida who developed an abscess of the knee after an injury
sustained at a beach. Two fatal infections in American soldiers
in Vietnam were reported by Ognibene and Thomas;[157] however, no
portal of entry of C. violaceum was noted. The case reported by
Johnson et al.[155] was in a child in South Carolina who died of septicemia 3 weeks after a splinter injury of her foot. Whether or not
this injury represented the portal of entry of C. violaceum could
not be established; however, the interval between onset of illness
and death has been reported to range from 7 days to 15 months.[155]
Johnson et al.[155] were able to isolate C. violaceum subsequently
from several natural bodies of water in South Carolina.

Mycobacteria. Post-traumatic infections with mycobacteria have
been due to M. fortuitum, M. abscessus (M. chelonei), and M. marinum.
The sources of the first two of these species, in addition to soil
and lower animals,[158-160] may be dirty skin, foreign bodies, contaminated needles or syringes, or contaminated injectable materials.[161]   The type species of M. fortuitum was described by Cruz[162]
and was isolated from a postinjection abscess. Wells et al.[163]
described isolation of this species from enlarged cervical lymph
nodes of a woman after she had extracted her own carious molar
tooth with a pair of scissors. Beck[164] isolated M. fortuitum from
subcutaneous abscesses in the arms of two patients who had received
BCG inoculations and from a buttock abscess of a third patient who
had received a course of iron-dextran injections for anemia. Additional cases of postinjection abscesses have been reported by Owen
et al.,[165] Canilang and Armstrong,[166] Clapper and Whitcomb,[167] and
Hand and Sanford.[161] The cases reported by Hand and Sanford[161]
included one in a patient with a laceration of the right pretibial
area due to a lawn edger, one in a soldier in Vietnam with multiple
gunshot wounds, and one in a patient with a thorn injury of her left
pretibial area. Another infection occurring after a leg injury by
a lawn edger was reported by Offer et al.[168] A number of post-traumatic ophthalmic infections have been reported;[169-174] all but one[171]
of these infections were corneal.

Inasmuch as M. fortuitum and M. abscessus (M. chelonei) resemble one another in some respects and can be easily confused with one another, it is quite possible that in some instances M. fortuitum was actually M. abscessus (M. chelonei).[168] The two species produce identical disease.[161] A case of infection with M. abscessus (M. chelonei) which was established 48 years after the initial trauma was reported by Moore and Frerichs.[175] Postinjection abscesses due to this species have also been reported,[176,177] as has infection in a deep puncture wound.[178]

Swimming pool granulomas due to M. marinum were reported in 290 cases in Colorado in 1962 by Philpott et al.[179] Such granulomas were initially called to the attention of physicians by Linell and Norden [180] in 1954 when they described an epidemic of skin lesions, particularly on the elbows, following scratches and abrasions received in a specific swimming pool in Orebro, Sweden. The organism in both instances was isolated from the swimming pools. Other such cases have been encountered in Hawaii[181,182] and England.[183] Cott et al.[184] and Zeligman[185] have reported cutaneous infections due to M. marinum acquired in tributaries of the Chesapeake Bay. Skin infections with M. marinum have been observed after abrasions received in tropical fish aquariums.[186,187]

Fungi. Members of the general Nocardia, Actinomyces, Mycobacterium, and Streptomyces have been placed in the family Actinomycetaceae and in the order Actinomycetales. Their classification as higher bacteria, rather than as fungi, is based on the absence of chitin or cellulose in their cell walls. However, because Nocardia, Actinomyces, and Streptomyces commonly form mycelia in tissue and in culture, they are usually included in textbooks of mycology and will be so considered in this discussion.

Actinomycosis has occurred after human bite wounds.[87] Klaber[188] reviewed 67 reported cases of cutaneous actinomycosis, of which 31 involved antecedent "vegetation trauma." Davies[189] reported 1 case and reviewed 13 previously reported cases of primary actinomycosis of the breast. Cullen and Sharp[190] described three cases of mixed war wound infections due to anaerobic Actinomyces. Post-traumatic actinomycotic infection of the eye has been reported by Makley et al.[191]

Isolation of Nocardia from soil was described in 1936 by Gordon and Hagan [192] and by others since.[193,194] Cutaneous and subcutaneous nocardiosis mayoccur after traumatic introduction of the organism; however, this interpretation of a localized cutaneous or subcutaneous lesion should be viewed with caution because localized lesions have occurred at the site of a bruise without evidence of any break in the skin surface.[195-197] Cases with localized lesions related to an abrasion with a thorn and to hand injury were cited

by Pizzolato.[196] Cullen and Sharp[190] described two cases of severe
war wounds of the hip in which mixed infections with aerobic actinomycetes developed and one case of postinjection abscess with a
pure culture of an aerobic actinomycete. Post-traumatic ophthalmic
infections with N. asteroides have also been reported.[198-201] No-
cardial mycetoma occurs frequently in the lower extremities and on
the back; the organism presumably is introduced through puncture
wounds and abrasions of the skin.[196] The disease occurs predomi-
nately in tropical and temperate regions and is sporadic and infrequent elsewhere.

The commonest form of sporotrichosis follows cutaneous inoculation of the spores of Sporothrix schenkii in a penetrating wound.
Infection may be classified as epidermal, dermal, or subcutaneous.[202]
Typically, the infection involves the lymphatics which drain the
primary lesion; the lymph nodes enlarge and may suppurate, and the
connecting lymphatics become indurated.

The fungus is widely distributed in nature and has been recovered from soil, plants, and insects.[202] Lurie[202] cited cases of infections after trauma involving wire, nails, woods, gardening, farm-
ing, and a variety of occupations or avocations involving close contact with vegetation and soil. In addition, he cited many cases related to insect bites. Dahl et al.[203] reported an outbreak of lymph-
ocutaneous sporotrichosis in children related to their playing in
stacks of baled prairie hay from which the fungus was recovered.
Although ingestion of alcohol has not been cited as a primary cause
of sporotrichosis, its influence on the activities of its imbibers
has been.[204,205] These last three reports inspired a poem in a
JAMA editorial entitled "Beware, the Sporothrix":[206]

> When drinking wine amongst the roses
> Or guzzling beer while throwing bricks
> Or playing games in bales of hay
> Where lurks the tricky sporothrix,
> Beware, the price you pay for play
> When you get struck by dread mycoses.

Infections acquired in the laboratory, after injections with
contaminated syringes, and even by way of the chin rest of a violin
are also cited by Lurie.[202] Extracutaneous forms of sporotrichosis
are generally attributed to hematogenous dissemination of the fungus,
although cases of post-traumatic intraocular sporotrichosis have been
cited by Francois and Rysselaere[207] and reported by Levy.[208] Extracutaneous forms may involve any tissue or organ of the body, including the mucous membranes, musculoskeletal system, viscera, central
nervous system, and organs of special senses;[202] however, in many
cases the primary lesion is unknown.

Primary mucocutaneous histoplasmosis is extremely rare, inasmuch as most such lesions are secondary to disseminated disease.[209] Primary cutaneous infections have resulted from accidental inoculation of Histoplasma capsulatum into the skin.[210,211] Similarly rare is primary cutaneous coccidioidomycosis, which resembles the primary "chancre" phase of several other granulomatous diseases, which occurs within 3 weeks after inoculation through a break in the skin, and which heals spontaneously within a few weeks.[212] The original report of such an infection was made by Wilson et al.[212] in 1953 and was followed by several reports by other authors.[213-219]

The first reported case of primary cutaneous blastomycosis was by Evans[220] in 1903; a second case was reported in 1951 by Schwarz and Baum;[221] and three other cases were reported in 1955 by Wilson et al.[222] All but one of the patients were physicians who had accidentally inoculated themselves with culture material or during autopsy of patients known to have had blastomycosis; the remaining patient was an autopsy room attendant. A chancriform lesion developed in these cases, similar to that which may be observed in tuberculosis, yaws, syphilis, espundia, sporotrichosis, blastomycosis, nocardiosis, or coccidioidomycosis.[216] Lymphangiitis and lymphadenopathy develop but are limited to the affected extremity.

The subject of primary cutaneous cryptococcosis was recently discussed by Noble and Fajardo.[223] They reviewed five cases, including one of their own, in which the lesion, proved by biopsy and culture, was confined to the skin and in which no systemic involvement became evident for a minimum follow-up of 4 weeks. In no case, however, was cutaneous inoculation proved, and in no case was there any convincing evidence that systemic disease had not occurred.

If the frequent immersion of the hands in water is considered to represent a form of trauma (psychologic, physical, or both), candidiasis, or candidosis as some prefer to call it,[224] of the hands occurs in housewives, chefs, bartenders, and fishmongers. Maceration or damage by moisture is the most important predisposing factor.[224] Otherwise, serious infection with Candida is usually seen in debilitated patients, including those severely traumatized or who have received antimicrobial, corticosteroid, immunosuppressive, irradiation, or hyperalimentation therapy. Breaks in the normal mucocutaneous barriers and urethral and intravenous catheters are common portals of entry for this organism. The presence of underlying endocrine or malignant diseases, malnutrition of malabsorption syndromes, and bone marrow suppression are other predisposing factors.[224]

Eight reported cases of primary phycomycosis were cited by Baker,[225] three of which occurred in extensive thermal burns. Of 40 reported cases of disseminated phycomycosis also collected by Baker,

Table 8.--Fungi Most Frequently Isolated
in Keratomycosis*

Fungus	Cases (no.)
Aspergillus	39
Penicillium	11
Candida	11
Nocardia	9
Cephalosporium	9
Fusarium	7
Blastomyces	5
Sporothrix	5

*Based on data of Gingrich[201] and Zimmerman.[226]

Table 9.--Post-traumatic Intraocular Mycosis:
Endophthalmitis or Vitreous
Abscess*

Organism	Cases (no.)
Actinomyces	1
Aspergillus	8
Sporothrix	1
Cladosporium	1
Mycelial fungi	14
Unidentified fungi	2

*Collected by François and Rysselaere.[207]

2 followed burns and 1 followed septic abortion. Also cited by
Baker is one case of focal mucormycosis in a fracture site in a leg.

Other fungal diseases, excluding those caused by dermatophytes,
which may result from trauma, albeit frequently minor, of the skin
include paracoccidioidomycosis (Touth American blastomycosis), ke-
loidal blastomycosis (Lobo's disease), mycetoma, subcutaneous phycomycosis, and chromoblastomycosis. These diseases are rare on the
North American continent and will not be discussed in this review.
Those who are interested are referred to Roger D. Baker's excellent
book, Human Infection With Fungi, Actinomycetes, and Algae.[225]

There is one final important area of post-traumatic fungal disease which requires review. Oculomycoses are being recognized and
reported more often.[207] Keratomycosis is rarely a primary disease.
Most cases are secondary to trauma; however, other preexisting corneal conditions may also predispose to keratomycosis.[226] Corneal
erosion is usually caused by contaminated vegetable matter; however,
10 to 25% of normal eyes have been found to harbor fungi, and in
diseased eyes, particularly when treated with antimicrobials or
steroids, the incidence has been found to be much higher.[226] The
fungi most frequently isolated in cases of keratomycosis are listed
in Table 8; however, cases with Monosporium, Scopulariopsis, Fusi-
dium, Curvularia, Trichosporon, Gibberella, Absidia, and others
have been reported.[201,226,227] Francois and Rysselaere[207] collected
20 cases of post-traumatic intraocular mycoses (Table 9). The diagnosis of oculomycosis is made by demonstration of hyphae and yeasts
in scrapings and by culture.[228]

CONCLUSION

In this review of the microbiology of trauma, a deliberate effort
has been made to confine the discussion and bibliography to the subject of primary microbial contamination or infection and to avoid,
insofar as possible, the subject of secondary or nosocomial infection. Obviously, patients with serious, or occasionally minor, trauma are frequently hospitalized. Once hospitalized, they are subject
to microbial contamination and infection from an enormous variety
and number of sources. Hospital-acquired infection and its control
are exceedingly complex problems which are beyond the scope of this
review.

It is apparent that, with the exception of rabies, the primary
microbiology of trauma is bacterial (including mycobacterial) and
fungal. It is equally apparent that this microbiology includes an
almost infinite number of species of bacteria and fungi and that
primary contamination and infection are frequently due to mixtures
of microorganisms of either exogenous or endogenous origin. Many

new causes of post-traumatic infection are being recognized, with obvious implications for their management and control.

REFERENCES

1. Klainer, A. S., Beisel, W. R.: Opportunistic infection: a review. Am. J. Med. Sci., 258:431-456, 1969.

2. Ad Hoc Committe of the Committee on Trauma: Postoperative wound infections: the influence of ultraviolet irradiation of the operating room and of various other factors. Ann. Surg., 160 Suppl:1-192, 1964.

3. Dillon, M. L., Postlethwait, R. W., Bowling, K. A.: Operative wound cultures and wound infections: a study of 342 patients. Ann. Surg., 170:1029-1034, 1969.

4. Davidson, A. I. G., Smith, G., Smylie, H. G.: A bacteriological study of the immediate environment of a surgical wound. Br. J. Surg., 58:326-333, 1971.

5. Davidson, A. I. G., Clark, C., Smith, G.: Postoperative wound infection: a computer analysis. Br. J. Surg., 58:333-337, 1971.

6. Altemeier, W. A., Culbertson, W. R.: Surgical infections. In Surgery: Principles and Practice. Fourth edition. Edited by Rhoads, J. E., Allen, J. G., Harkins, H. N., Moyer, C. A. Philadelphia, J. B. Lippincott Company, 1970, pp. 48-51.

7. Burnet, F. M., White, D. O.: Natural History of Infectious Disease. Fourth edition. London, Cambridge University Press, 1972, pp. 70-87.

8. Zucker, T. F., Zucker, L. M.: Nutrition and natural resistance to infection. In Modern Nutrition in Health and Disease: Dietotherapy. Fourth edition. Edited by Wohl, M. G., Goodhart, R. S. Philadelphia, Lea & Febiger, 1968, pp. 600-611.

9. Axelrod, A. E.: Nutrition in relation to acquired immunity. In Modern Nutrition in Health and Disease: Dietotherapy. Fourth edition. Edited by Wohl, M. G., Goodhart, R. S. Philadelphia, Lea & Febiger, 1968, pp. 612-622.

10. Axelrod, A. C.: Immune processes in vitamin deficiency states. Am. J. Clin. Nutr., 24:265-271, 1971.

11. Selvaraj, R. J., Bhat, K. S.: Metabolic and bactericidal ac-
        tivities of leukocytes in protein-calorie malnutrition.
        Am. J. Clin. Nutr., 25:166-174, 1972.

12. Perillie, P. E., Nolan, J. P., Finch, S. C.: Studies of the
        resistance to infection in diabetes mellitus: local exu-
        dative cellular response. J. Lab. Clin. Med., 59:1008-
        1015, 1962.

13. Menkin, V.: Diabetes and inflammation. Science, 93:456-458,
        1941.

14. Cohn, Z. A.: Relation of cell metabolism to infection with
        rickettsial and bacterial agents. Bacteriol Rev., 24:96-
        105, 1960.

15. Cohn, Z. A., Morse, S. I.: Functional and metabolic properties
        of polymorphonuclear leucocytes. I. Observations on the
        requirements and consequences of particle ingestion. J.
        Exp. Med., 111:667-687, 1960.

16. Brayton, R. G., Stokes, P. E., Schwartz, M. S., et al: Effect
        of alcohol and various diseases on leukocyte mobilization,
        phagocytosis and intracellular bacterial killing. N. Engl.
        J. Med., 282:123-128, 1970.

17. Bybee, J. D., Rogers, D. E.: The phagocytic activity of poly-
        morphonuclear leukocytes obtained from patients with dia-
        betes mellitus. J. Lab. Clin. Med., 64:1-13, 1964.

18. Bryant, R. D., DesPrez, R. M. VanWay, M. H., et al: Studies on
        human leukocyte motility. I. Effects of alterations in
        pH, electrolyte concentration, and phagocytosis on leuko-
        cyte migration, adhesiveness, and aggregation. J. Exp.
        Med., 124:483-499, 1966.

19. Mowat, A. G., Baum, J: Chemotaxis of polymorphounuclear leuko-
        cytes from patients with diabetes mellitus. N. Engl. J.
        Med., 284:621-627, 1971.

20. Wilson, W. E. C., Kirkpatrick, C. H., Talmage, D. W.: Suppres-
        sion of immunologic responsiveness in uremia. Ann. Intern.
        Med., 62:1-14, 1965.

20a. Forscher, B. K., Cecil, H. C.: Some effects of alloxan diabetes
        on acute inflammation. J. Appl. Physiol., 13:278-282, 1958.

21. MacLennan, J. D.: The histotoxic clostridial infections of man.
        Bacteriol Rev., 26:177-274, 1962.

22. Dineen, P.: Infections following crushing or gross impact.
     In Infectious Diseases. Edited by Hoeprich, P. D. Hag-
     erstown, Maryland, Harper & Row, Publishers, 1972, pp.
     1215-1217.

23. Dineen, P.: Infections following penetrating injuries. In
     Infectious Diseases. Edited by Hoeprich, P. D. Hagers-
     town, Maryland, Harper & Row, Publishers, 1972, pp. 1219-
     1221.

24. Weed, L. A.: Technics for the isolation of fungi from tissues
     obtained at operation and necropsy. Am. J. Clin. Pathol.,
     29:496-502, 1958.

25. Attebery, H. R., Finegold, S. M.: Combined screw-cap and rub-
     ber-stopper closure for Hungate rubes (pre-reduced anae-
     robically sterilized roll tubes and liquid media). Appl.
     Microbiol., 18:558-561, 1969.

26. Sutter, V. L., Attebery, H. R., Rosenblatt, J. E., et al: Anae-
     robic Bacteriology Manual. Anaerobic Bacteriology Labora-
     tory, Wadsworth Hospital Center, Veterans Administration,
     Los Angeles, The Regents of the University of California,
     1972.

27. Holdeman, L. V., Moore, W. E. E.: Anaerobe Laboratory Manual.
     Blacksburg, Virginia, Virginia Polytechnic Institute and
     State University, 1972.

28. Fleming, A.: Bacteriology of wounds: nature of the infection
     in war wounds. In Surgery of Modern Warfare. Vol. 1.
     Third edition. Edited by Bailey, H. Baltimore, Williams
     & Wilkins Company, 1944, pp. 34-42.

29. Rosenblatt, J. E., Fallon, A., Finegold, S. M.: Comparison of
     methods for isolation of anaerobic bacteria from clinical
     specimens. App. Microbiol., 25:77-85, 1973.

30. Martin, W. J.: Practical method for isolation of anaerobic bac-
     teria in the clinical laboratory. Appl. Microbiol., 22:
     1168-1171, 1971.

31. MacLennan, J. D.: Anaerobic infections of war wounds in the
     Midle East. Lancet, 2:63-66; 94-99; 123-126, 1943.

32. Hall, H. E., Angelotti, R., Lewis, K. H., et al: Characteris-
     tics of Clostridium perfringens strains associated with
     food and food-borne disease. J. Bacteriol., 85:1094-1103,
     1963.

33. Qvist, G.: Anaerobic cellulitis and gas gangrene. Br. Med. J.,
        2:217-221, 1941.

34. Smith, L. D. S., Holdeman, L. V.: The Pathogenic Anaerobic
        Bacteria. Springfield, Illinois, Charles C. Thomas, Pub-
        lisher, 1968, pp. 203-205.

35. Williams, R. E. O., Miles, A. A.: Infection and sepsis in in-
        dustrial wounds of the hand: a bacteriological study of
        aetiology and prophylaxis. Med. Res. Counc. Spec. Rep.
        Ser. (Lond). Report #266:7-73, 1949.

36. Thadepalli, H., Gorbach, S. L., Broido, P., et al: A prospec-
        tive study of infections in penetrating abdominal trauma.
        Am. J. Clin. Nutr., 25:1405-1408, 1972.

37. Bornstein, D. L. Weinberg, A. N. Swartz, M. N., et al: Anae-
        robic infections -- review of current experience. Medi-
        cine (Baltimore), 43:207-232, 1954.

38. Rustigian, R., Cipriani, A.: The bacteriology of open wounds.
        JAMA, 133:224-229, 1947.

39. Pulaski, E. J., Meleney, F. L., Spaeth, W. I. C.: Bacterial
        flora of acute traumatic wounds. Surg. Gynecol. Obstet.,
        72:982-988, 1941.

40. Meleney, F. L.: Bacterial synergism in disease processes: with
        aconfirmation of the synergistic bacterial etiology of a
        certain type of progressive gangrene of the abdominal wall.
        Ann. Surg., 94:961-981, 1931.

41. Hite, K. E., Locke, M., Hesseltine, H. C.: Synergism in exper-
        imental infections with nonsporulating anaerobic bacteria.
        J. Infect. Dis., 84:1-9, 1949.

42. Pirogoff, N.: Cited by Walter, C. W: The Aseptic Treatment
        of Wounds. New York, The Macmillan Company, 1948, p. 1.

43. Altemeier, W. A.: The bacteriology of war wounds: collective
        review. Int. Abstr. Surg., 75:518-533, 1942.

44. Wangensteen, O. H., Wangensteen, S. D.: Military surgeons and
        surgery, old and new: an instructive chapter in management
        of contaminated wounds. Surgery, 62:1102-1124, 1967.

45. Robb-Smith, A. H. T.: Tissue changes induced by Cl. welchii
        type A filtrates. Lancet, 2:362-368, 1945.

46. Gye, R., Rountree, P. M., Loewenthal, J.: Infection of surgi-
    cal wounds with Clostridium welchii. Med. J. Aust., 1:761-
    764, 1961.

47. Wilson, T. S.: Significance of Clostridium welchii infections
    and their relationship to gas gangrene. Can. J. Surg.,
    4:35-42, 1960.

48. Altemeier, W. A., Fullen, W. D.: Prevention and treatment of
    gas gangrene. JAMA, 217:806-813, 1971.

49. Sandusky, W. R. Pulaski, E. J., Johnson, B. A., et al: Anae-
    robic nonhemolytic streptococci in surgical infections
    on general surgical service. Surg. Gynecol. Obstet., 75:
    145-156, 1942.

50. MacLennan, J. D.: Streptococcal infection of muscle. Lancet,
    1:582-584, 1943.

51. Anderson, C. B., Marr, J. J., Jaffe, B.M.: Anaerobic strepto-
    coccal infections simulating gas gangrene. Arch. Surg.,
    104:186-189, 1972.

52. Beathard, G. A., Guckian, J. D.: Necrotizing fasciitis due to
    group A β-hemolytic streptococci. Arch. Intern. Med.,
    120:63-67, 1967.

53. Quintiliani, R. Engh, G. A.: Overwhelming sepsis associated
    with group A beta hemolytic streptococci. J. Bone Joint
    Surg. (Am), 53:1391-1399, 1971.

54. Altemeier, W. A., Culbertson, W. R.: Acute non-clostridial
    crepitatnt cellulitis. Surg. Gynecol. Obstet., 87:206-
    212, 1948.

55. Rein, J. M., Cosman, B.: Bacteroides necrotizing fasciitis of
    the upper extremity: case report. Plast. Reconstr. Surg.,
    48:592-594, 1971.

56. Veillon, A., Zuber, A.: Sur quelques microbes strictement anae-
    robies et leur role dans la pathologie humaine. C R Soc.
    Biol. (Paris), 49:253-255, 1897.

57. Finegold, S. M., Rosenblatt, J. E. Sutter, V. L., et al:
    Scope(R) Monograph on Anaerobic Infections. Kalamazoo,
    Michigan, Upjohn Company, 1972.

58. McCloskey, R. V.: Scarlet fever and necrotizing fasciitis
    caused by coagulase-positive hemolytic Staphylococcus
    aureus, phage type 85. Ann. Intern. Med., 78:85-87, 1973.

59. Levin, M. J., Gardner, P., Waldvogel, F. A.: "Tropical" pyo-
myositis: an unusual infection due to Staphylococcus
aureus. N. Engl. J. Med., 284:196-198, 1971.

60. Altrocchi, P. H.: Spontaneous bacterial myositis. JAMA, 217:
819-820, 1971.

61. Smith, L. D. S., Holdeman, L. V.: The Pathogenic Anaerobic
Bacteria. Springfield, Illinois, Charles C. Thomas, Publisher, 1968, pp. 245-246.

62. Pritchard, J. A., Whalley, P. J.: Abortion complicated by
Clostridium perfringens infection. Am. J. Obstet. Gynecol.
111:484-490, 1971.

63. Berggren, R. B., Batterton, T. D., McArdle, G., et al: Clostridial myositis after parenteral injections. JAMA, 188:
1044-1048, 1964.

64. Smith, L. D. S., Holdeman, L. V.: The Pathogenic Anaerobic
Bacteria. Springfield, Illinois, Charles C. Thomas, Publisher, 1968, pp. 220-238.

65. Willis, A. T.: Anaerobic Bacteriology in Clinical Medicine.
Second edition. London, Butterworth & Co., Ltd., 1964,
pp. 143-145.

66. Smith, L. D. S., George, R. L.: The anaerobic bacterial flora
of clostridial myositis. J. Bacteriol., 51:271-279, 1946.

67. Dhayagude, R. G., Purandare, N. M.: Studies on anaerobic infection. Indian. J. Med. Res., 37:283-292, 1949.

68. DeHaven, K. E., Evarts, C. M.: The continuing problem of gas
gangrene: a review and report of illustrative cases.
J. Trauma, 11:983-991, 1971.

69. Smith, L. D. S., Holdeman, L. V.: The Pathogenic Anaerobic
Bacteria. Springfield, Illinois, Charles C. Thomas, Publisher, 1968, pp. 374-382.

70. Hall, I. C.: The occurrence of Bacillus botulinus, types A and
B, in accidental wounds. J. Bacteriol., 50:213-217, 1945.

71. Thomas, C. G., Jr., Keleher, M. F., McKee, A. P.: Botulism, a
complication of Clostridicum botulinum wound infection.
Arch. Pathol., 51:623-628, 1951.

72. Hampson, C. R.: A case of probably botulism due to wound infection. J. Bacteriol., 61:647, 1951.

73. Davis, J. B., Mattman, L. H., Wiley, M.: Clostridium botulinum in a fatal wound infection. JAMA, 146:646-648, 1951.

74. United States Public Health Service, Center for Disease Control: Type A botulism from wound infection. Morbidity Mortality Weekly Rep., 20:183-184, May 22, 1971.

75. United States Public Health Service, Center for Disease Control: Wound botulism. Morbidity Mortality Weekly Rep., 20:453; 458, December 18, 1971.

76. Cherington, M. Merson, M. H.: Wound Botulism and Treatment of Botulism. International Conference on Anaerobid Bacteria: Bacteriological and Clinical Considerations. United States Department of Health, Education, and Welfare, Public Health Service, Center for Disease Control, Atlanta, Georgia, 1972

77. Smith, L. D. S., Holdeman, L. V.: The Pathogenic Anaerobic Bacteria. Springfield, Illinois, Charles C. Thomas, Publisher, 1968, pp. 308-313.

78. Smith, L. D. S., Holdeman, L. V.: The Pathogenic Anaerobic Bacteria. Springfield, Illinois, Charles C. Thomas, Publisher, 1968, pp. 256-281.

79. LaForce, F. M., Young, L. S., Bennett, J. V.: Tetanus in the United States (1965-1966): epidemiologic and clinical features. N. Engl. J. Med., 280:569-574, 1969.

80. Christensen, N. A., Thurber, D. L.: Clinical experience with tetanus: 91 cases. Proc. Staff Meet Mayo Clin., 32:146-158, 1957.

81. Condit, P. K.: Tetanus in California: epidemiology and a review of 232 cases. Calif. Med., 90:318-321, 1959.

82. De Waal, H. L.: Wound infection: a preliminary note on a combined clinical and bacteriological investigation of 708 wounds. Edinburgh Med. J., 50:577-588, 1943.

83. Gunn, A. A.: Bacteroides septicaemia. J. R. Coll. Surg. Edinb. 2:41-50, 1956.

84. Bodner, S. F., Koenig, M. G., Goodman, J. S.: Bacteremic Bacteroides infections. Ann. Intern. Med., 73:537-544, 1970.

85. Marcoux, J. A., Zabransky, R. J., Washington, J. S. II, et al: Bacteroides bacteremia. Minn. Med., 53:1169-1176, 1970.

86. Felner, J. M., Dowell, V. R., Jr.: "Bacteroides" bacteremia.
     Am. J. Med., 50:787-796, 1971.

87. Guidry, D. J.: Actinomycosis. In Human Infection With Fungi,
     Actinomycetes, and Algae. Edited by Baker, R. D. New
     York, Springer-Verlag, 1971, pp. 1019-1058.

88. Altemeier, W. A., Culbertson, W. R., Fullen, W. D., et al:
     Intraabdominal abscesses. Am. J. Surg., 125:70-78, 1973.

89. Beerens, H., Tahon-Castel, M.: Infections humaines a bacteries
     anaerobies non toxigenes. Brussels, Presses Academiques
     Europeennes, 1965, pp. 136-137.

90. Nettles, J. L., Kelly, P. J., Martin, W. J., et al: Musculo-
     skeletal infections due to Bacteroides: a study of eleven
     cases. J Bone Joint Surg. (Am), 51:230-238, 1969.

91. Ziment, I., Miller, L. G., Finegold, S. M.: Nonsporulating
     anaerobic bacteria in osteomyelitis. Antimicrob Agents
     Chemother, 1967, pp. 77-85, 1968.

92. Pearson, H. E., Harvey, J. P., Jr.: Bacteroides infections in
     orthopedic conditions. Surg. Gynecol. Obstet., 132:876-
     880, 1971.

93. Leake, D. L.: Bacteroides osteomyelitis of the mandible: a
     report of two cases. Oral Surg., 34:585-588, 1972.

94. Rotheram, E. B., Schick, S. F.: Nonclostridial anaerobic bac-
     teria in septic abortion. Am. J. Med., 46:80-89, 1969.

95. Smith, J. S., Southern, P. M., Jr., Lehmann, J. D.: Bacteremia
     in septic abortion: complications and treatment. Obstet.
     Gynecol., 35:704-708, 1970.

96. Gelb, A. F., Seligman, S. J.: Bacteroidaceae bacteremia: ef-
     fect of age and focus of infection upon clinical course.
     JAMA, 212:1038-1041, 1970.

97. Miles, A. A., Schwabacher, H., Cunlifee, A. C., et al: Hospital
     infection of war wounds. Br. Med. J., 2:855-859; 895-900,
     1940.

98. Lindberg, R. B. Wetzler, T. F., Marshall, J. D., et al: The
     bacterial flora of battle wounds at the time of primary
     debridement: a study of the Korean battle casualty. Ann.
     Surg., 141:369-374, 1955.

99. Pulaski, E. J.: War wounds. N. Engl. J. Med., 249:932-938, 1953.

100. Kovaric, J. J., Matsumoto, T.,Dobek, A. S., et al: Bacterial flora of one hundred and twelve combat wounds. Milit. Med., 133:622-624, 1968.

101. Heggers, J. P., Barnes, S. T., Robson, M. C., et al: Microbial flora of orthopaedic war wounds. Milit. Med., 134: 602-603, 1969.

102. Matsumoto, T.,Wyte, S. R., Moseley, R. V., et al: Combat surgery in communication zone. I. War wound and bacteriology (preliminary report). Milit. Med., 134:655-665, 1969

103. Tong, M. J.: Septic complications of war wounds. JAMA, 219: 1044-1047, 1972.

104. Feltis, J. M., Jr.: Surgical experience in a combat zone. Am. J. Surg., 119:275-278, 1970.

105. Nicholls, R. J.: Initial choice of antibiotic treatment for pyogenic hand infections. Lancet,1:225-226, 1973.

106. Fitzgerald, R. H., Jr., Linscheid, R. L., Washington JA II: Unpublished data.

107. Stanley, N. F., Stanley, E. R., Stanley, F., et al: The brolga bites back. Med. J. Aust., 2:1461-1462, 1972.

108. Altemeier, W. A.: The significance of infection in trauma. Bull Am. Coll. Surg., Feb., 1972, pp. 7-16.

109. Allen, H. F., Jr.: Current status of prevention, diagnosis and management of bacterial corneal ulcers. Ann. Ophthalmol., 3:235-246, 1971.

110. Golden, B., Meek, E. S.: Corneal infections. In Infectious Diseases. Edited by Hoeprich, P. D. Hagerstown, Maryland, Harper & Row, Publishers, 1972, pp. 1253-1257.

111. Golden, B. Meek, E. S.: Intraocular infections. In Infectious Diseases. Edited by Hoeprich, P. D. Hagerstown, Maryland, Harper & Row, Publishers, 1972, pp. 1259-1262.

112. Liebow, A. A., Mac Lean, P. D., Bumstead, J. H., et al: Tropical ulcers and cutaneous diphtheria. Arch. Intern. Med., 78:255-295, 1946.

113. Riddell, G. S.: Cutaneous diphtheria: epidemiological and dermatological aspects of 365 cases amongst British prisoners of war in Far East. J. R. Army Med. Corps, 95:64-87, 1950.

114. Belsey, M. A., Sinclair, M., Roder, M. R., et al: Corynebacterium diphtheriae skin infections in Alabama and Louisiana: a factor in the epidemiology of diphtheria. N. Engl. J. Med., 280:135-141, 1969

115. Dixon, J. M. S.,Thorsteinson, S.: Diphtheria bacilli isolated in Alberta in 1967 from the throat, nose, ears and skin. Can. Med. Assoc. J., 101:204-207, 1969.

116. Flor-Henry, P.: Cutaneous diphtheria: a brief historical review and discussion of recent literature, with presentation of two cases. Med. Serv. J. Canada, 17:823-830, 1961.

117. Cockcroft, W. H., Boyko, W. J., Allen, D. E.: Cutaneous infections due to Corynebacterium diphtheriae. Can.Med. Assoc. J., 108:329-331, 1973.

118. Adler, J. L., Burke, J. P., Finland, M.: Infection and antibiotic usage at Boston City Hospital, January, 1970. Arch. Intern. Med., 127:460-465, 1971.

119. Waldvogel, F. A., Medoff, G., Swartz, M. N.: Osteomyelitis: a review of clinical features, therapeutic considerations and unusual aspects. N. Engl. J. Med., 282:260-266, 1970.

120. Duncan, D. W., Razzell, W. E.: Klebsiella biotypes among coliforms isolated from forest environments and farm produce. Appl. Microbiol.,24:933-938, 1972.

121. Washington, J. A. II, Birk, R. J., Ritts, R. E., Jr.: Bacteriologic and epidemiologic characteristics of Enterobacter hafniae and Enterobacter liquefaciens. J. Infect. Dis., 124:379-386, 1971.

122. Gilardi, G. L., Bottone, E., Birnbaum, M.: Unusual fermentative, gram-negative bacilli isolated from clinical specimens. I. Characterization of Erwinia strains of the "lathyri-Herbicola group." Appl. Microbiol.,20:151-155, 1970.

123. Von Graevenitz, A.: Erwinia species isolates. Ann. N. Y. Sci., 174:436-443, 1970.

124. Pien, F. D., Martin, W. J., Hermans, P. E., et al: Clinical
     and bacteriologic observations on the proposed species,
     Enterobacter agglomerans (the Herbicola-Lathyri bacteria).
     Mayo Clin. Proc., 47:739-745, 1972.

125. Komagata, K., Tamagawa, Y., Iizuka, H.: Characteristics of
     Erwinia herbicola. J. Gen. Appl. Microbiol. (Tokyo),14:
     19-37, 1968.

126. Starr, M. P., Chatterjee, A. K.: The genus Erwinia: entero-
     bacteria pathogenic to plants and animals. Annu. Rev.
     Microbiol., 26:389-426, 1972.

127. Gaughran, E. R. L.: From superstition to science: the history
     of a bacterium. Trans N. Y. Acad. Sci., 31:3-24, 1969.

128. Wilkowske, C. J., Washington, J. A. II, Martin, W. J., et al:
     Serratia marcescens: biochemical characteristics, anti-
     biotic susceptibility patterns, and clinical significance.
     JAMA, 214:2157-2162, 1970.

129. Crowder, J. G., Gilkey, G. H., White, A. C.: Serratia marces-
     cens bacteremia: clinical observations and studies of
     precipitin reactions. Arch. Intern. Med., 128:247-253,
     1971.

130. Henjyoji, E. Y., Whitson, T. C., Ohashi, D. K., et al: Bac-
     teremia due to Serratia marcescens. J. Trauma, 11:417-
     421, 1971.

131. Kelly, P. J., Washington, J. A. II, Wilko-ske, C. J.: Unpub-
     lished data.

132. Bigger, J. F., Meltzer, G., Mandell, A., et al: Serratia mar-
     cescens endophthalmitis. Am. J. Ophthalmol., 72:1102-1105,
     1971.

133. Ewing, W. H., Davis, B. R.: Biochemical Characterization of
     Citrobacter freundii and Citrobacter diversus. United
     States Department of Health, Education, and Welfare, Pub-
     lic Health Service, Center for Disease Control, Atlanta,
     Georgia, 1971, pp. 2-3.

134. Gonzalez, A. B., Ruffolo, E. H.: Edwardsiella tarda: etio-
     logic agent in a post-traumatic subgaleal abscess. South.
     Med. J., 59:340-346, 1966.

135. Jordan, G. W., Hadley, W. K.: Human infection with Edwardsi-
     ella tarda. Ann. Intern. Med., 70:283-288, 1969.

136. Kominos, S. D., Copeland, C. E., Grosiak, B., et al: Intro-
     duction of Pseudomonas aeruginosa into a hospital via vege-
     tables. Appl. Microbiol.,24:567-570, 1972.

137. Moncrief, J. A.: Burns. N. Engl. J. Med., 288:444-454, 1973.

138. Gilardi, G. L.: Infrequently encountered Pseudomonas species
     causing infection in humans. Ann. Intern. Med., 77:211-
     215, 1972.

139. Taplin, D., Bassett, D. C. J., Mertz, P. M.: Foot lesions
     associated with Pseudomonas cepacia. Lancet, 2:568-571,
     1971.

140. Ederer, G. M., Matsen, J. M.: Colonization and infection with
     Pseudomonas cepacia. J. Infect. Dis., 125:613-618, 1972.

141. Whitmore, A., Krishnaswami, C. S.: An account of the discovery
     of a hitherto undescribed infective disease occurring among
     the population of Rangoon. Indian. Med. Gaz.,47:262-267,
     1912.

142. Howe, C., Sampath, A., Spotnitz, M.: The pseudomallei group:
     a review. J. Infect. Dis.,124:598-606, 1971.

143. Washington, J. A. II: Aeromonas hydrophila in clinical bacter-
     iologic specimens. Ann. Intern. Med., 76:611-614, 1972.

144. Taplin, D., Rebell, G., Zaias, N.: The human skin as a source
     of Mima-Herellea infections. JAMA, 186:952-955, 1963.

145. Gardner, P., Griffin, W. B., Swartz, M. N., et al: Nonfermen-
     tative gram-negative bacilli of nosocomial interest. Am.
     J. Med., 48:735-749, 1970.

146. Reynolds, R. C., Cluff, L. E.: Infection of man with Mimeae.
     Ann. Intern. Med., 58:759-767, 1963.

147. Daly, A. K., Postic, B., Kass, E. H.: Infections due to organ-
     isms of the genus Herellea: B5W and B. Anitratum. Arch.
     Intern. Med., 110:580-591, 1962.

148. Epidemiology: Pasteurella and Yersinia infections. Br. Med.
     J., 1:320, 1972.

149. Tindall, J. P., Harrison, C. M.: Pasteurella multocida infec-
     following animal injuries, especially cat bites. Arch.
     Dermatol., 105:412-416, 1972.

150. Holmes, M. A., Brandon, G.: Pasteurella multocida infections
     in 16 persons in Oregon. Public Health Rep.,80:1107-
     1112, 1965.

151. Barth, W. F., Healey, L. A., Decker, J. L.: Septic arthritis
     due to Pasteurella multocida complicating rheumatoid arth-
     ritis. Arthritis Rheum.,11:394-399, 1968.

152. Bell, D. B., Marks, M. I., Eickhoff, T. C.: Pasteurella mul-
     tocida arthritis and osteomyelitis. JAMA, 210:343-345,
     1969.

153. Olson, J. R., Meadows, T. R.: Pasteurella pneumotropica in-
     fection resulting from a cat bite. Am. J. Clin. Pathol.
     51:709-710, 1969.

154. Miller, J. K.: Human pasteurellosis in New York State. N. Y.
     State J. Med., 66:2527-2531, 1966.

155. Johnson, W. M., DiSalvo, A. F., Steuer, R. R.: Fatal Chromo-
     bacterium violaceum septicemia. Am. J. Clin. Pathol., 56:
     400-406, 1971.

156. Dauphinais, R. M., Robben, G. G.: Fatal infection due to
     Chromobacterium violaceum in Vietnam. Am. J. Clin. Pathol.
     50:592-597, 1968.

157. Ognibene, A. J., Thomas, E.: Fatal infection due to Chromo-
     bacterium violaceum in Vietnam. Am. J. Clin. Pathol., 54:
     607-610, 1970.

158. Kushner, D. S., McMillen, S., Senderi, M.: Atypical acid-fast
     bacilli. II. Mycobacterium fortuitum: bacteriologic
     characteristics and pathogenicity for laboratory animals.
     Am. Rev. Tuberc., 75:108-122, 1957.

159. Gordon, R. D., Smith, M. M.: Rapidly growing, acid-fast bac-
     teria. II. Species' description of Mycobacterium fortui-
     tum Cruz. J. Bacteriol., 69:502-507, 1955.

160. Wolinsky, E., Rynearson, T.: Mycobacterial flora of soil (ab-
     stract). Am. Rev. Respir. Dis., 94:478-479, 1966.

161. Hand, W. L., Sanford, J. P.: Mycobacterium fortuitum - a human
     pathogen. Ann. Intern. Med., 73:971-977, 1970.

162. Cruz, J. Da. C.: "Mycobacterium fortuitum," um novo bacilo
     acido-resistente patogenico para o homen. Acta Med Rio de
     Janerio, 1:297-301, 1938.

163. Wells, A. Q., Agius, E., Smith, N.: Mycobacterium fortuitum.
      Am. Rev. Tuberc., 72:53-63, 1955.

164. Beck, A.: Mycobacterium fortuitum in abscesses of man. J.
      Clin. Pathol., 18:307-313, 1965.

165. Owen, M., Smith, A., Coultras, J.: Granulomatous lesions
      occurring at site of injections of vaccines and antibio-
      tics. South. Med. J., 56:949-952, 1963.

166. Canilang, B., Armstrong, D.: Subcutaneous abscesses due to
      Mycobacterium fortuitum: report of a case. Am. Rev. Resp.
      Dis., 97:451-454, 1968.

167. Clapper, W. E., Whitcomb, J.: Mycobacterium fortuitum abscess
      at injection site (letter to the editor). JAMA, 202:550,
      1967.

168. Offer, R. C., Karlson, A. G., Spittell, J. A.: Infection
      caused by Mycobacterium fortuitum. Mayo Clin. Proc., 46:
      747-750, 1971.

169. Turner, L., Stinson, I.: Mycobacterium fortuitum as a cause
      of corneal ulcer. Am. J. Ophthalmol., 60:329-331, 1965.

170. Levenson, D. S., Harrison, C. H.: Mycobacterium fortuitum
      corneal ulcer. Arch. Ophthalmol., 75:189-191, 1966.

171. Lauring, L. M., Wergeland, F. L., Sack, G. E.: Anonymous Myco-
      bacterium keratitis. Am. J. Ophthalmol., 67:130-133, 1969.

172. Wunsh, E. E., Boyle, G. L., Leopold, I. H., et al: Mycobac-
      terium fortuitum infection of corneal graft. Arch. Oph-
      thalmol., 82:602-607, 1969.

173. Zimmerman, L. E., Turner, L., McTigue, J. W.: Mycobacterium
      fortuitum infection of the cornea: a report of two cases.
      Arch. Ophthalmol., 82:596-601, 1969.

174. Willis, W. E., Laibson, P. R.: Intractable Mycobacterium
      fortuitum corneal ulcer in man. Am. J. Ophthalmol., 71:
      500-504, 1971.

175. Moore, M., Frerichs, J. B.: An unusual acid-fast infection of
      the knee with subcutaneous, abscess-like lesions of the
      gluteal region: report of a case with a study of the or-
      ganism, Mycobacterium abscessus, n. sp. J. Invest. Derma-
      tol., 20:133-168, 1953.

176. Inman, P. M., Beck, A., Brown, A. E., et al: Outbreak of in-
      jection abscesses due to Mycobacterium abscessus. Arch.
      Dermatol., 100:141-147, 1969.

177. Borghans, J. G. A., Stanford, J. L.: Mycobacterium chelonei
      in abscesses after injection of diphtheria-pertussis-teta-
      nus-polio vaccine. Am. Rev. Respir. Dis., 107:1-8, 1973.

178. Gangadharam, P. R., Hsu, K. H. K.: Mycobacterium abscessus
      infection in a puncture wound. Am. Rev. Respir. Dis.,
      106:275-277, 1972.

179. Philpott, J. A., Jr., Woodburne, A. R., Philpott, O. S., et al:
      Swimming pool granuloma: a study of 290 cases. Arch. Der-
      matol., 88:158-162, 1963.

180. Linell, F., Norden, A.: Mycobacterium balnei: a new acid-fast
      bacillus occurring in swimming pools and capable of pro-
      ducing skin lesions in humans. Acta Tuberc. Scand. Suppl.
      33:1-84, 1954.

181. Sommer, A. F. J., Williams, R. M., Mandel, A. D.: Mycobacterium
      balnei infection: report of two cases. Arch. Dermatol.
      86:316-323, 1962.

182. Walker, H. H., Shinn, M. F., Higaki, M., et al: Some charac-
      teristics of "swimming pool" disease in Hawaii. Hawaii
      Med. J., 21:403-409, 1962.

183. Thomas, D. T.: Swimming-pool granuloma. Br. Med. J., 1:437,
      1967.

184. Cott, R. E., Carter, D. M., Sall, T.: Cutaneous disease caused
      by atypical mycobacteria: report of two chromogen infec-
      tions and review of the subject. Arch. Dermatol.,95:259-
      268, 1967.

185. Zeligman, I.: Mycobacterium marinum granuloma: a disease ac-
      quired in the tributaries of Chesapeake Bay. Arch. Derma-
      tol.,106:26-31, 1972.

186. Swift, S., Cohen, H.: Granulomas of the skin due to Mycobac-
      terium balnei after abrasions from a fish tank. N. Engl.
      J. Med., 267:1244-1246, 1962.

187. Adams, R. M., Remington, J. S., Steinberg, J., et al: Tropical
      fish aquariums: a source of Mycobacterium marinum infec-
      tions resembling sporotrichosis. JAMA, 211:457-461, 1970.

188. Klaber, R.: Primary cutaneous actinomycosis: with a note on
        the bacillus actinomycetem comitans. Br. J. Dermatol., 46:
        12-19, 1934

189. Davies, J. A. L.: Primary actinomycosis of the breast, Br. J.
        Surg., 38:378-381, 1951.

190. Cullen, C. H. Sharp, M. E.: Infection of wounds with Actinomy-
        ces. J. Bone Joint Surg. (Br), 33:221-227, 1951.

191. Makley, T. A., Kissen, A. T., Suie, T.: Mycotic infections of
        the eye (report of six cases). Ohio State Med. J., 57:45-
        47, 1961.

192. Gordon, R. D., Hagan, W. A.: A study of some acid-fast acti-
        nomycetes from soil with special reference to pathogenicity
        for animals. J. Infect. Dis.,59:200-206, 1936.

193. McClung, N. M.: Isolation of Nocardia asteroides from soils.
        Mycologia, 52:154-156, 1960.

194. Gonzalez-Ochoa, A.: Mycetomas caused by Nocardia brasiliensis;
        with a note on the isolation of the causative organism from
        soil. Lab. Invest.,11:1118-1123, 1962.

195. Emmons, C. W., Binford, C. H., Utz, J. P.: Medical Mycology.
        Second edition. Philadelphia, Lea & Febiger, 1970, pp. 94-
        108.

196. Pizzolato, P.: Nocardiosis. In Human Infection With Fungi,
        Actinomycetes, and Algae. Edited by Baker, R. D., New York
        Springer-Verlag, 1971, pp. 1059-1080.

197. Goodman, J. S., Koenig, M. G.: Nocardia infections in a gener-
        al hospital. Ann. N. Y. Acad. Sci., 174:552-567, 1970.

198. Schardt, W. M., Unsworth, A. C., Hayes, C. V.: Corneal ulcer
        due to Nocardia asteroides. Am. J. Ophthalmol., 42: 303-
        305, 1956.

199. Henderson, J. S., Wellman, W. E., Weed, L. A.: Nocardiosis of
        the eye: report of case. Proc. Staff Meet Mayo Clin., 35:
        614-618, 1960.

200. Newmark, E., Polack, F. M., Ellison, A. C.: Report of a case
        of Nocardia asteroides keratitis. Am. J. Ophthalmol., 72:
        813-815, 1971.

201. Gingrich, W. D.: Fungus diseases -- etiology. In Symposium on

Infectious Diseases of the Conjunctiva and Cornea, New
Orleans Academy of Ophthalmology. St. Louis, C. V. Mosby
Company, 1963, pp. 148-156.

202. Lurie, H. I.: Sporotrichosis. In Human Infection With Fungi,
Actinomycetes, and Algae. Edited by Baker, R. D. New
York, Springer-Verlag, 1971, pp. 614-675.

203. Dahl, B. A., Silberfarb, P. M., Sarosi, G. A., et al: Sporotrichosis in children: report of an epidemic. JAMA, 215:
1980-1982, 1971.

204. Kedes, L. H., Siemienski, J., Braude, A. I.: The syndrome of
the alcoholic rose gardener: sporotrichosis of the radial
tendon sheath; report of a case cured with amphotericin B.
Ann. Intern. Med., 61:1139-1141, 1964.

205. Sanders, E.: Cutaneous sporotrichosis: beer, bricks, and
bumps. Arch. Intern. Med., 127:482-483, 1971.

206. Editorial: Beware, the Sporothrix. JAMA, 215:1976, 1971.

207. Francois, E. J., Rysselaere, M.: Oculomycoses. Springfield,
Illinois, Charles C. Thomas, Publisher, 1972, pp. 48-107.

208. Levy, J. H.: Intraocular sporotrichosis: report of a case.
Arch. Ophthalmol.,85:574-579, 1971.

209. Curtis, A. C., Grekin, J. N.: Histoplasmosis: a review of
the cutaneous and adjacent mucous membrane manifestations
with a report of three cases. JAMA, 134:1217-1223, 1947.

210. Tosh, F. E., Balhuizen, J., Yates, J. L., et al: Primary cutaneous histoplasmosis. Arch. Intern. Med., 114:118-119,
1964.

211. Tesh, R. B., Schneidau, J. D., Jr.: Primary cutaneous histoplasmosis. N. Engl. J. Med.,275:597-599, 1966.

212. Wilson, J. W., Smith, C. E., Plunkett, O. A.: Primary cutaneous coccidioidomycosis: the criteria for diagnosis and a
report of a case. Calif. Med.,79:233-239, 1953.

213. Trimble, J. R., Doucette, J.: Primary cutaneous coccidioidomycosis: report of a case of a laboratory infection. Arch.
Dermatol., 74:405-410, 1956.

214. Wright, E. T., Newcomer, V. D., Nelson, N. H.: Primary inoculation coccidioidomycosis (abstract). Arch. Dermatol.,
79:118-119, 1959.

215. Wilson, J. W., Wright, E. T.: Cured primary cutaneous cocci-
     dioidomycosis (abstract). Arch. Dermatol., 81:300, 1960.

216. Wilson, J. W.: Cutaneous (chancriform) syndrome in deep my-
     coses. Arch. Dermatol., 87:81-85, 1963.

217. Winn, W. A.: Primary cutaneous coccidioidomycosis. Arch. Der-
     matol., 92:221-228, 1965.

218. Levan, N. E., Huntington, R. W., Jr.: Primary cutaneous cocci-
     dioidomycosis in agricultural workers. Arch. Dermatol.,
     92:215-220, 1965.

219. Huntington, R. W., Jr.: Coccidioidomycosis. In Human Infec-
     tion With Fungi, Actinomycetes, and Algae. Edited by
     Baker, R. D. New York, Springer-Verlag, 1971, pp. 147-210.

220. Evans, N.: A clinical report of a case of blastomycosis of the
     skin from accidental inoculation. JAMA, 40:1772-1775, 1903

221. Schwarz, J., Baum, G. L.: Blastomycosis. Am. J. Clin. Pathol.
     21:999-1029, 1951.

222. Wilson, J. W., Cawley, E. P., Weidman, F. D., et al: Primary
     cutaneous North American blastomycosis. Arch. Dermatol.,
     71:39-45, 1955.

223. Noble, R. C. Fajardo, L. F.: Primary cutaneous cryptococcosis:
     review and morphologic study. Am. J. Clin. Pathol. 57:13-
     22, 1972.

224. Winner, H. I.: Candidosis. In Human Infection With Fungi,
     Actinomycetes, and Algae. Edited by Baker, R. D. New
     York, Springer-Verlag, 1971, pp. 832-918.

225. Baker, R. D.: Mycormycosis. In Human Infection With Fungi,
     Actinomycetes, and Algae. Edited by Baker, R. D. New
     York, Springer-Verlag, 1971, pp. 832-918.

226. Zimmerman, L. E.: Keratomycosis. Survey Ophthalmol., 8:1-
     25, 1963.

227. Jones, B. R., Richards, A. B., Morgan, G.: Direct fungal
     infection of the eye in Britain. Trans Ophthalmol. Soc.
     UK, 89:727-721, 1970.

228, Francois, E. J., Rysselaere, M.: Oculomycoses. Springfield,
     Illinois, Charles C. Thomas, Publisher, 1972, pp. 17-34.

PHYSIOLOGIC EFFECTS OF ENDOTOXEMIA

Hugh C. Gilbert

Gerald S. Moss

Abraham Lincoln School of Medicine

INTRODUCTION

Since the recognition of the specific shock syndrome associated with gram negative septicemia, the pathophysiology of endotoxemia has received world-wide attention. Numerous investigators have contributed to the understanding of the pathophysiologic interplay felt to be a part of the endotoxic syndrome. This effort has spanned the spectrum of the life sciences and has resulted in a wealth of information focusing on various possible pathophysiologic mechanisms.

Concurrent with this scientific attack, there has been a 5-8 fold increase in the reported incidence of gram negative septicemia. McCabe and his associates[28] have indicated that as many as 300,000 patients are hospitalized because of gram negative infections yearly and that more than 100,000 deaths result from gram negative bacteremia each year in the United States. In the last 2½ years, at the University of Illinois Hospitals, 342 patients were treated for gram negative septicemia with a 20% mortality rate. Even though the impetus has been to find specific pathophysiologic mechanisms amenable to treatment, the mechanism or mechanisms by which endotoxins induce lethality have not been fully clarified.

This communication summarizes some of the important aspects of the pathophysiology of endotoxemia. Because of species variability, emphasis will be placed on sub-human primate models.

BIOCHEMISTRY OF ENDOTOXIN

The spectrum of activity of endotoxin is astounding. "They can effect structure and function of numerous organs and cells, change tissue and blood levels of many enzymes, modify carbohydrate, fat, and protein metabolism, raise or lower body temperature, increase or decrease resistence to bacterial and viral infections and other noxious stimuli (including themselves), cause hemorrhage and increase coagulation of blood, modify hemodynamics in every accessible ana- tomical site, cause or prevent shock, modify gastric secretion, destroy tumors, and effect the function of several endocrine glands."[3]

The designation "endotoxin" originates from the viewpoint that this material was an internal constituent of gram-negative bacteria. Early in its history, it was felt that disruption of bacterial cells was essential for release of endotoxin. Endotoxin can be extracted from intact bacteria and today a variety of agents are used to ex- tract endotoxin. Hundreds of enterobacterial lipopolysaccharides have been extracted each having a specific antigenicity as well as toxicity. The antigencity of endotoxin resides in the carbohydrate moiety. These are, for the most part, identical to the O somatic antigens.

The specific chemical arrangement of the phospholipid, protein and carbohydrate moieties which together represent the lipopolysac- charides termed endotoxins is largely unknown.

EFFECTS OF ENDOTOXIN ON CELLULAR METABOLISM

Endotoxin exerts profound effects on cellular metabolism. In vitro metabolic studies demonstrate that endotoxin has a glycolytic effect comparable to the addition of exogenous insulin. In vivo effects reflect not only the direct effects of lipopolysaccharides, but also the profound effects of cellular hypoperfusion which results from endotoxic shock. When cellular hypoperfusion occurs, substrates such as oxygen, glucose, amino-acids and fatty acids are available for cellular metabolism in reduced amounts. This occurs as a result of decreased cellular fluxes. The reduction in oxygen availability produces a constellation of cellular changes. It has been demon- strated that pyruvate, not being able to be transformed into citrate or oxaloacetate, is diverted toward lactic acid. This switch from aerobic to an anerobic glycolytic cycle has a profound effect on the availability of energy rich phosphates to maintain the active trans- port of glucose in amino-acids and fatty acids across the cell mem- brane. The effect on specific enzyme systems is currently an active area of investigation. These studies focus attention to the mito- chondria as a probable site where endotoxin exerts its influence on cellular metabolism.

The effect of endotoxemia on rat liver mitochondrial metabolism of succinate has been extensively studied. An inhibition of succinic dehydrogenase has been identified in endotoxemia. This biochemical change has been correlated with conformational abnormalities in morphology. White and his co-workers[43] have found that mitochondria become swollen when exposed to endotoxin suggesting an alteration in membrane permeability. In addition, the functional ability of the mitochondria to alter its structure in the presence of substrate was impaired. This swelling of mitochondria after exposure to endotoxin has been described by others and seems to be a consistent morphological abnormality.

The energy deficit secondary to endotoxin challenge results in a decrease in the production of energy rich organic phosphate by the uncoupling of oxidative phosphorylation. The mechanisms that result in this uncoupling of oxidation and respiration are not fully known. Tryptophan metabolism is diminished because of inhibition of the liver enzymes, tryptophan pyrolase. This reduces the production of nicotinic acid which is a necessary substrate for nicotinamide adenine dinucleotide (NAD, a substance necessary for electron transport). Malfunction of membrane transport after endotoxin challenge has been described. It is unknown whether this results from a primary effect on the functional integrity of the membrane or is an indirect effect of decreased ATP.

The fragility of the lysosome in low flow states has been demonstrated. Increases in acid hydrolases during shock, induced by an endotoxic challenge have been described. It has been speculated that lysosomal fragility and rupture causing release of acid hydrolases produces an intracellular autodigestion of hypoperfused cells. This could result in alteration in cellular metabolism and effect membrane permeability. This has been offered as a second mechanism by which endotoxin effects cellular metabolism.

DETECTION OF ENDOTOXIN

A necessary condition for the study of the early effects of endotoxemia is detecting its presence in a sufficiently small concentration early in the course of the disease. Biological assays have been unsatisfactory because they take excessive time and require the use of intact laboratory animals. A new technique has been devised to provide early detection of endotoxemia. This method relies on the finding that very small concentrations of gram-negative endotoxin causes gelation of lysates of the amebocytes of the horseshoe crab (limulus polyphemus). Levin and his co-workers[26] have demonstrated that there is a good correlation between the presence of

endotoxemia and a positive gelation of limulus lysates. The limulus assay appears to be an important adjunct to the diagnosed gram-
negative endotoxemia. The assay is sufficiently sensitive (.005 -
.0005 g/ml). It can be performed in two hours and does not depend upon the presence of viable organisms. False-negatives, secondary to massive antibiotic treatment, have not been experienced.
For these reasons the limulus assay has been heralded as a sensitive, reliable and practical method for the detection of endotoxemia.

EFFECT OF ENDOTOXIN ON CARDIAC FUNCTION

The underlying mechanisms by which endotoxin produces its shock
syndrome are largely unknown. The apparent disparity between tissue
perfusion and cell function has lead to the study of central effects
which may be important in septicemia.

In 1962, Alican, Doltan and Hardy[1] compared the contractile
force of the right ventricle of open chested dogs subjected to
endotoxin with control dog hearts. They reported an initial increase
in the contractile force followed by a decline. This pattern reflected the fluctuation of the arterial pressure. Cardiac output
decreased with venous return. They concluded that peripheral pooling
may account for these changes.

Solis and Downing[40] reported on the effect of endotoxemia on
Frank-Starling Function Curves in isolated cat hearts. Afterload
(aortic distolic pressure), heart rate, and preload (left ventricular
end diastolic pressure) could be controlled. The infusion of 5 mg/kg
of E. coli endotoxin shifted the Frank-Starling Function Curves to
the right indicating a decreased "ventricular function." These
function curves remained depressed and right shifted throughout the
two hour study period. These investigators concluded that endotoxemia has a primary effect on the myocardium as manifested by the
depressed ventricular function curves. The addition of isoproterenol
tended to normalize ventricular function.

In the same year, Kutner and Cohen[23] reported on the effect of
endotoxemia on isolated cat papillary muscle. The effect of E. coli
endotoxin was assessed by adding 0.1 to 9.2 mgm of purified endotoxin
to test bathing solutions. No measurable effect on inotropy was
noted following the addition of endotoxin.

Siegel and Fabian[38] reported in 1967 on the effect of endotoxin
on denervated dog hearts. "Myocardial contractility" was assessed
using the isometric time-tension index. This index is calculated
by dividing the maximum rate of pressure development (dp/dt) by the
area under the rising portion of the isometric pressure curve, de-

termined by the r wave of the electrocardiogram to the peak dp/dt. These investigators suggest that endotoxin decreased myocardial contractility. Isoproterenol was found to augment the isometric time-tension index in endotoxemia.

Priano, Wilson, and Traber[32] studied the effect of endotoxemia on isolated dog heart lung preparations using the maximum rate rise of the intraventricular pressure (dp/dt) as an index of myocardial contractility. Preload and afterload were carefully monitored and heart rate controlled. High fidelity tracings of the ventricular pressure trace were obtained with a catheter tip transducer. Negative inotropic effects were absent in heart lung preparations studied for three hours.

The work of Lefer and his co-workers[25] has called attention to the possible role of toxic factors developing during a variety of low flow states. A peptide of 800-1000 M. W. has been identified which depresses the developed tension of isolated cat papillary muscles. This peptide termed "the myocardial depressant factor" (MDF) has been implicated in endotoxemia. MDF has been found to depress the developed tension and prolong the duration of the action potential in cat papillary muscle. MDF has been postulated to exert its negative inotropic influence by effecting excitation-contraction coupling. This concept has been suggested because the addition of calcium ions normalizes the developed tension in muscles subjected to MDF.

This provocative finding stimulated a number of studies aimed at evaluating "myocardial performance" in endotoxemia. Early studies by Hinshaw, Archer, Greenfield, and Guenter[16] failed to detect any significant impairment of cardiac performance in isolated dog hearts subjected to three hours of endotoxemia. Hinshaw's model provides for isolating a dog's left ventricle and delivering blood flow from a roller-pump to the left ventricle of the denervated preparation. The aortic outflow of the test heart is returned to the support dog's femoral vein via a reservoir and a second roller-pump. Adjusting the resistence of the connecting tubes alters preload and afterload. The intrinsic heart rate of the test heart is not controlled. Ventricular pressures are measured via transducers attached to a fluid filled canula inserted into the apex of the test left ventricle. Myocardial performance was unchanged after three hours of endotoxemia. However, alterations in "cardiac performance" were reported in hearts taken from animals subjected to endotoxemia for five to seven hours. A reduction in "cardiac performance" was demonstrated by the inability of these hearts to perform normally at afterloads exceeding 100 mm Hg. dp/dt was diminished in hearts harvested from animals subjected to five to seven hours of endotoxemia.

CRITERIA FOR ASSESSING MYOCARDIAL PERFORMANCE

The criteria for assessing myocardial performance has been controversial for a number of reasons. Since the heart's prime function is delivery of sufficient oxygenated blood to meet metabolic requirements, the determination of cardiac output has been considered as the prime determinant of myocardial performance. Brunwald[5] has pointed out that there may be a disparity between the output of the heart and its "contractile state." He suggests that the cardiac output is dependent on three separate factors: 1) the preload, which determines the heart muscle fibers and diastolic length 2) the afterload, which is closely related to the intermyocardial systolic tension and 3) the contractile state of the heart. It follows that only when afterload and preload are constant will the extent of myocardial fiber shortening depend directly on the heart's contractile state. Heart rate is the 4th major determinant and its effect on this delicate system is complex since sympathetic stimulation produces alteration in both heart rate and contractile state. Therefore, the depressed cardiac output seen in hemorrhagic shock may be explained solely on the diminution of preload and while the ability of the heart to perform as a muscle may remain unimpaired. The same sort of effect has been demonstrated with the cardiac glycosides which exert powerful positive inotropic influences, yet do not elevate the cardiac output in normal subjects. For the above reasons, methods were devised to consider how the heart performs as a muscle.

MYOCARDIAL CONTRACTILITY

The ability of the heart to contract is dependent on the inherent arrangement of the contractile elements within each sarcomere. Quantification of this relationship has been studied extensively. By contractility we mean the ability of the heart to develop force independent of the existing loading conditions. A variety of indices felt to correlate with changes in the inotropic state have been proposed. Central to most of these indices has been the measurement of the ventricular pressure trace and its first derivative (dp/dt). Serious errors in the estimation of the first derivative can result if fluid filled catheters are used for the recording of the ventricular pressure trace. Catheter tipped transducers with linear frequency responses beyond 100 Hz coupled to accurate electronic recording devices permit high fidelity measurements of the ventricular pressure trace.

The use of dp/dt alone as an index of contractility has been questioned because it has been found to change with alteration of loading conditions alone. Variants of dp/dt as the dp/dt rise time, and the acceleration index ($dp/dt_{max}/dp/dt$ rise time) have the same inherent errors.

The time tension index developed by Siegel and Sonneblick[37] is effected by afterload. This limitation precludes its use in the study of endotoxemia because any alteration in aortic pressure could on its own, independent of changes in the contractile state of the heart, effect the time tension index.

Recently, concepts of skeletal muscle mechanisms have been applied to the intact heart. The force velocity relationship of skeletal muscle has allowed quantification of the contractile state by extrapolating the velocity of the contractile element to zero load. This number, V_{max} has been found to change when the contractile state of the muscle has been altered. The equation relating the force contraction to the velocity of shortening in skeletal muscle happens to be represented by a hyperbola. Hill's model[15] of skeletal muscle provide for a contractile element, CE, linked in series with an elastic element, SE. This formulation has been extended to cardiac muscle and tested in papillary muscles as well as in the intact heart. A third element, the parallel elastic component was added to account for resting tension in heart muscle. It is believed that during isovolumic contraction the rate of stretch of the series elastic element is equal to the rate of shortening of the contractile element. The velocity of shortening (V_{ce}) therefore can be estimated by the following equation:

$$V_{ce} = V_{se} = \frac{dp/dt}{K} \cdot \text{Developed Pressure}$$

where K is the series elastic constant estimated from papillary muscle experiments. It has been suggested that when V_{ce} is plotted against developed pressure a hyperbolic force velocity relationship is described. Extrapolation to zero developed pressure could permit estimation of the maximum velocity of the contractile element, V_{max}, for the intact ventricle. V_{max} should be independent of loading conditions and reflect solely on the dynamic changes in inotropy alone.

There is growing concern that application of the force velocity relationship described for skeletal muscle is not valid in the intact heart. In order to use the V_{ce}-developed pressure relationship, one has to assume that the isovolumic phase is truly a constant volume, that ventricular geometry and wall thickness are constant during the isovolumic phase and that the K constant derived from cat papillary muscles applies to all ventricles.

We have been using the estimate of the velocity of the contractile element at 50 mm Hg developed pressure ($V_{ce}50$) as an index of the contractile state of the intact heart. This index has been shown in our laboratory to be sensitive to alterations in the inotropic state while remaining essentially unchanged when loading conditions are altered. This index has an added advantage because it does not require extrapolation to zero developed pressure. We have noted that as the developed pressure approached zero, the variability of the estimate to V_{ce} increases making extrapolation to zero load inconstant from beat to beat.

The Effect of Endotoxemia on the Myocardial Contractility of the Awake, Intact Baboon

Continuing studies are underway investigating the effect of endotoxin on the myocardial contractility of the awake, intact baboon. Access to the left ventricle is made possible by a chronically implanted polyethylene catheter. On the day of the study, the catheter is retrieved from a subcutaneous pocket and free flow of ventricular blood established. A catheter tipped transducer is threaded into the left ventricle via the polyethylene catheter allowing for a high fidelity recording of the ventricular pressure trace. This pressure signal is differentiated electronically and recorded on magnetic tape. Baboons are placed in a restraining chair and subjected to endotoxemia. In some recently completed studies, endotoxemia was produced with the administration of a 4.55 mg/kg bolus of endotoxin followed by a constant infusion of 1 mg/kg/hr. Myocardial contractility was assessed at two and four hours into endotoxemia. No significant change in dp/dt_{50} was found to occur during the period studied. Current studies are underway to consider the late effects on myocardial contractility. Preliminary analysis of data suggests that no appreciable differences are noted in V_{ce50} when comparing animals subjected to 10 hours of endotoxemia to control animals restrained in a chair for 10 hours.

EFFECT OF ENDOTOXIN ON THE LUNGS

Included among the prominent physiologic disturbances studied in endotoxemia are alterations on lung structure and function. The following morphologic changes have been ascribed to occur in rodent lung after endotoxin:

Clumping of polymorphonuclear leukocytes.
Deposition of fibrin.
Rupture of capillary walls.
Platelet agglutination within capillaries.

This constellation of morphologic changes has focused attention on the lung as an important "shock" organ in endotoxemia. Two prevailing view points have developed, each of which may play an important role in the pathophysiology of endotoxemia.

Stein and Thomas[41] focused attention on the possibility that humoral factors may effect pulmonary function in endotoxemia. They found that during endotoxemia, airway constriction could be demonstrated in dogs. Serotonin antagonists prevented this alteration in airway resistance suggesting that serotonin may be important in this pathophysiologic alteration.

Davis, Meeker and McQuarrie[8] found serum serotonin levels in the pulmonary artery of dogs exceeded significantly those in the femoral artery after endotoxin injection. This group has also reported increases in plasma serotonin and histamine levels following endotoxin administration in rabbits. In this series of experiments, a fall in platelet count paralleled the increases in plasma serotonin levels suggesting that platelets could be the source of the increase in circulating serotonin levels.

The role of leukocytes and platelets in the pulmonary response to endotoxin is controversial. Some believe they are important only as sources of humoral factors. Others, suggest that microemboli and disseminated intravascular coagulation (DIC) play important roles in the pathophysiology of endotoxemia.

Kux, Coalson, Massion and Gunter[24] have presented evidence suggesting that the increases in pulmonary vascular resistance seen in perfused dog lungs following the addition of endotoxin is not due to mechanical obstruction of the vascular bed by leukocytes and platelets.

Robb, Marqulis and Jabs[34] demonstrated showers of microemboli developing after endotoxemia in rabbit lung.

McKay et al. [30] reported on ultrastructural changes in monkeys subjected to endotoxemia. After 15 minutes, leukocytes were observed distending the pulmonary capillary; at one hour, fibrin strands were demonstrated. These authors have placed emphasis on the possibility of DIC occurring as well as leukocyte sequestration and platelet agglutination. In contrast, Coalson, Hinshaw and Guenter[7] could not demonstrate any evidence of fibrin strands in lungs harvested from monkeys subjected to endotoxemia.

Species variability is an important probable cause for this divergence of opinion. It is not unexpected that rabbits with a limited fibrinolytic system demonstrate microemboli and intravascular coagulation following endotoxin administration (Schwartzman Phenomenon). The increase in pulmonary artery pressure seen in monkeys and cats is not a prominent feature of baboon or dog endotoxemia. For this reason, a unified concept to the pathophysiology of endotoxemia and pulmonary function cannot be determined.

Clinical data has placed emphasis on A-V shunting, congestive

atelectasis and interstitial edema as pulmonary manifestations of
sepsis. These alterations are non-specific and occur in a variety
of conditions other than endotoxemia. Microemboli, with or without
DIC or release of vasoactive peptides could account for the clinical
picture described above.

OXYGEN DELIVERY IN ENDOTOXEMIA

The affinity of hemoglobin for oxygen has been shown to respond
to a variety of influences. These changes are reflected in shifts
of the oxyhemoglobin dissociation curve. When the oxyhemoglobin
dissociation curve shifts to the left, hemoglobin increases its
affinity for oxygen and decreases its off-loading of oxygen. A
right shifted curve indicates off-loading is enhanced. The sigmoid
shape of the dissociation curve indicates that shifts in the location
of the oxyhemoglobin curve will have greater effects on the steeply
sloped central portion than on the upper portion. This central region defines the state of hemoglobin's affinity at the tissue capillary while the upper flat portion relates to oxygen uptake in the
pulmonary capillary.

The partial pressure of oxygen at which 50% of the hemoglobin
is saturated (P_{50}) defines the position of the oxyhemoglobin dissociation curve. The normal P_{50} of freshly drawn volunteer blood is
approximately 26 mm Hg. There are a constellation of factors which
change in endotoxemia that are known to effect the P_{50}. Increases
in temperature decreases the affinity of hemoglobin for oxygen and
increases the P_{50}. Increases in pCO_2 decrease hemoglobin's affinity
for oxygen and result in an increase in P_{50}. The inverse relationship between 2,3 diphosphoglyceric acid and ATP to oxygen affinity
of hemoglobin has been an exciting area of study. The effect of
pH on the oxyhemoglobin dissociation curve has been defined by Bohr.[8]
Increases in pH shifts the oxyhemoglobin curve to the left

$$\frac{(\Delta \log P_{50})}{\Delta \log pH}$$

while a decrease in pH causes the opposite effect. The dynamic
inter-relationship of these factors determines the ability of hemoglobin to off-load oxygen. Alteration of off-loading has recently
been stressed as an important "chemical shunt" which may play a significant role in acutely ill patients.

McCoon and Del Guercio[29] have reported alterations in the P_{50}
of 15 septic patients. In contrast to the hypothetical rightward
shift expected in the face of acidosis (Bohr effect) they reported
that the presence of acidosis was usually associated with a left

shifted oxyhemoglobin dissociation curve. This apparent conflict has been resolved by implicating other factors, the so-called non-Bohr metabolic opposing effects. One of these factors is the influence of pH on the synthesis of 2,3 diphosphoglyceric acid. In the presence of acidosis there is an inhibition of hexokinase resulting in a reduction of 2,3 diphosphoglyceric acid production. This would tend to shift the oxyhemoglobin dissociation curve to the left.

The degree by which a left shifted oxygen dissociation curve effects the oxygen flux at the cellular level is still unknown. Chance[36] has suggested that it takes approximately a 1 mm oxygen gradient to allow the mitochondria to be adequately perfused.

Current studies in our laboratory are underway to determine the effect of endotoxin on P_{50} of whole blood and stroma free hemoglobin solution. In vitro incubation of 3.5 mg of purified endotoxin with 6 ml of heparinized blood obtained from normal volunteers has not effected the P_{50} when compared to controls. This data suggests that the effect of endotoxin on the functional integrity of the erythrocyte with respect to oxygen kinetics is not reflected in changes in the oxyhemoglobin dissociation curve.

EFFECT OF ENDOTOXIN ON RENAL FUNCTION

The study of the effect of endotoxin on kidney function is difficult for a number of reasons. The hypotensive response following endotoxin injection is felt to reduce the glomerular filtration rate, renal blood flow, and diminish urine production. Reflex renovascular constriction has been observed in low flow states which reduces the effective renal blood flow even more. Finally, experimental endotoxemia, regardless of animal model, is not a steady state condition. This prevents quantification of clearance studies.

The essential questions which could determine the effect of endotoxin on renal function are:

1. To what degree is the distribution of renal blood flow effected after endotoxemia?

2. Does this alteration in renal blood effect the glomerular filtration rate?

3. Does endotoxin effect tubular function?

Injecting endotoxin directly into the renal artery of a dog produces a prompt increase in renal vascular resistance. This observation has been confirmed by Cavanagh[6] in the baboon and Siegel[38] and Hinshaw[18] in canines. Hermreck and Thal,[13] however, found that the production of sepsis in the hind limb of canines reduced renal vascular resistance and increased renal blood flow by 60%. They have suggested that powerful vasodilatory substances are produced in the septic limb that causes an increase in renal blood flow.

Hinshaw, Bradley, and Carlson[19] designed a model which approximated a steady state. They perfused isolated dog kidneys controlling flow and resistance. After measuring clearance of creatinine and extraction of PAH, these investigators suggested that tubular function was preserved during endotoxemia.

Hinshaw et al. [18] repeated these measurements in intact canines. In this study, renal function was found to be dependent on the administered dose of endotoxin. These investigators concluded that direct vasoconstriction occurs followed by further vasoconstriction secondary to vasoactive substances transported to the kidney. The combined action of systemic hypotension and chemical mediators was felt to decrease renal blood flow. The decrease in PAH extraction demonstrated reduced tubular function which was felt to reflect on hypoxia at the tubular level.

A prevailing attitude has been that the functional integrity of the kidneys is preserved in endotoxemia if renal blood flow remains adequate. This concept is underscored by numerous clinical reports. Strauch et al.[42] found that lethal endotoxemia was associated with the absence of a measurable osmolar clearance.

Analysis of data from the Vietnam conflict has suggested that resuscitative measures have helped preserve renal function in critically wounded patients. The creatinine and osmolar clearances collected in 18 septic patients by Fletcher, Hirsch and Lucas[9] demonstrates that renal function can be normal in sepsis.

However, the recent study of Lucas et al.[27] suggests that renal function is altered in sepsis. Re-emphasizing the importance of vasodilation and the hyperdynamic state of sepsis, Lucas found that both glomerular filtration and tubular function were depressed despite an increased renal blood flow. The cited studies reflect on the need for further investigation of the effect of endotoxin on renal function.

EFFECT OF ENDOTOXIN ON CLOTTING

The relationship between endotoxemia and activation of coagulation has been intensely studied. There is no doubt that the syndrome of DIC occurs in sepsis.

It has been shown that bacterial endotoxins are strong thrombo-

plastic agents capable of producing DIC in animals.   Coagulation
changes have been demonstrated in septic patients that typify DIC.
Hardaway [11,12] believes that DIC is important in the pathogenesis of
shock.   Microinfarction and focal tissue necrosis are thought to be
the consequence of DIC.   It is postulated that a reduction in capillary flow produces a "stagnant hypoxia."   Regional acidosis resulting from the reduction in capillary flow is felt to inactivate
endogenous heparin producing a locus of hypercoagulable blood.   The
triad of hypercoagulability, diminished flow and release of thromboplastin from tissue injury accounts for the activation of the
clotting sequence.

Ratnoff[33] has demonstrated the pivotal role that Hageman Factor
(XII) activation plays in initiating clot formation and clot dissociation.   The activation of fibrinolysis (plasmin activation) in
turn initiates activation of the complement sequence and results in
kinin production.

Clotting defects have been described in patients with gram-negative septicemia, gram-positive septicemia, fungal and Richettsiel
infections.   The association of thrombocytopenia, diminished factor
analysis and the presence of fibrin split products have been found
to be reliable laboratory guides to the presence of consumptive
coagulopathy.   The presence of DIC in sepsis activates fibrinolysis
and the interplay between clot formation and dissolution will determine the hemostatic state of capillary beds.   The functional state
of the capillary is effected not only by microemboli retarding flow,
but also by the release of vasoactive substances initiated by the
production of clot dissolution.

THE EFFECT OF ENDOTOXEMIA ON SUB-HUMAN PRIMATES

Because the response of mammals to endotoxin challenge varies
from species to species, a search for an appropriate animal model
applicable to clinical sepsis has been undertaken.   Experience with
sub-human primates as hemorrhagic shock models and their similarities
to hypotensive shock in humans has pointed to primates as feasible
models for endotoxin research.   As experience with primate models
increased, it became apparent that the individual variability in
response to endotoxin of known potency played an important role in
the overall measured effect.   For this reason, some investigators
have begun using live bacteria rather than purified endotoxin in the
hope to achieve narrower confidence limits for parameters measured.

Guenter, Fiorica and Hinshaw[10] found that the hemodynamic and
metabolic changes induced by E. coli organisms was similar to that
of the purified endotoxin in Rhesus Monkeys.   In general, the effects
of live organisms paralleled the effect of the endotoxin.   Cardiac

output fell, mean arterial blood pressure fell, heart rate increased. Blood gases demonstrated reductions in pH and pCO_2, and blood lactate levels increased after two hours of hypotension. These changes were correlated with elevations in plasma catechols.

Cavanagh et al.[6] induced endotoxin shock in baboons. A reduction in mean arterial blood pressure was associated with a steady increase in heart rate. Cardiac output fell throughout the 180 minute study period. This was associated with an increase in total peripheral resistance. A precipitous fall in renal blood flow was noted for the entire study and correlated with anuria. Evidence for a consumptive coagulopathy was described.

Horwitz, Moquin and Herman[21] induced sepsis in baboons using live E. coli organisms. The hemodynamic effect of live organisms correlated with Cavanagh's description for endotoxin. Arterial blood gases demonstrated acidosis, hyperventilation and a fall in pO_2. The progressive increase in prothrombin time and partial thromboplastin times linked with the presence of fibrin split products demonstrated that a consumptive coagulopathy occurs in septic baboons. Fibrin split products were present in every septic animal.

In our baboon studies using purified endotoxin, the same general changes are noted. Arterial blood gases are hall-marked by significant reductions in pCO_2. Acidosis occurs but pH changes are not significantly different from controls. We have found that cardiac output and total peripheral resistance were not significantly changed after four hours of endotoxemia.

THE FATE OF INJECTED ENDOTOXIN

Host defence mechanisms have been implicated as important factors which influence the individual response to injected endotoxin. A number of techniques have been devised for the investigation of the fate of endotoxin. Radiolabelled endotoxin has been used to assess the degree of organ concentration of endotoxin. The consensus of numerous studies suggest that endotoxin accumulates within reticulum cell rich organs. Spleen, liver and the buffy coat of blood are particularly important areas of incorporation of radiolabelled endotoxin while erythrocytes do not fix endotoxin in vivo.

Rubenstein, Fine and Coons[35] used indirect immunofluorescent techniques to study the localization of endotoxin. This technique demonstrated the uptake of endotoxin by endothelium of arteries, capillaries and veins in many organs.

The functional integrity of the reticuloendothelial system (RES) has become an active area of investigation. The cytotoxic effects of endotoxin on mouse peritoneal leukocytes was observed by Wiener, Beck and Shilo.[44] Beeson[2] showed that blockade of the RES was associated with enhanced lethality when animals were challenged with minimal does of endotoxin.

Holper, Trejo, Brettschneider and Di Luzio[20] recently emphasized the importance of a functionally intact RES. They demonstrated that the addition of lead acetate would enhance the lethality of a minimale bolus of endotoxin (LDo) in baboons. This effect has been previously demonstrated in lower animals and is felt to result from a RES blockade. The increased lethality was correlated with liver function tests. In the face of RES blockade, a LDo bolus of endotoxin became lethal. This was associated with an increase in BSP retention and an elevation in specific liver enzymes. The alteration in liver function after the lead acetate-endotoxin injection was similar to changes effected by LD_{100} doses of endotoxin alone.

Lead acetate, an inoculum of Pertussis, Brucellosis inoculation and BCG vaccination have been used to enhance the lethal effect of endotoxin.

Reduced lethality has been described after zymosan (an extract of yeast cell wall), cationic polypeptide antibiotics and unsaturated fatty cells.

It has become apparent that detoxification of endotoxin is, by and large, cell mediated and plays an important role in the pathophysiology of sepsis. In vivo biochemical detoxification involving serum enzymes as well as immunochemical detoxification have also been considered important host mechanisms. Skarnes[39] has isolated two serum enzymes which effect endotoxin. One apparently degrades endotoxin into smaller units without effecting its biological potency. The second enzyme digests these subunits into a non-toxic substance.

The role of humoral immunity on enteric sepsis has not been completely defined. Kim and Watson[22] passively transferred endotoxin tolerance using 19 S immunoglobulins produced in rabbits. Nowotny et al.[31] reported that O antiserum neutralizes toxic reactions of endotoxin when incubated in vitro with the lipopolysaccharide. A number of reports suggest that in some cases specific O antigens protect laboratory animals. It is felt that specific O antigens do not afford protection in gram-negative septicemia in humans.

McCabe[28] has demonstrated that antibodies to rough mutant strains (Rd and Re) provide protection to mice challenged with heterologous gram-negative bacilli. In a recent clinical study antibody titers to O specific and two shared cross-reactive antigens (Ca and Re)

were measured. Septic shock with mortality was less frequent in
patients with O antibody titers of 1:640 or greater. However, the
incidence of shock and death was not significantly different in
patients with low titers. Likewise, the presence of Ca antibody
had little influence on the course of bacteremia. A significant
decrease in the incidence of shock with mortality was found in
patients having the Re antibody titers greater than 1:80. These
observations emphasize the importance of humoral antibody in host
mechanisms to gram-negative organisms. The effect of these antibodies on the detoxification of endotoxin remains to be elucidated.

CONCLUSION

The diverse effects of endotoxin on a variety of animal models
has made a unified approach to the pathophysiology of endotoxemia
difficult. The addition of variability of host responses to a
standardized endotoxin preparation increases this difficulty.

There are several factors which are probably important in the
overall response of particular species challenged with endotoxin.
Those species in which the enzymatic machinery for fibrinolysis is
diminished or absent display intravascular coagulation as a prominent feature of endotoxemia.

There is growing interest in a unified approach to low flow
states. This concept has merit from a biochemical standpoint. It
has pointed to the pivotal role of mitochondria, lysosomes and cell
membrane transport systems as areas principally effected during
hypoperfusion. It has been difficult to separate the direct effects
of endotoxin on cell machinery from the effect secondary to hypoperfusion. The importance of hypoperfusion cannot be over-empha-
sized, but there is sufficient in vitro evidence to support the
concept that endotoxemia has a direct effect on cell structure and
function.

REFERENCES

1. Alican, F. M., Dolton, L., and Hardy, J. D.: Experimental
        Endotoxin Shock: Circulatory Changes with emphasis upon
        cardiac function. Amer. J. Surg., 193:702, 1962.

2. Beeson, P. B.: Effect of reticulo-endothelial blockade on
        immunity to the Schwartzman Phenomenon. Proc. Soc. Exp.
        Biol. Med., 64:146, 1947.

3. Bennett, I. L.: Approaches to the Mechanism of Endotoxin Action.
        (Eds.) Landy, M., and Braum, W. In: Bacterial Endotoxins.
        Quinn and Boden Company, Inc., Rahway, New Jersey, p. XIV,
        1964.

4. Bohr, C., Hasselbalch, K., and Krogh, A.: Ueber einem in
    biologischer Beziehung wichtigen Einfluss, den die Kohlen-
    säurespannung des Blutes auf dessen Sauerstoffbindung übt.
    Skand. Archiv. F. Physiol., XVI:401, 1904.

5. Braunwald, E.: Editorial: On the Difference Between the Heart's
    Output and its Contractile State. Circulation, 43:171,
    1971.

6. Cavanagh, R., Papineni, S. R., Sutton, D. M. C., Bhagat, C., and
    Bachman, B. D.: Pathophysiology of endotoxin shock in
    the primate. Amer. J. Obstet. Gynecol., 108:705, 1970.

7. Coalson, J. J., Hinshaw, L. B., and Guenter, C. A.: The pulmo-
    nary ultrastructure in septic shock. Exp. Mol. Pathol.,
    12:84, 1970.

8. Davis, R. B., Meeker, W. R., and Bailey, W. L.: Serotonin re-
    lease by bacterial endotoxin. Proc. Soc. Exp. Biol. Med.,
    108:774, 1961.

9. Fletcher, J. R., Hirsh, E. F., and Lucas, S.: Renal function
    after sepsis in combat casualties. Surg. Gynec. Obstet.,
    133:237, 1971.

10. Guenter, C. A., Fiorica, V., and Hinshaw, L. B.: Cardiorespira-
    tory and metabolic responses to live E. coli endotoxin in
    the monkey. J. Appl. Physiol., 26:789, 1967.

11. Hardaway, R. M.: Intravascular Coagulation in Shock. (Eds.)
    Malinin, T. I., Zeppa, R., Gollan, F., and Callahan, A. B.
    In: Reversibility of Cellular Injury Due to Inadequate
    Perfusion. C. C. Thomas Company, Springfield, Illinois,
    pp. 373-386, 1972.

12. Hardaway, R. M., Hiesni, E. A., Guber, E. F., Noyes, H. E., and
    Burns, J. W.: Endotoxin Shock: A manifestation of intra-
    vascular coagulation. Ann. Surg., 154:791, 1961.

13. Hermreck, A. S., and Thal, A. P.: Mechanisms for the high cir-
    culatory requirements in sepsis and septic shock. Ann.
    Surg., 170:677, 1969.

14. Herring, W. B., Herion, J. C., Walker, R. I., and Palmer, J. G.:
    Distribution and clearance of circulating endotoxin. J.
    Clin. Invest., 42:79, 1963.

15. Hill, A. V.: Heat of shortening and dynamic constants of
          muscle. Proc. Royal Soc.- London, S. B., 126:135, 1938.

16. Hinshaw, L. B., Archer, L. T., Greenfield. L. J., and Guenter,
          C. A.: Effects of endotoxemia on myocardial hemodynamics,
          performance, and metabolism. Amer. J. Physiol.,221:504,
          1971.

17. Hinshaw, L. B., Greenfield, L. S., Owen, S. E., Black, M. R.,
          and Guenter, C. A.: Precipitation of cardiac failure
          in endotoxin shock. Surg. Gynec. Obstet., 135:41, 1972.

18. Hinshaw, L. B., Spink, W. W., Vich, J. A., Mallet, E. and Fin-
          stad, J.: The effect of endotoxin on kidney function and
          renal hemodynamics in the dog. Amer. J. Physiol., 201:
          144, 1961.

19. Hinshaw, L. B., Bradley, G. M., and Carlson, C. H.: Effect of
          endotoxemia on renal function in the dog. Amer. J.
          Physiol., 196:1127, 1959.

20. Holper, K., Trejo, R. A., Brettschneider, L., and Di Inzio,
          N. R.: Enhancement of endotoxin shock in the lead-sensi-
          tized sub-human primate. Surg. Gynec. Obstet., 136:593,
          1972.

21. Horwitz, D. L., Moquin, R. B., and Herman, C. M.: Coagulation
          changes of septic shock in the sub-human primate and their
          relationship to demodynamic change. Ann. Surg., 175:417,
          1972.

22. Kim, Y. B., and Watson, D. W.: Modification of host responses
          to bacterial endotoxin. II. Passive transfer to immunity
          to bacterial endotoxin with factions containing 195 anti-
          bodies. J. Exp. Med., 121:751, 1965.

23. Kutner, F. R., and Cohen, J.: Effect of endotoxin on isolated
          cat paillary muscle. J. Surg. Res., 6:83, 1966.

24. Kux, M., Coalson, J. J., Jacqueline, J., Massion, W. H., and
          Guenter, C. A.: Pulmonary effects of E. coli endotoxin
          role of leukocytes and platelets. Ann. Surg., 175:26,
          1972.

27. Lucas, C. E., Rector, F. E., Werner, M., and Rosen berg, I. K.:
          Altered renal homeostasis with acute sepsis. Arch. Surg.,
          106:444, 1973.

28. McCabe, W. R., Kreger, B. E., and Johns, M.: Type-specific
          and cross reactive antibodies in gram-negative bacteria.

N. Eng. J. Med., 287:261, 1972.

29. McCoon, R., and Del Guercio, L. R. M.: Respiratory function
    of blood in the acutely ill patient and the effect of
    steroids. Ann. Surg., 174:436, 1971.

30. McKay, D. G., Margaretten, W., and Csavossy, I.: An electron
    microscope study of endotoxin shock in Rhesus Monkeys.
    Surg. Gynec. Obstet., 125:825, 1967.

31. Nowotny, A.: Molecular aspects of endotoxin reactions. Bac-
    teriological Reviews, 33:72, 1969.

32. Priano, L. O., Wilson, R. D., and Traber, D. L.: The direct
    effects of endotoxin on the heart. Proc. Soc. Exp. Biol.
    Med., 135:495, 1970.

33. Ratnoff, O. D.: The interrelationship of clotting and immuno-
    logic mechanism. Hosp. Prac., 6:119, 1971.

34. Robb, H. J., Marqulis, R. R., and Jabs, C. M.: Role of pulmo-
    nary microembolism in the hemodynamics of endotoxin shock.
    Surg. Gynec. Obstet., 135:777, 1972.

35. Rubenstein, H. S., Fine, J., and Coons, A. H.: Localization
    of endotoxin in the walls of the peripheral vascular sys-
    tem during lethal endotoxemia. Proc. Soc. Exp. Biol. Med.,
    111:458, 1962.

36. Scheetz, W. L.: Discussion of: Respiratory function of blood
    in the acutely ill patient and the effect of steroids. By:
    McCoon, R., and Del Guercio. In: Ann. Surg., 174:449,
    1972.

37. Siegel, J. H., and Sonnenblick, E. H.: Isometric time-tension
    relationships as an index of myocardial contractility.
    Circ. Resch., 12:597, 1963.

38. Siegel, J. H. and Fabian, M.: Therapeutic advantages of an
    inotropic vasodilator in endotoxin shock. JAMA, 200:696,
    1967.

39. Skarnes, R. C.: The inactivation of endotoxin after interaction
    with certain proteins of normal serum. Ann. N. Y. Acad.
    Science, 133:644, 1966.

40. Solis, R. T., and Downing, S. E.: Effects of E. coli endotox-
    emia on ventricular performance. Amer. J. Physiol., 211:
    307, 1966.

41.   Stein, M., and Thomas, D. P.:   Role of platelets in the acute
      pulmonary responses to endotoxin.   J. Appl. Physiol., 23:
      47, 1967.

42.   Strauch, M., McLaughlon, S. S., Mansherger, A., Young, J.,
      Mendonca, P., Gray, K., and Cowly, R. A.:   Effects of
      septic shock on renal function in humans.   Ann. Surg.,
      165:539, 1967.

43.   White, R. R., Mela, L., Miller, L. D., and Berwich, L.:   Effect
      of E. coli endotoxin on mitochondrial form and function:
      Inability to complete succinate-induced condensed-to-
      orthodox conformational changes.   Ann. Surg., 174:983,
      1971.

44.   Weiner, E., Beck, A., and Shilo, M.:   Effect of bacterial lipo-
      poly saccharides on mouse peritoneal leukocytes.   Lab.
      Invest., 14:475, 1965.

THE ADULT RESPIRATORY DISTRESS SYNDROME

Edward W. Humphrey, Michael L. Schwartz,

William F. Northrup, Charles A. Murray, III

University of Minnesota

Acute respiratory insufficiency, although previously recognized in certain types of extra-thoracic trauma prior to and during World War II, has recently been shown to accompany a wide variety of conditions not primarily involving the chest. In addition to direct damage by penetrating injuries of the chest, burns involving the airway, contustion of the lung, and lung transplantation or reimplantation, pulmonary function may be compromised by fat embolism, high concentrations of oxygen, long term positive ressure respiratory support, intracranial injury, extra-thoracic trauma, and shock of several etiologies. The clinical picture and the histology of the lung changes in each of these conditions is quite similar reflecting the probability that the lung is limited in the response it can produce to damage. Although in a given patient, several of these factors may contribute to the clinical condition, they will be discussed separately.

A. SHOCK AND EXTRATHORACIC TRAUMA

The onset of the respiratory distress syndrome is typically delayed for 12 to 24 hours after trauma or resuscitation from shock. It is unusual for it to occur in a patient who dies without regaining an adequate cardiac output, but it is very common in those who are resuscitated from shock yet die later. Moon[144] reported from the AFIP intense pulmonary hyperemia in 80 percent of those who died in shock. During the first 12 to 24 hours the patient may have no obvious respiratory difficulties. He may be alert, his vital signs stable and the urine output adquate, but there is usually some hyperventilation.

At the end of this period, the chest x-ray begins to show a bilateral patchy infiltrate that resembles early pulmonary edema. While the patient is breathing room air the p_{aO_2} is low, usually under 60 torr. The $paCO_2$ remains low. The p_{aO_2} response to increased concentrations of inspired O_2 is less than expected indicating increased venous admixture and an increased A-a gradient. Over the ensuing hours or days, the pulmonary function gradually decreases, the arterial lactate is increased and the chest x-ray shows a markedly increased infiltrate. Even at this time, the process may occasionally be reversed, but usually it ends in death. CO_2 retention occurs and the p_{aO_2} will fall below 40 with the patient breathing 100 percent O_2. This may all happen with a normal or subnormal central venous pressure. The pulmonary compliance falls until very high inspiratory pressures are needed, pressures which may of themselves cause a fall in cardiac output or a pneumothorax.

The pathology of this syndrome has been extensively discussed by several authors[134,47,34]. Early in its course there are petechial hemorrhages on the pleural surface and scattered areas of atelectasis. The histology shows capillary and venous engorgement, minimal interstitial edema, and small areas of atelectasis. In the fully developed syndrome, the lungs are solid and resemble liver. There are gross areas of hemorrhage throughout the lung. Microscopically, there is a marked infiltrate of polymorphonuclear leukocytes. There is peribronchial and perivascular hemorrhage as well as intra alveolar hemorrhage and hyaline membranes.

The etiology of the respiratory distress syndrome following shock or extrathoracic trauma is unknown. The three most widely held hypotheses at present are: 1) Pulmonary venospasm, caused by the extrinsic or intrinsic nerve supply of the lung, increases the hydrostatic pressure of the pulmonary vascular system and causes capillary disruption; 2) Circulating substances released elsewhere in the body, either directly or indirectly increase capillary permeability or cause capillary disruption; 3) Multiple microemboli to the lungs occur as a result of disseminated intravascular coagulation during shock.

It is known that increased intracranial pressure is associated with the development of pulmonary edema, so it is logical to assume that stimuli mediated by the autonomic nervous system could play a role in the etiology of this syndrome. Other data bearing on this are those of Daly and Daly,[53] who believe that carotid chemo or baroreceptor stimulation may cause pulmonary vasoconstriction. Keller[118] found a significant increase in the small pulmonary vein pressure in dogs after hemorrhage and reinfusion. Daniel and Cate[55] reported that bilateral cervical vagotomy markedly increased pulmonary edema from a saline infusion, while a dorsal sympathectomy offered protection. Sugg and his coworkers[204] found that reimplantation of one lung in the dog prevented the changes of hemorrhagic shock from appearing in that lung, but bilateral cervical and thoracic sympathectomy did not. It is possible, however, that blood flow through the reimplanted lung was not great enough to produce the characteristic lesions. Greenfield, et al[90] have reported that acute or chronic unilateral pulmonary denervation by hilar stripping prevented the alterations on the denervated side after 3 hours of hemorrhagic hypotension. One strong argument against this theory is that Ingram and Associates[111], as well as others, have found that sympathetic stimuli to the lungs do not influence its histology.

The possibility that blood borne substances may affect pulmonary capillary integrity is supported by several investigators. Jenkins[114] placed a tourniquet around an extremity after making a window in the thorax. Upon release of the tourniquet, he found spotty hemorrhagic areas appearing on the lung within a few minutes. Willwerth et al.[229] have shown that a lung, excluded from the circulation during shock, is spared the usual changes following hemorrhagic hypotension. Farrington and co-workers[74] reported septal edema and polymorphonuclear invasion when blood from a septic dog was used to perfuse the left lower lobe of a healthy dog. Similar changes were seen when only the plasma from a septic dog was used. They conclude that these changes are due to blood borne factors and may involve the kallikrein system. Kinins, histamine, and serotonin have all at one time been implicated in this condition, but with little supporting evidence. Studies in this author's laboratory have failed to show any abnormality in the kallikrein system of dogs after 2 hours of hemorrhagic hypotension at 50 mm Hg followed by a reinfusion. Wilson[231] has reported a significant sequestration of polymorphonuclear leukocytes in the lung following shock, and electron micrographs have shown them adherent to endothelial cells. He has postulated that alterations in these leukocytes and substances released from them could be responsible for the injuries seen in the lung. Circulating fat has also been implicated in this syndrome. Allardyce and Groves[8] found both fatty acids and triglycerides elevated after injecting endotoxin, but fat embolism occurred only with concurrent intravascular coagulation.

Microemboli as a cause of the respiratory distress syndrome following shock has been proposed by Blaisdell[34]. Autopsies done on 36 patients who died from pulmonary insufficiency after major soft tissue trauma showed pulmonary venous and capillary engorgement and thromboemboli in the small pulmonary vessels. He postulated that pulmonary insufficiency follows microemboli either from 1) Release of vasoconstrictor substances from the embolus; or 2) physical obstruction of the pulmonary circulation possibly by further intravascular coagulation. This obstruction of nutrient flow could cause endothelial and alveolar cell injury. In support of this theory, Pennington[161] after administering endotoxin and Wilson[230] after hemorrhagic hypotension have reported decreases in circulating platelets. Wilson has also reported a greater decrease in platelet counts from the left atrium than from the right; but he has been unable to demonstrate significant depositions of platelets in the lung by histological means. Kuida[121] first reported a pulmonary hypertensive response to endotoxin in the intact animal. This response was later shown[105] to depend on the formed elements of the blood so that a mechanism has been described which might cause vasospasm in the lung from a platelet embolus.

A principal reason for the slow progress in determining the etiology and treatment of the respiratory distress syndrome after shock has been the difficulty in developing a reproducible model in animals. The findings following hemorrhagic shock in dogs described by Wilson[231] of vascular congestion, perivascular and interstial edema and areas of atelectasis are patchy. They may be seen to a lesser extent after 4 hours of anesthesia and positive pressure anesthesia.

Sealy[182] for example, reported that one half of his animals developed significant morphological lung damage. In this author's laboratory, 2 hours of hemorrhagic hypotension at 50 mm Hg followed by
reinfusion of the blood, did not produce pulmonary changes in the
dog after 4 hours that could be distinguished from 4 hours of positive pressure ventilation without hypotension unless phenoxybenzamine 2 mg/kg was administered to the dogs just prior to the reinfusion of blood. Maintenance of the systemic pressure at 40 mm Hg
for 2 hours by hemorrhage will produce severe pulmonary pathology,
but more than 60 percent of these dogs die within 24 hours so that
the investigation of the late effect of shock on the lung becomes
difficult.

   In contrast to the increased venous admixture found with the
clinical syndrome, shock in experimental animals often causes a
rise in the $P_{a_{O_2}}$. Gerst[87] has demonstrated a decrease in pulmonary shunting and an increase in respiratory dead space during shock
in dogs. Cahill[43] has reported an increase in compliance during
shock in animals, rather than the decrease seen in the human respiratory distress syndrome.

   One of the early histological changes described is interstitial
edema. Gump[91] has reported an increase in the pulmonary extravascular water, measured with a double indicator-dilution technique,
of a series of patients with acute pulmonary failure. This increase
is difficult to demonstrate in the experimental animal. In Table I,
the volume of total lung water, determined by the difference between
wet and dry weights after sacrifice, is compared to the volume of
extravascular fluid (PEVF) calculated from indicator-dilution curves
for: 1) dogs sacrificed immediately; 2) those ventilated for 4 hours
with room air; and 3) those in hemorrhagic hypotension for 2 hours,
then followed for 2 hours after the reinfusion of blood. It is apparent that the weighed water did not increase following shock in
spite of the increase in calculated extravascular fluid. Because
the indicator-dilution technique measures only that volume contained
within perfused tissue, an increase in PEVF can indicate either an
actual increase in water or an increase in the perfused capillary
bed of the lung. In this instance, the latter of these possibilities
is the more likely. Further evidence that an acute change in the
calculated PEVF in vivo represents a change in the area of the perfused capillary bed is found in Figure 1[179a]. In this experiment the
cardiac index was varied by opening or closing chronic bilateral
femoral A-V fistulae or by acute hemorrhage. On the ordinate, the
fraction of pulmonary capillaries perfused is the ratio of calculated PEVF to weighed lung water. At and above a cardiac index of
about 5L/min., this ratio approximates 1.0, and it decreases to a
minimum value of 0.5 at a cardiac index of 1 to 2 L/min.

   Ventilation-perfusion abnormaliaties may be the principal cause

TABLE I

	4th Hour Calculated PEVF (cc/kg)	4th Hour Weighed Lung Water (cc/kg)	$\dfrac{\text{Calculated water}}{\text{Weighed Water}} \times 100$
Normal Initial Value (0 time)	(4.0)	(6.2)	64%
Group 2	5.2	6.3	81%
Group 3	4.7	5.9	79%
Group 4	3.9	5.7	68%
Group 5	3.9	6.2	64%
Group 6	5.0	5.6	88%

Group 2 = 4 hours of anesthesia
Group 3 = 2 hours of hypotension with reinfusion of blood
Group 4 = 2 hours of hypotension with reinfusion + isoproteranol
Group 5 = 2 hours of hypotension with reinfusion + methylprednisolone
Group 6 = 2 hours of hypotension with reinfusion + phenoxybenzamine

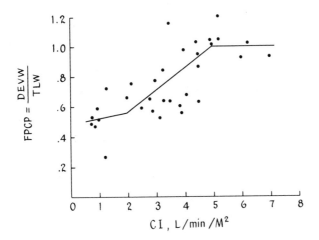

Figure 1

of the low P_{aO_2} in the respiratory distress syndrome. The question
arises why the blood flow to these areas of poor ventilation does
not decrease as it does in normal animals with an atelectatic lobe.
A partial answer to this question may be available from current
data. Wahrenbrock[220] has reported that although the pulmonary shunt
flow in an animal with normal lungs may decrease during shock, in an
animal with an atelectatic lobe, the fractional shunt flow through
that lobe is increased during shock. A further answer may be found
from the data in Table I. The increase in the calculated PEVF following 2 hours of hypotension plus reinfusion must represent a recruitment of pulmonary capillaries by a yet undetermined mechanism.
The data from Groups 4 and 5 indicate that this recruitment can be
prevented by methylprednisolone or by isoproteranol.

Further data bearing on this question are given in Table 2.
Schwartz has shown (179) that shortly after the production of left
lower lobe atelectasis in an otherwise normal animal, the fractional
shunt flow increases to 0.21 from a control value of 0.14. However,
within the subsequent 2 hours this abnormality is almost completely
corrected by decreasing the blood flow to the atelectatic lobe. However, in an animal that is pretreated with methylprednisolone,
30 mg/kg, the shunt flow through the atelectatic lobe is essentially
unchanged after 2 hours. It is possible that some substance with a
similar action prevents the correction of the abnormal ventilation-
perfusion ratios in the respiratory distress syndrome.

TABLE 2

| Group | Control | Shunt flow / Cardiac Output | |
		Atelectasis + 15 min.	Atelectasis + 2 hours
1--L.L.L. atelectasis	0.14	0.21	0.15
2--L.L.L. atelectasis + methylprednisolone	0.15	0.26	0.25
3--4 hours of anesthesia	0.14	0.12	0.10

(179)

B. NEUROGENIC PULMONARY EDEMA

A variety of central nervous system disorders have been associated with pulmonary abnormalities, including trauma[22,98,168,189,60] hemorrhage,[60,223] neoplasm,[23] seizure disorders,[35,60] infection,[96] and the experimental introduction of a variety of substances into the central cerebrospinal fluid.[23,176,190,191,223]

In adult autopsy series, the association of pulmonary edema with CNS pathology is roughly 20 percent.[23,44] In 1939, Weisman[223] made an exciting observation when he found lung edema and congestion in approximately two-thirds of 686 cases of traumatic and spontaneous intracranial hemorrhage dying within 30 minutes to 1 hour after injury. Marked pulmonary edema and congestion are almost constant findings in adolescents who sustain fatal head injuries in civilian traffic accidents.[22,168] In a series of 56 young soldiers who died at various times following blunt or penetrating head trauma, only 2 had combined lung weights under 800 grams. Pulmonary edema was found in 18 of 20 young soldiers who died within minutes of intracranial trauma. It was also noted that pulmonary edema did not occur if cervical cord transection or massive hemorrhage accompanied the head injury.[189] It is thus quite clear that considerable pulmonary congestion and edema are almost constant findings after significant head injury and that the pulmonary edema can occur suddenly, even within seconds or minutes of injury.

In the experimental animal, head injury models[132,63,61,188,23,44,22] or the injection into the central cerebral spinal fluid of veratrine[223], thrombin and fibrinogen,[23] fibrin[176] or endotoxin[64,190,191] can rapidly produce pulmonary edema. The basic mechanism for the production of pulmonary edema in all of these experimental models appears to be a reflex involving a massive sympathetic discharge. The reflex appears to be the same as described by Cushing[52] in 1901. He demonstrated that increases in intracranial pressure are followed by an increase in systemic arterial pressure, and that the diastolic pressure exceeds the intracranial pressure by several millimeters of mercury. This response is apparently universal among all animals studied and presumbably represents a mechanism by which the cerebral perfusion can be maintained despite the increased pressure on the brain.[188]

As the intracranial pressure is elevated in animals, a coordinated cardiovascular hemodynamic response follows which consists of an increase in cardiac output due to an increase in venous return and a positive inotropic effect on the heart. These changes are associated with a moderate rise in both systemic and pulmonary vascular pressures. Further increases in intracranial pressure produce a rise in calculated total peripheral resistance with an attendant increase in diastolic pressure slightly higher than intracranial pressure. In 15-20 percent of the animals the total peripheral resistance may rise to extreme heights, greater than 11,000 dyne-sec/cm^5. Within minutes of this increase in calculated total peripheral resistance, cardiac output falls, the left atrium distends with pressures exceeding 40 mm Hg and pulmonary edema occurs.[63,98,61,188] Cervical vagotomy does not alter this hemodynamic response.[63] However, cervical cordotomy prevents the entire sequence of events. These data support the proposal that the entire sequence of hemodynamic events which finally results in passive pulmonary congestion and edema, results from massive sympathetic discharge.

The earliest hemodynamic change appears to be constriction of the venous capacitance bed[40] which increases the volume of blood returning to the heart. Total peripheral resistance drops in this phase while the central blood volume increases.[61,188]

One of the earliest functional changes is a decrease in p_{aO_2}. If the elevation of intracranial pressure is gradual, almost immediate significant decreases in p_{aO_2} occur in the absence of gross or microscopic evidence of pulmonary edema. It has been assumed that this early hypoxemia is due to pulmonary arteriovenous shunting. Of great interest is the fact that cervical cordotomy and alpha adrenergic blockade prevent the hypoxemic response, whereas bilateral adrenalectomy, vagotomy, atropine, or beta-adrenergic blockade do not. The mechanism of this hypoxemic response has been interpreted as being due to direct (but unspecified) effects of mass sympathetic discharge on the pulmonary vascular bed or to passive (but also unspecified) changes in pulmonary blood flow from the systemic neurovascular response. The authors concluded, somewhat vaguely, that "it is probable that both peripheral and pulmonary alpha-adrenergic mechanisms are involved in the hypoxemic response to increased intracranial pressure."[31]

Moderate acidosis is another result of increased intracranial pressure.[31] How much this has to do with the hypoxemic response is unknown. Further studies on the role of the sympathetic nervous system in the gross pulmonary pathology occurring with head injury have been done in rats and monkeys. It was first noted in rats that the gross pulmonary pathology could be prevented or ameliorated by pretreatment with sympatholytic, antiepinephrine, or general anesthetic agents, again suggesting that the sympathetic nervous system is involved in the development of the pulmonary pathology.[44] Significant decreases in pulmonary compliance have been observed in rats and monkeys with head injury.[44,22] These compliance changes can occur even in the absence of any gross lung changes.[44] The observation that pretreatment with an alpha-adrenergic blocking agent prevents this compliance change supports the proposal that the sympathetic nervous system in some way may be responsible, even in the apparent absence of pulmonary edema. This loss of compliance has been assumed to be due to loss of surface tension forces. However, the specific nature of this "sympathoneuroendocrinogenic influence on surface tension factors" is not known.[44]

It has been suggested that the massive sympathetic discharge occurring with increases in intracranial pressure is associated with a failure of the heart to relax during diastole producing an elevated left ventricular end diastolic pressure and a consequent failure to accomodate the increased cardiac return.[131,223] However, while it is true that the left ventricular end diastolic pressure and the left atrial pressure do indeed rise leading to passive pulmonary congestion, other investigators have not found any evidence for loss of left ventricular compliance in response to sympathetic stimulation.[141]

Some experimental work suggests that a persistent vagal discharge secondary to increased intracranial pressure may produce pulmonary edema by bradycardia.[45] However, significant bradycardia has not been the usual result of increased intracranial pressure. It is usually a transient phenomenon and is not associated with a decrease in cardiac output.[188,63,61] Furthermore, vagotomy affords no protection against pulmonary edema whereas cervical cordotomy and alpha adrenergic blocking agents can prevent it.[61,62]

Stellate ganglionectomy alone has been shown to prevent pulmonary edema resulting from the intracisternal injection of veratrine,[223] suggesting that some of the pathology may result from the pulmonary vasoconstriction per se. Berman and Ducker[30] anticipate hypoxemia if constriction of the arterial side of the pulmonary circuit occurs (but do not explain how). They also suggest that pulmonary venous constriction would be expected to favor the transudation of fluid from the pulmonary capillary bed. Some experimental work however, has shown no benefit with stellate ganglionectomy.[176]

The syndrome of pulmonary pathology resulting from intracranial infections[96] seems to follow a slightly different course. In the experimental animal, a minute quantity of sterile gram negative endotoxin within the central cerebral spinal fluid system produces a gradually developing pulmonary edema.[64,190,191] The pulmonary pathologic changes, which are mainly those of edema, resemble the changes seen in patients with gram negative meningitis[96] and in animals with increased intracranial pressure. But, although massive pulmonary edema occurs with both intracerebral endotoxin injection and with cerebral trauma mechanisms, the hemodynamic and respiratory findings are different. In the first place, neither hypoxemia nor acidosis is a feature of intraventricular endotoxin injection[191] as it is with head injury.[31] There is also no increase in pulmonary artery or systemic artery pressure. Although the cardiac output gradually falls and the systemic vascular resistance gradually rises, there is no indication that the lungs become passively congested as in the head injury models.[191]

The clinical application of these observations of abnormal pulmonary findings in a variety of CNS disorders lies in the potential manipulation of the sympathetic nervous system either with alpha-adrenergic or ganglionic blocking drugs or with bilateral stellate ganglion blocks.[189,60] Berman and Ducker[30] have reported transient reversal of arterial hypoxemia with bilaterial stellate ganglion blockade in a patient with acutely increased intracranial pressure and progressive respiratory insufficiency.

Stellate ganglionectomy, however, would not be expected to be of primary benefit in patients with neurogenic pulmonary edema since the primary pathogenic mechanism does not appear to lie within the pulmonary vasculature, but rather within the systemic vasculature.[176] On the other hand, if the early hypoxemia and pulmonary compliance changes are not due to passive pulmonary congestion and edema, pulmonary sympathetic interruption by stellate ganglion blockade does have potential usefulness. Stellate ganglionectomy might also decrease the left ventricular end diastolic pressure and thereby lessen the passive congestion if compliance changes are a significant factor in this type of pulmonary injury.[223] Systemic ganglionic blockade has been shown experimentally to prevent neurogenic pulmonary edema,[176] presumably by its action on the systemic vasculature.

C. FAT EMBOLISM

This entity was first recognized clinically by Ernst von Bergmann in 1873.[29] His patient developed hemoptysis, respiratory distress, audible bubbling rales and cyanosis 2½ days after trauma.

The incidence of fat embolism has been reported as 80 to 100 percent in autopsy series of patients dying after a fracture[184,84,159,181], and it is a major or contributing cause of death in about 15 percent of patients dying after trauma.[81]

The pathologic changes in the lungs of patients who do not die acutely of pulmonary fat emboli are hyperemia, petechial hemorrhages and edema.[180] The brain has petechiae in the white matter of the cerebral and cerebellar hemispheres.[181] Microscopically fat emboli, microinfarcts, perivascular edema and hemorrhages can be seen. Fat emboli may be seen in virtually any organ. They are particularly well seen in the glomerular capillaries and the myocardial sinusoids.

Fat embolism occurs most frequently after a fracture. The time of onset of symptoms is variable. It can occur within minutes after

trauma when massive amounts of neutral fat embolize to the lungs or
may be delayed several days when neutral fats are being degraded to
FFA. In Sevitts' series[81], 25 percent had the onset of symptoms by
12 hours, 60 percent by 24 hours, and 84 percent by 48 hours.

Classically the onset of symptoms of pulmonary fat embolism is
abrupt. Tachypnea, tachycardia, pyrexia, and neurologic symptoms
appear simultaneously. Tachypnea can change to respiratory distress
and failure in a matter of hours. The tachycardia is usually over
130 beats/minute. Hyperpyrexia is an almost universal finding
(119,102) and is unresponsive to aspirin. The neurologic manifestations are generally disorientation and delerium[73] but can range from
headache, restlessness and irritability to stupor and coma. Ascu-
lation of the chest generally reveals bubbling rales. Petechiae
can occur in up to 50 percent[102,181,157], but they can fade in 4-6
hours and are easily missed. They occur most frequently on the
neck, axilla and anterior chest. Fundiscopic examination can show
petechial hemorrhages and edematous patches.[181]

There are increased bronchovascular markings and fluffy densities in both lung fields[101,102] on chest x-ray. The ECG can show
a pattern of right ventricular strain or less commonly one of myocardial injury.[157] Electroencephalogram shows generalized dysrhyth-
mia without localized findings.[181,102]

Sprule[201] was the first to describe the remarkable decrease in
P_{aO_2} which occurs in fat embolism. Interestingly, skin color is
often not indicative of the degree of arterial desaturation even
though tachypnea or respiratory distress may be present. Often
with the onset of symptoms there is marked fall in hemoglobin
(3-5 gm.%) without detectable source of blood loss.[102,101,73] The
platelet count is invariably decreased.[16] Peltier has stated[157],
"if there is a thrombocytopenia below 150,000 and the arterial P_{O_2}
is below 60 mm Hg., the diagnosis of fat embolism can be positively
made if any of the criteria for the clinical diagnosis are present."
If the urine supernatant is stained for fat, it will be positive in
50 percent of patients with fractures[157] indicating its questionable
significance as a diagnostic test in PFE. Serum and sputum tests
for fat and serum lipase determinations are even less reliable[73,16,
209,101].

There are at least two popular schools of thought as to the
origin of fat emboli. The mechanical theory, formed by Gauss in
1924[85] is that the emboli are formed by intravisation of marrow fat.
The other or physiochemical theory, first stated by Lehman and Moore
in 1927[125] is that emboli are formed from alterations in the stability of the emulsion of fat in the circulatory system giving rise to
large droplets (10 to 50 μ) which act as emboli.

There is considerable evidence to support that part of the mechanical theory which holds that noncollapsed osseous veins remain patent to admit fat droplets as soon as the hemorrhage and exudation raise marrow pressure above venous pressure. Several authors[10] have demonstrated bone marrow elements in the lungs after fractures. Pulmonary fat emboli can be traced to marrow fat previously stained with dye.[88] A tourniquet applied to a traumatized extremity above a fracture will sequester intravasated fat droplets until the tourniquet is removed.[160] Increasing the pressure in the marrow cavity[226] or above a fracture[99] will increase the amount of fat embolization. Meer[140] demonstrated the origin of embolic fat in humans to be from marrow by sampling femoral venous blood from an extremity with a fracture of the femur. Sevitt[184] has pointed out that there is a close correlation between the number of pulmonary fat emboli and the severity of bone injury. Contrary to early studies[119], there is sufficient fat in the marrow to account for pulmonary fat emboli. Finally, analysis of the FFA content of pulmonary fat emboli shows a great similarity to bone marrow fat but not circulating lipids.[170,94]

Equally as impressive are the following observations which favor the physiochemical theory. First, there are numerous conditions in which fat emboli have been documented without trauma or fractures. These conditions include steroid induced and alcoholic fatty liver, renal transplantation, blood transfusion, diabetes mellitus, burns, high altitude flights, severe infection, neoplasma, poisoning, child birth, sickle cell disease, and heart lung bypass. Second, several authors[126,201,104] have noted cholesterol crystals in fat emboli in concentrations ranging from 5 to 30 percent. When marrow fat is examined it is consistently less than 1 percent cholesterol. Szabo[207] concluded, by studying the rate of absorption of 131_I triolein injected into the ends of fractures, that marrow fat quantitatively could not be the only source of fat emboli. Other authors[86] have found the fatty acid content of embolic fat to be between marrow triglycerides and circulating triglycerides. Ashbough (SGO '66) has argued that in view of the great filtering capacity of lungs there would be no systemic fat emboli if all emboli originated in marrow fat.

Certainly one of the factors which has stimulated interest in the physiochemical theory is the rapid mobilization of lipids that occurs during trauma. Baker[16] has shown an increase in catecholamines with trauma. Epinephrine and norepinephrine are known to be potent mobilizers of fat from adipose tissue[107] by stimulating the hydrolysis of neutral fat to FFA and glycerol.[107]. The degree of elevation of FFA has been related to the degree of trauma.[133] Several triglycerides are also elevated after trauma, particularly in patients who develop subsequent fat emboli.[133] In this same study, total lipids were stable while cholesterol increased in patients who developed subsequent fat emboli. The magnitude of mobilization of lipids led Bergentz[27] to speculate that injury results in alterations in lipid suspension stability that could result in fat emboli.

The evidence seems to indicate that marrow fat is at least partly responsible for pulmonary fat emboli but under some circumstances fat emboli can form without trauma. Even after fractures, plasma lipids (such as cholesterol) may combine in some way with marrow fat to form fat emboli. Less speculative is the fate of embolic fat in the pulmonary capillaries. Acutely, there may be a mechanical blockage of blood flow through the pulmonary circulation by the neutral fat. Neutral fats have been shown to cause adherence to platelets on the surface.[158] These platelets break down in the lungs releasing vasoactive amines[206] which cause broncho-constriction[210] and pulmonary venous constriction.[236] Experimentally, reserpine depletion of platelet amines has been shown to lessen the shunting that occurs in experimental fat emboli.[146]

If survival occurs, lung lipase, a facultative enzyme in pneumocytes promotes hydrolysis of neutral fats to FFA and glycerol. FFA is far more toxic to pulmonary capillaries and alveoli than neutral fat.[158] Lung surfactant activity is decreased[95] and there is associated atelectasis. The role that the clotting mechanism plays in the pathobiology of pulmonary fat embolism is yet unsettled. Although an increase in coagulability, changes in platelets[161,211,102], and aggregation of RBC's[102] do occur, there is no evidence that they are related to the pathophysiology of PFE causally. Fractures should be splinted early and movement of the fractured extremity minimized. Manipulation, when necessary, should be performed gently. Tight casts and dressings which increase pressure in the medullary cavity must not be used.

Equally agreed upon is the need for respiratory therapy. Arterial desaturation out of proportion to the clinical appearance of the patient is the rule and not the exception. Arterial blood gases should be monitored frequently and respiratory support begun promptly. In a series reported by Fischer[79] over half the patients required tracheostomy and respiratory support. Careful attention to adjustments of inspired oxygen content and ventilation are essential. Also essential is frequent clearing of the tracheobronchial tree by whatever means are necessary.

Steroids were first advocated by Ashbough[11] after dramatic results were achieved in 2 cases. Now most authors advocate their use.[79,158,73,102,25] Experimentally, when given after fat embolism (142,225) steroids do not affect survival, but when given prior to fat embolism by Wertzberger[225] it reduced the mortality in rats from 47 to 5 percent. There are no controlled clinical series that demonstrate the value of steroids. There does seem to be a temporal relationship between steroids and clinical improvement. Fischer[79] feels that arterial P_{O_2} improves in 6 hours, tachycardia is decreased in 24 hours, the chest x-ray improves within 48 hours and neurologic picture improves as does pulmonary compliance by 72 hours.

Herman introduced ethanol as an emulsifying agent to aid in clearing fat from the pulmonary circulation.[210] It is now agreed[158,4,119,25] that it is a lipase inhibitor both in vivo and in vitro Experimentally it has been shown in rabbits to decrease the mortality from neutral fat emboli from 42 to 9 percent. Alder showed that it supressed the rise in serum lipase[4]; however, no patient benefit was shown. Other authors believe its mild sedative and analgesic properties and weak vasodilating action are more important than its specific activity as a lipase inhibitor.[101,181,25]

Heparin has been advocated in PFE but its role is controversial.[101,73] Its actions are threefold: 1) as an anticoagulant heparin decreases platelet adhesiveness and blocks the release of platelet vasoactive amines; 2) it is a lipolytic agent; 3) it is a lipase stimulator. Peltier believes that if heparin is helpful it is as an anticoagulant.[156] Experientally there is evidence to support Peltier. Sachdiva[136] showed a decrease in mortality of rabbits given large doses of heparin after neutral fat emboli. Control animals died of acute right heart failure. Other investigators[142] have shown no benefit in experimental aleic acid emboli and Ross[174] showed an increased mortality with use of heparin on a rat-neutral fat model. There is no evidence that lipase stimulation would be beneficial, and some possibility that it might be detrimental. The use of heparin also involves the risk of bleeding in a patient with recent fractures and trauma to other organs. The hemorrhagic lesions in the lung and brain in fat emboli should also discourage the use of heparin in full anticoagulant doses.

Low molecular weight dextran has been advocated in the treatment of fat emboli because of its properties of preventing aggregation of the formed elements of blood[73,25], but its use did not improve the mortality in a neutral fat[127] or fatty acid[142] model in rabbits.

D. PULMONARY CONTUSION

Pulmonary contusion is a bruise of the lung. The magnitude of the problem of pulmonary contusion is realized from the data that there are over 13 million auto accidents yearly with over 50,000 deaths, half of which are directly caused by or contributed to by thoracic trauma.[48] Pulmonary contusion is the most common type of chest injury. It is present in 70 percent of significant chest trauma.[228,212]

Automobile accidents are the commonest cause of pulmonary contusion; falls from heights are a distant second. Burford[42] described patients with pulmonary contusion as dyspneic with paroxysms of a painful cough usually ineffective in clearing the lungs. Often patients are only slightly symptomatic immediately post trauma and only after 24 hours become dyspneic and cyanotic.[57,5] In only half of the patients is there evidence of trauma to the bony thorax.[173] Indeed, the pulmonary injury may be far out of proportion to the chest wall injury.[83] Especially in young patients, where the thorax is highly elastic, there may be no evidence of chest wall trauma.[5] Associated trauma, especially to head and abdomen, occurs in over half the cases.[83] A Flail chest may be present and associated pathology such as hemopneumothorax may mask the contusion. Fulton has stressed the progressive pathologic and physiologic nature of this condition.[82] Ten of the 12 patients that he studied clinically needed respiratory assist and 8 had progressive hypoxemia.

The roentgen picture of pulmonary contusion is generally a patchy, irregular, mottled radiodensity which does not conform to an anatomic segment of the lung.[173,48,57,212] However, early after injury the chest film may be perfectly clear. Erickson[72] showed that a pulmonary contusion was present on x-ray by 30 minutes in 33 percent and by 4 hours in 78 percent of the animals tested. Eleven percent developed densities after 24 hours. This time sequence of appearance agrees with the clinical studies cited above. The radiodensity may increase in size and density for up to 48 hours, but after this further increase in size or density is most likely due to secondary infection or atelectasis. Resolution of a contusion generally occurs in 1 to 2 weeks without secondary infection.

The basic pathology is disruption of the vasculature and alveoli of the lung with edema and erythrocytes[57,5,228,54] in the interstitial space and alveoli. This can occur locally or throughout the whole lung. Experimentally Fulton[82] has inflicted trauma to the lung and noted initial destruction of alveoli and hemorrhage followed in 1 to 2 hours by edema and cellular infiltrate. After 24 hours the architecture of the lung is obscured.

Air hammers[82] falling weights[152,108,54] and sling shots have been used to cause pulmonary contusion. All the methods cited cause a very similar type of injury. After injury the venous effluent from the affected lobe has a lower PO_2 and higher PCO_2 than normal lobes[82,152]. The compliance progressively falls[152],[108]. By 24 to 48 hours, surfactant can be shown to be decreased.[152] Until greater than 50 percent of the lung is affected by the contusion there is no correlation between the amount of injury and the PO_2. If over half the lung is injured, the PO_2 is reduced.[72] Daniel and Cate[54] have shown an increase in the susceptibility to edema of lungs adjacent to the area of contusion. This phenomenon seems to be preventible by prior pulmonary sympathectomy.

The first steps in therapy include resuscitation of the patient and diagnosis of other injuries. Flail chest and hemopneumothorax must be treated. The recommended therapy for an isolated contusion include: 1) nerve block, 2) antibiotics, 3) steroids, 4) tracheostomy and assisted ventilation, and 5)oxygen. Nerve block seems to be particularly helpful when there are multiple rib fractures with an underlying contusion. Burford reported numerous cases of immediate benefit with productive cough and clearing of the tracheobronchial tree.

Antibiotics are recommended to prevent secondary infection of the contused lung but there is no evidence that this is beneficial. Steroids have also not been shown to be beneficial, but clearing the airway of secretions is most helpful.

Tracheostomy and assisted vhntilation is necessary in many of these patients. Due to the progressive nature of this lesion, these patients should be watched very closely with frequent arterial blood gas determinations over the first 48 hours. The choice of a respirator ideally would include one with a periodic hyperinflation and a positive end expiratory pressure. A volume controlled respiration is often necessary because of the very low compliance in some of these patients. Oxygen should be administered as needed.

E. PENETRATING INJURY OF THE LUNG

Penetrating injuries to the lung were recorded as early as 3000 BC. However, the modern concepts of therapy have evolved since World War I. Subsequent wars and weapons have radically changed our concepts of management. Today, the mortality of isolated penetrating lung injuries is less than 2 percent.

The pathology and extent of tissue injury depend on the offending weapon. In general, missiles with a velocity of greater than 2500 feet/sec are referred to as high velocity and cause extensive tissue damage through cavitation.[222,199.100]

Conventional low velocity missiles, knives, ice picks, etc. cause a central tract of penetration with a surrounding area of contusion. In the lung this contusion appears as blood in alveolar spaces. The laceration itself can be filled with air and/or blood. All penetrating injuries to the lung are contaminated wounds.[28] In combination with the devitalized tissue and blood, this contamination is a potential source of infection. Foreign bodies of clothing, bone, and the missile itself are frequently present in the lung after such injury.

Penetrating injury from a high velocity missile is a very common occurence now both in war and civilian casualties.[222,199,56,28] As a high velocity missile enters tissue all the cells close to the projectile become secondary missiles which move rapidly outward to give rise to a temporary cavity. The size of this cavity is directly related to the energy dissipated in the expanding tissue. The energy of destruction increases with the kinetic energy of the missile and so to the square of the velocity of the missile.[56] This explosive destruction gives rise to extensive tissue necrosis, larger cavities, massive chest wall trauma, and multiple in-driven foreign bodies (199,100). Intrabrochial bleeding occurs frequently.

Clinical Picture

In trying to establish the extent of lung injury from a penetrating wound several points in the history are important. First, what caused the wound? The length of a knife or the calibre of a rifle can help in estimating the amount of lung trauma. Second, in what direction was the wound inflicted? A ricochet in the thorax is rare, thus the projectile can be expected to damage organs in its direct line of flight. Third, has the patient coughed blood since injury? The presence of hemoptysis will definitely confirm puncture of the lung.

During the physical examination one should look initially at blood pressure, skin color, the amount of dyspnea and the wounds. The position of the trachea and apex impulse should be noted to determine the position of the mediastinum. Subcutaneous emphysema should be sought. The direction of the projectile should be established and other injuries investigated such as those to the heart, great vessels, diaphragm and abdominal viscera. Finally the chest is examined for the presence of a pneumothorax or a hemothorax.

Due to the low pressure in the pulmonary vasculature, bleeding from the pulmonary parenchyma generally ceases promptly. Thus a patient in shock will probably have blood loss from an additional source such as the heart, great vessels or an intraabdominal organ. Extreme dyspnea and cyanosis indicates additional pathology such as a tension pneumothorax, flail chest or an open thoracic wound.

When the patient's condition permits, a PA and a lateral upright chest film should be obtained. The upright position will allow one to accurately establish the presence and amount of air and blood in the thorax. The x-ray appearance of a laceration of the lung is in general obscured by surrounding pulmonary contusion. If there is little contusion, the laceration may be seen as a round blood or air filled space. Generally it is several days after injury before resolution of the surrounding contusion makes a laceration obvious. These cavities look like pneumatoceles or if partially filled with blood, they may appear to be a lung abscess.

Therapy

Initial therapy should be directed at resusitation of the patient. The airway should be cleared, the chest evacuated of blood and air, volume replacement begun and other injuries diagnosed. Foreign bodies impaled in the chest should not be removed until one is ready to perform definitive therapy. The pneumothorax and/or hemothorax should be treated immediately with placement of a thoracostomy tube. The patient is then observed. A thoracotomy is necessary in less than 10 percent of puncture wounds of the chest in civilian practice .[36,20,200,103,195,163] Pulmonary laceration alone is not an indication for thoracotomy. With high velocity injuries up to 20 percent of patients may need a thoracotomy.[103,222,100,37] The indications for thoracotomy are: 1) Massive bleeding from the chest tube causing hypotension or shock; 2)Continued bleeding of 250 cc/hr over 3 hours or more; 3) Massive air leak; 4) Injury or suspected injury to the heart, diaphragm, esophagus, or intra-abdominal organs. When a thoracotomy is necessary, a Carlens tube should be considered for endotracheal intubation. This will be especially valuable for large air leaks and when intra-bronchial bleeding is present. Massive air leaks are often due to bronchial injuries and not to simple pulmonary lacerations, and these should be explored immediately if there is no contraindication. Massive or continued hemorrhage should be explored, as in general these are due to injuries to other organs.[36,20,195.163] In a series by Shefts[103] only 4 of 200 patients had bleeding from a pulmonary laceration great enough to warrant exploration.

Burford[41] and Brewer[38] observed the great tendency of pulmonary lacerations to resolve. Due to the poor general condition of their patients undergoing thoracotomy for chest injuries, badly contused, hemorrhagic, boggy lungs were not resected, however, simple suture of these lacerations led to resolution, often complete, by 6-8 weeks. Resection of a lobe was reserved for[41]: 1) Irrepairable damage to a major bronchus; 2) Partial traumatic amputation of a lobe in which the bronchus had been transected and little or no blood supply remained; 3) Damage to a major pulmonary vessel requiring ligation. In over 2000 penetrating lung injuries only one lobectomy was performed. Today the incidence of pulmonary resection is reported to be higher in war injuries due to high velocity missiles. In a recent series the incidence of lobectomy or penumectomy was 40 percent of those patients requiring thoracotomy (8 percent of all patients with penetrating chest injury[222]).

The likelihood of secondary infections of pulmonary lacerations can be decreased by removal of foreign bodies. This is especially indicated when the foreign body measures greater than $1\frac{1}{2}$ cm. If thoracotomy is performed for other reasons, foreign bodies that can be easily removed should be.

The mortality of isolated pulmonary laceration is low (2%).
Most fatalities are due to other chest injuries, such as to the
heart, chest wall, or great vessel. The mortality for a penetrating
chest injury in general depends on the etiology of the wound, location, associated intra and extra thoracic trauma [103,49,89,195,74,
100,222]. In civilian practice at present it is 3-5 percent and in
the Viet-Nam conflict it is 9 percent.

F. RESPIRATORY TRACT DAMAGE IN BURNS

There are 2 types of respiratory tract damage in burns (RTDB)
related to the time of injury of the respiratory tract and the
time of injury elsewhere to the body. The first, referred to as
inhalation injury, is that injury to the respiratory tract which
occurs simultaneously with thermal injury elsewhere to the body.
(Actually, inhalation injury may occur without any thermal injury
to the body at all.) The term "pulmonary burn" will not be used
since it is clear that excessive heat is not always the cause of
respiratory tract damage and that it is rarely, if ever, the direct
cause of injury to the lung itself. The second kind of RTDB is
that which occurs in sequence, after the thermal injury elsewhere
to the body. It should be noted, however, that not all RTDB which
appears late to the clinician is in fact unrelated to original
injury to the respiratory tract. In fact, much if not the vast
majority, of the late-appearing respiratory tract pathology is related to subclinical inhalation injury sustained at the time of
the original burn.

The association of pulmonary injury with burns has been noted
for many years. Hepatization and engorgement of the lung with
pleural effusion was noted in association with burn injuries as early
as 1840.[130] In 1942, 491 people died in the Cocoanut Grove Nite
Club fire in Boston, Massachusetts. One hundred fourteen people
arrived at the Massachusetts General Hospital, of whom 75 were
either dead on arrival or died within the first few minutes of[76]
"uncontrollable anoxia."[76] This holocaust, which was extensively
analyzed and reported in the surgical literature,[51,24,13,177,133]
seems to have made surgeons aware for the first time on a large
scale of the serious problem of respiratory tract damage occurring
with burns and resulting from the inhalation of toxic products of
combustion.[133] As death from burns are becoming delayed, due primarily to more adequate management of burn shock and better control
of burn wound sepsis, more patients are now dying from pulmonary
complications.[169,164,186,165]

Clinical Setting of the Patient with Inhalation Injury

Incidence: The overall incidence of RTDB among large series of

patients is remarkably constant varying from 15-22.4 percent.[143,1,
202,186,166] The specific incidence of inhalation injury however,
is quite variable from series to series when reported as a separate
category. Of the 39 Cocoanut Grove victims who lived long enough to
be treated at the Massachusetts General Hospital, only 3 were free
of respiratory symptoms.[13] There appears to be a regional variation
in the incidence of inhalation injury with more inhalation injury
occurring where people are likely to be indoors more.[165,129] The
exhaustive study by Phillips and Cope[166] of 1,140 consecutive burn
patients cared for at the Massachusetts General Hospital in Boston
from 1939 to 1957 revealed an incidence of inhalation injury of
16.1 percent, whereas the review of 2, 297 burn patients cared for
at the Brooke Army Medical Center in Texas revealed an incidence of
only 2.9 percent.[58]

   Criteria for Diagnosis of Inhalation Injury: Because of the
delay in appearance of symptoms, physical findings and x-ray changes
in patients with inhalation injury, indirect evidence is necessary
to establish the diagnosis. Stone and Martin[202] have delineated 3
basic criteria: 1) Facial burns from flame injury involving the
mouth and nose; 2) Burns of the oral mucosa and nasal hairs; and
3) History of flame burns sustained in a closed space.

   Definite respiratory tract injury is assumed if all 3 criteria
are present. Probable respiratory tract injury is assumed if 2 of
the 3 criteria are present. In addition respiratory tract injury
is "almost certain" if wheezes and/or respiratory distress occur
before significant fluid replacement. Although most authors accept
Stone's criteria, some add the necessity of having specific symptoms
of wheezing, hoarseness, or dyspnea.[58] Cases with a clear history
of smoke inhalation should be included.[58]

   While it is clear that immediately fatal pulmonary injury can
occur in the absence of any thermal injury to the body,[13,133] there
does appear to be a rough correlation between the extent of body
burn and the incidence of respiratory tract damage.[80,1] Advanced
age may also be associated with a higher incidence of respiratory
tract injury.[1] Facial burns have long been considered by most to
be one of the prime indications of associated respiratory tract
damage.[166,202] Virtually all patients dying from tracheobronchitis
or pneumonia in a series of 817 burn victims from the Brooke Army
Hospital in Texas had facial burns[80], but in a series of 187 patients from Brooke Army Hospital with facial burns at autopsy 149
(80 percent) died from complications other than tracheobronchitis
or pneumonia. If the facial burn was associated with more extensive thermal injury elsewhere a greater incidence of respiratory
tract injury became evident. All the deaths in this series of
patients attributed to tracheobronchitis or pneumonia in associations with facial burns occurred in patients with greater than 39

percent total body burns. Whereas none of 59 patients with facial
burns in association with less than 15 percent total body surface
burn suffered either a fatal or a non-fatal respiratory complication.

Phillips and Cope,[166] concluded that "burns of the respiratory
area go hand in hand with respiratory difficulty" unless the patient is out of doors. In their series, the incidence of respiratory difficulties in patients with burns of the respiratory area
of the face was 59 percent, greater than twice the incidence occurring in patients whose faces were spared. First degree facial burns
had the same incidence of respiratory tract damage as facial sparing
but 2° and 3° facial burns were associated with a 63 percent incidence of respiratory tract damage.[166]

Confinement in enclosed spaces increased the incidence of respiratory difficulties in the MGH series from 13 percent in those
burned outdoors to about 50 percent in those burned indoors, approximately a fourfold increase.[166]

The nature of the burning agent is a significant predictor of
respiratory tract injury. Flame burns produced respiratory difficulties in 1/3 of the previously healthy patients in the MGH series,
more than 10 times the incidence with other agents.[166] The history
of smoke inhalation is a very important predictor of respiratory
tract damage. The patients in the MGH series without facial injury
who developed respiratory tract damage generally had been overcome
by smoke.[166] Of greatest value in predicting respiratory tract
injury is the combination of severe flame burns of the respiratory
area of the face sustained in a closed space (indoors) which resulted in respiratory tract damage in 88 percent of the patients
in the MGH series, an eightfold increase in incidence over patients
with similar facial burns in an unenclosed space.[166]

Symptoms and Signs: In a series of 66 patients with inhalation
injury symptoms were frequently absent on admission and appeared
anytime from the first to the fifteenth day. The second post-burn
day was the usual onset. Symptoms lasted from 1 day to 4 months
depending on the severity of injury.[58] An outstanding characteristic
of the immediate survivors of the Cocoanut Grove fire was hyperactivity, even to the extent of mania in some cases.[24] Evaluation of
the large MGH series has resulted in the recommendation that burned
patients with deep flame burns of the face sustained indoors who
manifest restlessness, irrational behavior, confusion, stupor, or
coma should be considered to be anoxic from respiratory tract injury
until proven otherwise.[167]

The most common observable symptom in the MGH series was labored
respirations, occurring in 98 percent of the fatalities and almost

1/2 the survivors.[167]   Actual wheezing or even stridor are common.
[58,167,13]   The second most common observable symptom in the MGH
series was cough.[167]   Hemoptysis is not infrequent,[167,58] but is
apparently more common among the fatalities.[167]   Carbonaceous sputum
is quite common,[167,168,58,13] but usually clears by the 2nd day.[58]
Bronchial casts are coughed up or suctioned through tracheal cannulas as early as the 1st post-burn day.[167,1,13]   Purulent sputum may
be noted as early as the 3rd day post-burn.[167,58]   Hoarseness and
sore throat are present in approximately 1/4 of the patients with
inhalation injury.[167,58]   Two patients in the MGH series permanently
lost their voices.[167]   Larynogoscopy generally shows edema and
reddening of the hypopharynx, larynx, tongue, and epiglottis occasionally with definite burned areas which occasionally extend beyond
the vocal cords.[167,13]   Cyanosis on admission was present in 50 percent of the fatal cases and 22 percent of the survivors in the
MGH series, but it may be masked by burn shock, or carbonmonoxyhemo
globinemia.   Fifteen percent of the survivors in the MGH series were
cherry red in color on admission from carbonmonoxyhemoglobinemia.[167]

   Audible chest findings are typically delayed for 24-36 hours.
[202,58]   The characteristic transient and migratory nature of the
auscultatory findings of localized pulmonary collapse and over-in-
flation was first noted among the survivors of the Cocoanut Grove
fire.   These abnormal findings were noted to frequently disappear
after coughing up mucus or bronchial casts,[13] but it has been emphasized that the disappearance or diminution of adventitial sounds
may be the result of total collapse of an area of lung rather than
improvement in aeration.[167]

Radiology of Inhalation Injury

   The development of x-ray abnormalities virtually always lags
behind the clinical and laboratory evidence of pulmonary problems.[1]
According to most authors, initial x-rays are usually normal.[58,237,
202,172]   Abnormal findings usually begin appearing by the 2nd and
3rd post-burn day.[58,237]   However, pulmonary edema, confirmed at
autopsy, was observed radiographically in 2 patients dead on arrival
from the Cocoanut Grove fire, demonstrating that abnormal findings
may be seen at an early time.

   Although the incidence of abnormal x-rays in patients with inhalation injury is not widely reported it seems clear that the majority had radiographic abnormalities.[167,177]   Abnormal chest x-rays
were observed in 1/2 of all the 1140 patients seen in the MGH series.
[167]   Among the 36 patients with inhalation injury from the Cocoanut
Grove fire slightly more than 1/3 of the patients had no abnormalities on chest x-ray at anytime.[177]

   The first abnormalities to show up on the chest x-ray are usually related to the 2 initial pathologic events occurring in the respiratory tract, namely, pseudomembranous necrotizing tracheo-bronchitis, and pulmonary edema,[58,237,1,133,167,196,80] and of these 2 pathologic events the most important in terms of radiographic abnormalities is the tracheobronchitis. The majority of the early x-ray changes are explained by various degrees of atelectasis and emphysema due to complete or partial obstruction of small bronchi and alveoli with sloughed mucosal debris.[177,167] Of 7 patients autopsied during the first 4 days post-burn with the findings of patchy pulmonary densities on chest x-ray, all showed exudate and necrotic debris filling bronchi and alveoli and pulmonary edema with focal atelectasis in 5 of them.[167] These pathologic events explain the frequent radiographic findings of: 1) Patchy densities or linear densities with pulmonary volume loss secondary to complete bronchial occlusion; and 2) Patchy emphysemia from partial bronchial occlusion.[167,177] These x-ray findings may shift from lobe to lobe or vanish with effective coughing.[167,177] Furthermore, the bronchial obstruction not infrequently causes massive collapse of 1 or more lobes, and has even resulted in collapse of an entire lung.[167] Compensatory emphysema is often seen when a significant amount of pulmonary collapse is present.[167]

Laboratory Findings

The Pa_{O_2} is frequently decreased in burn patients with and with out inhalation injury.[58,1,167,70,169] In the series of patients with inhalation injury from Brooke Army Hospital where the Pa_{O_2} was measured, it averaged 65 torr with a range from 35-97 torr. The Pa_{O_2} returned to normal on the average at 3 weeks with a range from 4 days to 4 months post-burn.[58] In another series of patients from the same hospital factors were looked for which were associated with hypoxemia. The Pa_{O_2} was decreased in 17 of 28 previously healthy young adult burned patients in the early post-burn period even though no abnormalities were present on physical examination or x-ray. It is interesting that the mere presence of facial burn, the presence or absence of enclosure, or the presence of singed nasal hairs were not associated with a decreased Pa_{O_2}. Factors with positive correlation included: 1) Higher burn index; 2) Severe facial burn; and 3) Pharyngitis.[70] In still another series from the Brooke Army Hospital 43 patients with greater than 50 percent burns had Pa_{O_2} determinations in the first 5 days. Two-thirds of the patients had a Pa_{O_2} less than 80 torr with a mean lowest value of 75.3 torr and a mean time of return to normal at the 7th day post-burn.[169] This series supports the contention that major thermal injury is associated with hypoxemia in the absence of direct respiratory tract damage.

Respiratory alkalosis seems to be the rule in the patient with a major burn.[97,169,70] The mean pH was 7.47 at the time of lowest pO_2

in the Brooke Army Hospital series of 62 patients with greater than
50 percent burns.   The pH characteristically was elevated even after
the P_{aO_2} returned to normal and the pH did not decrease with the
administration of oxygen during the hypoxemic period.[169] Patients
with severe inhalation injury may have respiratory acidosis.[202]

The importance of carbon monoxide poisoning in inhalation injuries was first emphasized in the reporting of the Cocoanut Grove
fire. Many of the patients dead on arrival at the hospital demonstrated signs of carbon monoxide poisoning. Fifteen percent of the
survivors of inhalation injury in the MGH series were cherry red
in color and oxygen saturations as low as 8.6 vols % secondary to
carbon monoxide were observed. Achauer, et al.[1] performed carbonmonoxyhemoglobin determinations on a series of 100 consecutive burn
patients, of whom 22 percent developed "clinically significant
pulmonary complications." The carbonmonoxyhemoglobin percentage
was less than 15 percent in all but 2 patients. The test was not
considered helpful in predicting pulmonary problems. The variable
incidences of significant carbonmonoxyhemoglobinemia between series
is probably related to differences in exposure times to the products
of combustion and/or to differences in the concentrations of the
products of combustion.

Abnormal spirometry is the rule in all major burn patients.   Except for patients with severe respiratory insufficiency,[202] the
minute volume is generally increased due to an increase in the respiratory rate. Minute volumes as high as 22L/min have been observed.
However, the tidal volume is apparently generally reduced.[97] The
vital capacity, which may be difficult to do on patients with severe
respiratory embarrassment,[167] is generally decreased in patients with
inhalation injury and seems to correspond inversely to the amount
of atelectasis present.[13,167]

Pulmonary Sepsis

Airborne pneumonia is usually manifest in the 2nd week post-burn
whereas hematogenous pneumonia begins in the 3rd week post-burn[169]
but it is clear from experimental work that airborne pneumonia can
begin as early as 8 hours after inhalation injury.[147] Organisms
have been observed in the lungs of patients as early as 17 hours
after burning and bronchopneumonia as early as the second and third
day.[167,113]

The incidence of pneumonia among all burn patients at Brooke Army
Hospital during 1962-63 was 23 percent. The incidence among all burn
patients from the same hospital during 1967-68 was 15 percent. Of
interest is the fact that the ratio of airborne/hematogenous pneumonia changed from 1/3 to 2/3 in 1962-63 to 2/3 to 1/3 in 1967-68.
Hematogenous pneumonia has become less of a problem largely because

of better control of burn wound sepsis with topical chemotherapy.[169]

Airborne pneumonia occurred in 58 percent of the patients with
inhalation injury from the Brooke Army Hospital series.[58] Although
the type of pneumonia was not specified, all but one patient in the
MGH series with inhalation injury developed superimposed pulmonary
sepsis if they lived more than 72 hours.[167] Stone and Martin,[202]
similarly reported that bacterial pneumonia (type unspecified) developed in 100 percent of patients with inhalation injury who
survived as long as the third day post-burn. These data support
the proposal that, given enough time, airborne pneumonia is virtually
an inevitable complication of inhalation injury.

Apart from the important association of tracheobronchitis with
airborne pneumonia in the patient with inhalation injury, the most
significant associations were with tracheostomies and mechanical
ventilators. Stone and Martin,[202] noted that the most severe
pneumonias occurred in patients who had been treated with positive
pressure respirators early in their hospital course. Of 16 patients
who received mechanical ventilator therapy for pulmonary edema and
who survived more than 2 days, 12 died of extensive necrotizing
bronchiopneumonia with innumerable intrapulmonary abscesses. They
noted a significant reduction in the incidence of severe and fatal
pneumonias when the ventilator equipment was properly and frequently
sterilized. Foley, et al.[80] from the Brooke Army Hospital have
emphasized that tracheobronchitis and pneumonia are often not related to the inhalation injury per se but rather to the tracheostomy.

The organisms which produce pulmonary sepsis are, for the most
part, the same organisms responsible for burn wound sepsis, regardless of whether the pneumonia is airborne or hematogenous.[169] The
spectrum of offending bacteria is virtually the same from series
to series and included staph aureus and the gram negative rods:
Pseudomonas, Klebsiella, Aerobacter, Proteus, and E. Coli.[169,186,
80,202]

Pathology and Pathophysiology of Inhalation Injury

Experimental: Stone et al.[203] modified the ambient air with 1)
increased temperature alone; 2) increased humidity with increased
temperature; and 3) smoke with increased temperature, and demonstrated in rats that mortality increased with each set of conditions
over a period of time up to 20 minutes exposure. It was interesting
that 100 percent mortality occurred by 9 minutes exposure to dry air
(10 percent humidity) at 100°C with interstitial pulmonary edema
and hemorrhage observed at autopsy. The addition of increasing
humidity increased the lethal nature of the environment but smoke
and heat together was the most lethal combination. Moritz, et al.[147]
were unable to produce any thermal injury below the pharynx in dogs

with hot dry air delivered directly into the pharynx.

All other attempts to produce respiratory tract damage directly in animals have involved the use of tracheal cannulas. Moritz, et al.[147] using hot dry air, flame, and steam via tracheal cannulas in dogs measured temperature gradients down the trachea and evaluated the respiratory tract damage pathologically with each agent. They demonstrated large decreases in temperature between the cannula opening in the upper trachea and the carina. Hot dry air at a temperature of about 300°C at the upper trachea was only 50°C at the lower trachea with no lower tracheal injury and no pulmonary injury. Direct flame exposure of the upper trachea at temperatures of about 500°C resulted in lower tracheal temperatures of only 50–135°C, but there was mild to moderate lower tracheal injury in all animals and mild pulmonary injury in 25 percent. The use of steam at a temperature of 100° in the upper trachea resulted in only a minimal temperature gradient with lower tracheal temperatures of 53–94°C. Lower tracheal injury was present in all animals and was severe in two thirds. The pathologic picture was one of destroyed tracheobronchial mucosa, hemorrhagic pulmonary edema, and bronchopneumonia as early as 8 hours post injury. Zikria, et al.[237] using steam delivered via tracheal cannula in dogs noted that the pathological events of respiratory tract damage passed through 3 phases. The first phase of coagulation necrosis and early reactivity occurred within the first hour and consisted of edema formation in the entire respiratory tract. The second phase of late reactivity lasted up to 24 hours and consisted of further sloughing of tracheobronchial mucosa, increased interstitial edema, perivascular edema, atelectasis, and hemorrhagic consolidation. The third phase was infection, which occurred after 24 hours and consisted of "bronchopneumonia behind respiratory tract obstruction secondary to mechanical or functional block and necrotizing tracheobronchitis." Aviado, et al.[14,15] using steam delivered by tracheal cannula in dogs demonstrated the rapid accumulation within minutes of protein-rich pulmonary edema. They also observed significant hemolysis of pulmonary blood with plasma hemoglobins as high as 1.0 gm/100cc. A rise in aortic blood temperature as high as 12.6°C was demonstrated and a transient reflex apnea and bradycardia occurred. The authors advanced the theory that thermal edema was caused by a combination of venous constriction and increased capillary permeability.

Rappaport, et al.[171] have presented evidence in rats that thermal injury not involving the respiratory tract may cause the release of fibrinplatelet microaggregates into the systemic circulation with subsequent embolization to the pulmonary vascular tree. In their experimental model the intravenous injection of colloidal carbon suspensions, which did not kill unburned rats, was rapidly fatal to 86 percent of the burned rats within 24 hours, if given between 1–4 hours post-burn. Death was associated with the rapid onset of

cyanosis, tachycardia, and respiratory distress. The lungs revealed
massive embolization of the carbon floccules which could be prevented
by pretreatment with heparin.

Matsuura, et al.[135] have demonstrated that thermal injury in dogs
causes a decrease in compliance and surfactant activity. Zikria, et
al.[237], have also demonstrated in patients with inhalation injury
and in dogs with steam inhalation injury that the surface tension in
the lungs is abnormally elevated. Cooper, et al.[50] in looking for a
mechanism of surfactant deterioration following thermal burn have
presented evidence that the "fundamental pathophysiologic mechanism
of pulmonary surfactant inhibition following thermal burn is due
to seepage of plasma across the alveolar capillary membrane."

Clinical: Two major problems arise in interpreting the pathologic findings of patients dying with thermal injury in general or
those specifically with inhalation injury. The first relates to
the specificity of respiratory tract pathology and its supposed
relationship to burn mortality. The second problem relates to how
much of the respiratory tract pathology is due to treatment. Since
much of the pathologic data involves treated burn patients, a search
for data from immediate fatalities provides some useful information
related to the specificity of inhalation injury pathology in the
absence of therapeutic invasion. Therefore, a division of patients
into categories of "dead on arrival" and "treated" is appropriate.

Victims Dead on Arrival: Eschar formation of the trachea, has
not been noted among DOA cases. This makes it "doubtful that flames
could have been inhaled far into the respiratory passages."[133] Specific findings have been those of a hemorrhagic tracheobronchitis
and have included submucosal hemorrhages, (often pin point) vascular
dilatation, marked edema of the mucosa, and frank sloughing of the
mucosa.[133,237] These findings have favored the theory that "some
physically or chemically irritating agent was inhaled."[133] In the
DOA cases the mechanism of death was assumed to be "anoxemia dependent in part upon inhalation of carbon monoxide and in all probability other gases, and in part upon edema of the lung."[133] Diffuse
pulmonary edema has been observed within the alveoli with indications
of high protein content as evidenced by its hematogeneous colloid
like precipitation. In addition, red blood cells were present within
alveoli indicating the presence of "significant capillary damage."[133]
Although the pulmonary edema can be impressive in promptly asphyxiated cases,[133] 7 patients examined by Zikria, et al[237] did not have
significant elevations of water content above normal.

Treated Patients: Bronchoscopic examinations of patients with
inhalation injury in the early post-burn period have revealed the
same findings as in the DOA cases.[237] Even in living patients with
tracheostomies, significant transverse ulcers have been seen through

the bronchoscope in the lower trachea and the main bronchi well beyond the reach of the tracheostomy tube.[237]   It is almost universally
believed that the basic pathologic feature of inhalation injury in
the airways is a diffuse shallow necrotizing pseudomembranous
tracheobronchitis.[133,167,237,196]

Severe necrotizing laryngitis has been observed with virtual
complete laryngeal stenosis,[133,167] and a greater involvement below
rather than above the vocal cords.[133]   In the Brooke Army Hospital
series of 188 autopsies of burn deaths, only 1 patient who did not
sustain steam burns demonstrated pathology in the trachea which could
be interpreted as thermal injury per se, and the specific pathology
only extended a few centimeters below the glottis.[169] Thus, clear
evidence for thermal injury to the tracheobronchial tree or lungs
is lacking. The tracheobronchitis is characteristically shallow
with necrosis never extending below 100 microns. The basement
membrane is sometimes partially destroyed in the larynx and trachea
but rarely in the bronchi. The similarity in appearance of the
airways in patients with inhalation injury and those who have been
exposed to several of the war gases, notably phosgene, mustard gas,
and chlorpicrin, and to nitrogen dioxide inhalation, supports the
theory that the tracheobronchitis is a result of the inhalation of
toxic products of combustion.[133] The importance of the necrotizing
tracheobronchitis is that the necrotic material sloughs and obstructs
the airway. It is in fact the obstructed upper airway, from edema
and the obstructed lower airway from sloughed necrotic epithelium
that is responsible for the majority of the clinical problems both
early and late following inhalation injury. The obstruction may
either be partial or complete, and may even be extensive enough to
involve all small bronchi.[167]   That tracheostomy contributes to the
tracheitis is undeniable but whether or not the "majority of laryngeal and tracheal lesions found at autopsy are due to repeated
trauma from tracheostomy tubes or endotracheal tubes,"[80] is uncertain. However, Foley, et al.[80] from the Brooke Army Hospital observed in 109 autopsied patients with tracheobronchitis, of whom
99 had had tracheotomy, that the laryngeal and tracheal mucous above
the tracheal stoma was grossly normal in the majority of patients,
supporting the proposal that the tracheostomy tube is the major
cause of tracheitis. The authors added, though, that the influence
of tracheostomy in patients with inhalation injury might be an
additive one since "lesions mechanically related to trauma from the
tracheostomy tube seem to occur more readily following suspected
inhalation injury."

The lungs usually demonstrate pulmonary edema or bronchopneumonia,
or both.[1,237,167,169,196,133,58] The pulmonary edema may be related
to the primary inhaled irritant[133] or probably more often to over
hydration or to existing heart disease.[145,202] The alveoli themselves have frequently been plugged by debris and often have them-

selves been partially destroyed. Alveolar hemorrhage is often
seen.[167,196] Scattered areas of localized pulmonary collapse or
over-expansion are the results of either complete or partial airway
obstructions. That the bronchopneumonia also results from the ob-
structing debris is clear from several pathologic series. Sochor and
Mallory[196], noted the "almost constant" association of bronchiolitis
and bronchopneumonia. Pruitt, et al.[169] from the Brooke Army Hospital noted the ability of bacteria, especially pseudomonas, to invade
areas of ulceration and contribute to the severity of the ulceration.
Hyaline membranes are occasionally seen.[169]

It is interesting that major bland pulmonary embolism is a rare
occurrence in the seriously burned patient and it is virtually
never the cause of death despite immobilization for long periods of
time.[169,185] This complication is not even mentioned in 2 of the
largest series of inhalation injuries.[58,202] Only 3 percent of the
patients in the MGH series died from pulmonary embolism[165] and less
than 1 percent of the patients in the Brooke Army Hospital series
of 1086 patients had it at all.[169] However, pulmonary microembolism
has been frequently reported.[133,69,80,196] Malloy and Brickley,[133]
in their report of the Cocoanut Grove fire fatalities, were apparently the first to report "the presence of multiple miliary emboli
throughout the pulmonary arterial tree" in 2 of the 3 treated patients which came to post-mortem examination. Socher and Mallory,
[196] noted fibrin thrombi in the small pulmonary arteries and arterioles in 1/2 of the patients with severe body burns autopsied at
Boston City Hospital during a 15 year period. Thus, it seems very
likely that thermal injury, as with other forms of trauma, results
in pulmonary microembolism which in turn may contribute to the
pulmonary damage which is often seen.[171,33,69]

Mortality of the Burn Patient

The mortality for patients with respiratory tract damage sustained at the time of burning have substantially greater mortality
than those without. It varies from 33-72 percent depending to some
extent on how many are immediately asphyxiated.[58,76,133,167.202]

Among more recent series which include more patients treated
with topical chemotherapy, the pulmonary pathology is the major
cause of death in the vast majority of cases. Shook, et al.[186]
reported that 84 percent of the deaths among 290 patients treated
between 1966 and 1968 were due to pulmonary pathology. Achauer,
et al.[1] reported that all the deaths in a series of 100 burn victims
treated between 1970 and 1971 occurred among patients with pulmonary
complications and that the lungs at post-mortem examination always
demonstrated either pulmonary edema or bronchopneumonia or both.
It is also clear among the large series, that the specific pulmonary
problem responsible for death is airborne pneumonia if the patient

survives beyond 3 days,[113,80,77,196,1,165] but it is not clear how many of these patients dying with pneumonia actually sustained inhalation injury.

The respiratory tract is responsible for death in about 75 percent of the patients dying with inhalation injury.[58,237,202] Immediate deaths are due to suffocation.[76] Early deaths are usually caused by respiratory insufficiency.[202,167,237,58] The later deaths are usually from pneumonia.[202,58,167]

Clinical Management of Patients
with Respiratory Tract Damage in Burns

Aviado[14] treated dogs subjected to steam inhalation injury with a combination of 1) antihistamines to combat the hyperhistaminemia; 2) pulmonary hypotensive drugs to "relieve pulmonary hypertension and edema"; and 3) oxygen inhalation for relief of anoxemia. He demonstrated an increase in survival from less than 1 hour to up to 8 hours. Zikria, et al,[237] demonstrated some survival improvement in dogs with steam inhalation injury treated with a combination of antibiotics and a positive pressure respirator. He concluded that in the acute phase of the burn "the respiratory insufficiency is secondary to edema and to patchy atelectasis which can be reduced by positive pressure respiration." He further suggested that since infection of the respiratory tract is such a substantial problem after 48 hours, "early antibiotic coverage may help to control it."

Any patient with criteria for inhalation injury present should be placed immediately in an environment of increased oxygen and humidity.[202,167] The obvious early problems with surfactant loss and obstruction of airways with sloughed necrotic epithelium make atelectasis prevention one of the primary aims of aggressive early management.[186,1,167] Hourly turning, coughing, and hyperventilation has obvious potential benefit.[167,186] The patient on a ventilator must have periodic hyperinflations.[186] Escarotomies should be done on patients with circumferential chest burns.[1] While it is important to keep up with the substantial fluid and electrolyte requirements of the severely burned patient, it is equally important not to cause pulmonary edema by over-hydrating the patient.[145,167,232,1] Frequent monitoring of arterial blood gases is helpful,[232] and the use of the flow-directed pulmonary artery catheter for measuring pulmonary artery pressures has advantages over the CVP.

The necessity of doing immediate laryngoscopy on all burn patients to assess the need for tracheal cannulation has been emphasized.[172] If severe edema is present, tracheal cannulation is advised either with an endotracheal tube if possible,[1] or with a tracheostomy.[97,172,232] Steroids should be given to lessen the problem of edema formation.[1] Some of the airway obstruction with inhalation injury is due to bronchospasm.[202] Bronchodilators have not been helpful but steroids have produced almost immediate reversal of bronchospasm in over 50 percent of the patients in Stone and Martin's series.[202] Pruitt, et al.,[169] have emphasized that because of the tendency of the burn patient to develop ulcerations of the G. I. tract, steroids should be used for as brief a period as possible. A vigorous cough is the most essential requirement for clearing the tracheobronchial tree of sloughed necrotic debris. Bronchoscopy is useful whenever the patient is unable to clear his obstructed tracheobronchial tree.[169,20] Humidity and mucolytic agents are probably useful in assuring adequate tracheobronchial toilet.[13,186,169,202] If tracheal cannulation is necessary early after burning[1], it seems desirable to use an endotracheal tube for a short time. Significant complications have been observed with the endotracheal tube as with the tracheostomy tube. Complications ranging from hoarseness to severe necrotizing laryngotracheobronchitis were observed in 4 of 10 patients treated with ventilatory assistance via a nasotracheal tube.[169] Tracheostomy, if needed to manage the patient with respiratory failure "must be considered cautiously, however, remembering the importance of sepsis from the operative procedure."[172] If a ventilator is used atelectasis prevention becomes a major management principle accomplished by periodic hyperinflation to total lung capacity.[186,1] Continuous positive pressure may be helpful.[1,20] Oxygen should be set as low as possible because of the potential danger of oxygen toxicity to the lung.

A follow-up study was done on 8 patients available for recall among 19 survivors of inhalation injury. Seven out of 8 patients had had marked symptoms of respiratory distress within 24 hours after injury and all were treated with tracheostomies, controlled ventilation, steroids, and antibiotics. Four of the patients developed pneumonia. All patients without pre-existing disease were symptom free at follow-up. Bronchoscopy and bronchograms were done on all patients between 3 months and 3 years after injury. Bronchoscopy revealed tracheobronchitis in 75 percent of the patients and left mainstem bronchial stenosis in 1 patient. Bronchography was normal except in the patient with bronchostenosis. Pulmonary function tests were normal. It was concluded that chronic inflammatory changes were not prevented by steroid administration.

G. PHYSIOLOGY OF PULMONARY REIMPLANTATION

Since the first successful orthotopic autotransplantation of the lung in 1951 by Juvenell, et al.[117] a voluminous surgical literature has arisen with a variety of conflicting reports about the results and even the feasibility of the procedure. It seems clear from this literature that much of the earlier data regarding abnormal post-reimplantation pulmonary physiology and the associated conclu-

sions must be disregarded, for pulmonary reimplantation, either for
basic physiological studies or for survival studies, can result in
normal or near-normal physiologic measurements regardless of the
parameter chosen to evaluate. However, it is equally clear that
near-perfect surgical technique is the key to success in experimental pulmonary reimplantation. Anastomotic artifacts have in
fact been the explanation for many of the early reports of abnormal
physiology.[39,65] Analysis of the work of the few investigators who
have worked with pulmonary autografting through the years and who
have developed technical expertise with this procedure has demonstrated fairly consistent patterns of normal respiratory and hemodynamic functions following reimplantation.

Serial roentgenography and histology have demonstrated a consistent pattern referred to by Siegleman, et al., as the "pulmonary
reimplantation response."[187a] The picture is one of pulmonary
edema developing in the immediate post-operative period, reaching
its peak during the third post-operative day, and completely clearing by 3-4 weeks. The edema is not uniformly distributed throughout the lung, but instead is located centrally, having a limited
peripheral extension in the shape of a flame or fan. This observation is an important one since peripheral lung biopsies are typically normal in the unilateral canine autograft without contralateral pulmonary artery ligation after the third post-operative
day, even though hilar edema is still obviously present on chest
x-ray. Traces of alveolar edema are the only histopathologic
finding.[187a]

The primate with a unilateral autograft and contralateral pulmonary artery ligation demonstrates a diffuse proteinaceous alveolar
exudate, interstitial edema, intravascular congestion, and areas of
local alveolar collapse during the first post-operative week. Extended ischemia has been shown to increase the severity of the alveolar and interstitial edema and to even make the picture "irreversible." At two weeks the lung biopsies demonstrate some alveolar exudate with continued interstitial edema and intravascular
congestion. At one month the alveolar exudate and interstitial
edema is almost completely resolved but mild congestion persists.
From 1-6 months the mild congestion persists and a diffuse interstitial leukocyte and macrophage infiltration develops with patchy
areas of bronchopneumonia. No significant areas of pneumonitis
are seen after 6 months. Gradually, by 9 months a uniform histologic picture is seen consisting of fibrous thickening and hypertrophy of the alveolar septa, characterized by a marked increase
in fibroblast proliferation, round cell infiltration, and fibrin
deposition in the extravascular spaces. Vascular pathology is not
present one year after pulmonary reimplantation in the primate with
contralateral pulmonary artery ligation even though all of the
cardiac output is going through the autograft.[115]

Survival studies have been done in dogs and baboons, but the only meaningful ones are those which immediately challenge the transplanted lung or lungs to provide sustained respiratory support. Unless one does cardio-pulmonary transplantation this can be accomplished by either unilateral pulmonary reimplantation and simultaneous contralateral pulmonary artery ligation, or by simultaneous bilateral pulmonary reimplantation. Meaningful success has been accomplished with both methods in the dog[6,218] and with unilateral reimplantation and immediate contralateral pulmonary artery ligation in baboons.[116] These studies have demonstrated that the immediately transplanted lung can indeed provide immediate and sustained respiratory support in the experimental animal.

For most physiologic studies unilateral reimplantation by itself is a satisfactory surgical procedure, and most of the data relating to separate physiologic parameters have been investigated in the animal with unilateral reimplantation. Although the baboon provides advantages of having fewer surgical complications and of being a primate,[93] the vast majority of the specific physiologic studies have been done in dogs.

The physiological characteristics of the vascular bed of the reimplanted lung are still variously interpreted. First, the mean pulmonary artery pressure measured anytime after unilateral reimplantation is normal, and no gradient is present across the anastomosis if the anastomosis is widely patent. This holds true even if the entire cardiac output is forced through the transplanted lung by occlusion of the contralateral pulmonary flow.[128,213,214,7] The data relating to resistance characteristics within the transplanted lung, however, have not been uniform. At low flows, mean pulmonary vascular resistance is not elevated.[213] However, if the resistance studies are done after contralateral pulmonary artery occlusion, either temporarily or permanently, variable results are reported. Lincoln, et al.[128] reported normal resistance in the transplanted lung immediately after reimplantation if the ischemia time is short. However, longer periods of ischemia up to 4 hours are associated with acute increases of pulmonary vascular resistance, probably related to damage within the pulmonary parenchyma. Most authors report an increase in pulmonary vascular resistance if the transplanted lung is forced to accept an increase in cardiac output by means of contralateral ventilation with 100 percent nitrogen via a tracheal divider or by means of temporary occlusion of the contralateral pulmonary artery.[213,120] These data have produced 2 different interpretations. Tisi, et al.[213] concluded that the "pulmonary vasculatures of the transplanted lung did not maintain normal elastance characteristics to increased blood flow." Koerner and Veith,[120] however, demonstrated identical curves of pulmonary vascular resistence and pulmonary blood flow in both normal and in autotransplanted lungs, and concluded that "the vasculature of

transplanted lungs does not have a fixed resistance and can dilate
normally with increased blood flow." The reimplanted lung does
respond to hypoxia by increasing its pulmonary vascular resistance.
[213,120]

The return towards normal of bronchial artery flow following
reimplantation has been well documented. Bronchial artery injection
techniques and bronchial artery flow studies using carbon monoxide
uptake in lungs with pulmonary artery occlusion have demonstrated
partial return of vascular channels as early as 6 days post-
operatively, and the systemic blood supply to the reimplanted lung
is re-established during the first month after complete interruption
of the bronchial circulation.[78,234]

The lymphatics have been studied with sky blue dye injected
into the reimplanted lung. Lymph flow was demonstrated by this
technique within 7 days and new lymphatic vessels were grossly
visible within 12 days. By the twentieth post-operative day, the
injected dye spread readily across the regenerated vessels, across
the bronchial suture line, into the hilar lymph nodes.[71] Thus, as
with the bronchial arteries, complete regeneration of the lymphatics
occurs within the first month.

Evidence of ischemia in the distal bronchial stump has long been
observed by bronchoscopy following reimplantation.[92] Pearson, et
al.[155] noted extensive mucosal edema and retention of secretions
distal to the level of division of the bronchus which gradually
disappeared within 6 weeks of the surgery. Transient mucosal edema
at the bronchial suture line has been demonstrated during the first
week in reimplanted lungs by deposition of a titanium aerosal
solution of human serum albumin tagged with technetium 99m or indium
113m during inhalation scanning.[112] The most comprehensive study
of the time course of functional return of the bronchi following
reimplantation has been done by Siegelman, et al.[187] They performed
plain chest x-rays and bronchography at weekly post-operative intervals following lung reimplantation in dogs. Transient bronchial
narrowing was observed, at the anastomosis, but also extending dis-
tally into all major lobar bronchi during the first month post-
operatively. By the end of the first month plain chest x-rays and
bronchograms were normal. Of interest is the observation that most
of the narrowing occurs in the first week, presumably from mucosal
edema, and that by the second week a progressive increase in width
of the bronchial lumen was observed.

That neural regeneration can occur following pulmonary reimplantation is well documented. The first analysis of the afferent
autonomics was done by Secrist and Trummer,[183] who demonstrated the
return of the pulmonary stretch-receptor (Hering-Breuer) reflex. In
a series of 14 dogs studied from 4 to 57 months after unilateral

reimplantation, both unilateral airway occlusion and inflation were produced with the use of a tracheal divider, and the response observed in the contralateral lung. Six dogs demonstrated return of the reflex in the reimplanted lung, the earliest occurring 7 months post-operatively. Eight dogs did not show any return of the reflex. Durvoisin, et al.[66] studied the same reflex in 44 dogs up to 17 months following unilateral reimplantation with the use of a tracheal divider by inflating the lung under study with various volumes to produce specific intrabronchial pressures. They demonstrated return of stretch-receptor activity in only 5 of the reimplanted lungs 9 to 12 months after surgery. Edmunds, et al.[68] also demonstrated the return of a week Hering-Breuer inflation and deflation reflex but in only one of 11 dogs studied 10-28 months after reimplantation. Further analysis of the afferent autonomics in these 11 dogs with ammonia inhalation, demonstrated weak ventilatory responses from the reimplanted lungs in only 2 of the dogs. These studies indicate that although afferent neural regeneration can occur, it does not return in the majority of animals and does not regenerate sufficiently to restore physiological activity.[66,68]

Regeneration of parasympathetic efferent nerves was studied by Edmunds, et al[68] using both electrical stimulation of the ipsilateral cervical vagus nerve and apnea to produce bronchoconstriction. In the 11 dogs studied 3-6 months after unilateral reimplantation, cervical stimulation caused diffuse bronchoconstriction in the reimplanted lungs of all dogs. In 5 of these dogs apnea caused similar bronchoconstriction in both the normal lung and in the reimplanted lung. These authors concluded that vagal bronchoconstrictor nerves regenerate rapidly.

Regeneration of sympathetic efferent nerves was studied by Edmunds, et al.[68] in 6 dogs 13-27 months afer unilateral lung reimplantation by supramaximal stimulation of the ipsilateral stellate ganglion and cardioaccelerator nerve to produce bronchodilatation. Bronchodilatation, however, was not observed, and it was concluded that the sympathetic efferent nerves do not regenerate sufficiently to restore physiological activity.

Ventilation and oxygen uptake has generally been shown in the dog to decrease following reimplantation and to return towards normal by the end of the first month if anastomotic artifacts are avoided and pneumonia isnot present.[65,110] In the baboon, both ventilation and oxygen consumption return to normal in the first month following reimplantation.[93] In both animals, the return of oxygen consumption lags behind the return of ventilation.[93,65,110] In the dog, normal preoperative values of ventilation and oxygen uptake are apparently not usually obtained.[110,213] Both Tisi, et al.[213] and Birch,[32] have demonstrated lower than normal lung volumes. Although Tisi, et al.[213] did not describe any histologic

abnormalities in their long term surviving dogs, Joseph and Morton
have described the sequence of morphologic alterations in the uni-
laterally reimplanted lung with contralateral pulmonary artery
ligation in the baboon. They described the disappearance of congestion and edema after the first month but also described the
gradual appearance in all survivors of fibroblastic proliferation
and hypertrophy of the alveolar septae after 6 months. Although
not a universal finding, pulmonary compliance is less than normal
in the long term dog survivors in the series by Tisi, et al.[213]
Overall airway resistance is normal if the bronchial anastomosis is
normal.[213]

The long term survivor of either bilateral simultaneous pulmonary reimplantation or of unilateral reimplantation with contralateral pulmonary artery ligation has normal arterial blood gases.[6]
In addition, diffusion capacity is normal.[213] Surfactant loss
occurs during the early post reimplantation period but returns rapidly to normal.[214,221] In the series reported by Waldhausen, et
al.[221] surfactant was normal by 8 days. Thus, it seems that most
of the physiological parameters either return to normal or near-
normal in the reimplanted lung, and that the time period of functional return of all significant parameters is within the first
month. It is clear that afferent autonomic and efferent sympathetic
nerves do not predictably return and that efferent parasympathetics
do return within the first 6 months. It is possible that some of
the abnormalities in lung size and in hemodynamics are related to
permanenent ischemia injury. It has been well shown that the lung is
very intolerant to ischemia.[128,219] Veith, et al[219] have shown that
in the absence of cooling or alveolar expansion lungs with even 1
hour of normothermic ischemia are unable to support life any longer
than 5 days at the most following injury.

H. OXYGEN INDUCED PULMONARY PATHOLOGY

Oxygen has been used for many decades as a clinical adjunct to
medical therapy and had been implicated as a pulmonary toxin for
almost as long. Smith,[194] in 1899, presented experimental evidence
that implicated oxygen in the production of a pathologic lung condition; however, even at the present time a pathopneumonic lesion
has not been definied. Experimental and clinical work have been
advanced to support and to deny oxygen implication in lung pathology.
However, the weight of evidence seems to favor the observation
that oxygen in concentrations clinically used can produce pulmonary
pathology.

Animal experimentation has generally confirmed the impressions
related to oxygen toxicity and has explained some of the probable
mechanisms. Species differences, noted by Lee[123] may be one cause

of the present confusion. Paine,[154] et al, noted mild pulmonary
edema and diffuse hyperemia in dogs exposed to oxygen concentrations at or below atmospheric pressure. In animals exposed to
85-90 percent oxygen concentration alveolar capillary congestion
and an increased number of leucocytes were noted. The animals
devloped symptoms of respiratory distress but did not die. A
second group of dogs exposed to 95-100 percent oxygen concentration
all died with a mean survival time of 39 hours. Their lungs showed
marked congestion, hyperemia and pulmonary edema. The lungs were
beefy red and weighed more than normal. Paine also observed that
the effects of exposure to 90-100 percent oxygen concentration could
be markedly reduced by brief intermittent periods of exposure to
room air.

Smith,[193] et al., in similar work, exposed dogs to 98 percent
oxygen concentration until death. (Average 54.6 hours.) The lungs
were noted to be beefy, edematous, congested, and hemorrhagic
grossly. The tidal volume increased while respiratory rate decreased
and late in the study compliance decreased. Left atrial pressure
was measured and was normal and the authors concluded that left
heart failure was not involved in the production of the pulmonary
changes observed. Laurenzi,[122] et al, studied the influence of
increased and decreased oxygen concentrations on trachial mucus
flow in cats and noted decreased mucus flow at both extremes of
concentration. He proposed a biochemical explantion that involved
inactivation of sulphydryl groups. This biochemical influence of
oxygen has been confirmed in vitro and he felt that the resultant
interference with carbohydrate metabolism could explain the abnormal
flow of tracheal mucus.

Soloway and co-authors felt that oxygen in excess concentration
inactivates sulphyhydryl dependant dehydrogenase and thereby inter-
fers with high energy phosphate bonding and carbohydrate metabolism.
Oxygen excess also inhibits proteolysis, stimulates histamine release
and causes decrease surfactant activity.

Experimental studies have generally confirmed the impression
that animal lungs are damaged by exposure to increased oxygen concentration. The pathologic changes involve alveolar capillary endothelial destruction, interstitial edema andhemorrhage and destruction of the type 1 pneumocyte.

Alterations in physiologic paramets include decreased p_{aO_2} right
to left shunting and late compliance changes. Changes in surfactant
content and function have not been conclusive. In each study a small
percentage of animals show a marked resistance to the development
of pulmonary oxygen toxicity. Lee,[124] et al., studied changes in
respiratory parameters in dogs exposed to increased oxygen concentrations (100%) at one atmosphere of pressure. He noted a progres-

sive decrease in the arterial oxygen concentration. He also measured
the presence of intrapulmonary shunting and at 24 hours observed
right to left shunts which averaged 30 percent of the cardiac output.
Pulmonary surfactant was decreased in all animals exposed to 100 percent oxygen but the percent reduction was quite variable.

Adamson,[2] et al, in a study using rats, studied pulmonary surfactant and electron microscopic changes. Ninety percent of the
rats died after a 7 day exposure to 90 percent oxygen concentration.
The survivors were returned to room air and 2 were sacrificed daily
during the recovery period. Rats in the surviving group appeared
normal after return to room air. Pathologic changes in the group of
rats which died after exposure to 90 percent oxygen were typical of
changes seen in the human respiratory distress syndrome. The lungs
were solid and hemorrhagic to gross inspection. The earliest electron microscopic changes were focal cytoplasmic swelling of the alveolar capillary endothelium and interstitial edema and these were
observed in both the animals dying and those which survived. The
animals which died developed extensive changes in the type one
pneumocyte with disintegration of the plasma membrane. Dilatation
of alveolar capillaries with occasional fibrin-platelet thrombi
also occurred. A hyaline membrane composed of damaged type one
pneumocytes and fibrin occurred in some animals. The type 2 pneumocyte was well preserved.

In the surviving animals the pathologic changes were quite different. Mild interstitial edema andmild swelling of the lamellar
bodies was observed. These changes resolved within 6 days and the
lungs were normal in appearance at 6 weeks. No fibrosis had
occurred as the result of oxygen exposure at 6 weeks. Surface tension measurements revealed no significant change from normal controls. Studies in healthy human volunteers have revealed no permanent disability from oxygen exposure. Welch,[224] et al, exposed
healthy males to 100 percent O_2 at 1 atmosphere of pressure and
noted substernal distress at 30 hours. The longest exposure was
110 hours and all volunteers recovered without incident.

Retrospective clinical pathologic studies in patients given
high concentrations of oxygen have uniformly shown pulmonary pathology. Nash,[150] et al, studied patients given varying concentrations
of oxygen with positive pressure ventilation. His autopsy study
using age, sex, and disease matched patients and controls revealed
four statistically significant pathologic lesions in the oxygen
therapy group. These were 1) Intra alveolar fibrin exudate, (hyaline membrane); 2) Alveolar septal and intralobular edema; 3) Mild
lymphocytic infiltration; 4) Early fibrosis.

The patients were divided into four groups based on oxygen concentration and length of exposure. Groups 1 and 2 were exposed to

no more than 90 percent O_2. Group 1 was exposed for less than 10 days and Group 2 for greater than 10 days. Groups 3 and 4 were exposed to 90-100 percent oxygen, Group 3 was exposed for less than 10 days and Group 4 for greater than 10 days. The four previously mentioned lesions were most common in Groups 3 and 4. The authors observed no correlation between time on the respirator and the appearance of pulmonary lesions unless oxygen was being administered and felt that this observation eliminated the respirator as a causative factor. Finally in the group of patients who lived long enough healing occurred by fibrosis.

Soloway[197] et al., in a clinical study observed similar pathologic lesions in 6 patients dying while on the respirator with oxygen support. Eeach of the patients in this study had been in shock and the author felt that only the hyaline membrane distinguished the lungs in this study from the lungs in any pathologic shock material.

Barbee[17] in a semi-controlled study of 10 patients with irreversible brain damage also noted significant pulmonary abnormalities after exposure to oxygen in increased concentration. All patients had normal heart and lung function.

One group of 5 patients was treated with intermittent positive pressure ventilation and room air and the other group was treated with IPPV and 100 percent oxygen. The $P_{a_{O_2}}$ was the most sensitive indicator of oxygen toxicity and began to drop rapidly at 40 hours in the 100 percent oxygen group. Also right to left shunting was observed in this group. The $P_{a_{CO_2}}$ lung-thorax compliance, and cardiac functions did not vary in either group. The dead space to tidal volume ratio which was increased at the start of the observation period in both groups increased significantly more after 30 hours in the 100 percent oxygen group.

Pulmonary edema and hemorrhage were observed in both groups as were numerous other lesions but significantly no hyaline membranes were observed in either group.

Clinical opinions are not uniform in the acceptance of oxygen as a toxic agent. Singer[192] et al., in a prospective study of patients undergoing open heart surgery could distinguish no pulmonary toxicity related to oxygen therapy as measured by intrapulmonary shunting, compliance or clinical course. The absence of any pulmonary changes may have been related to the fact that the aim of their therapy was to maintain a pO_2 of 80-120 mm. Hg. Eighteen patients were on 100 percent oxygen concentrations for a mean duration of 21 hours and it is possible that this exposure was not long enough to produce damage. Van De Water[215] et al., in normal human volunteers also noted no changes in physiologic

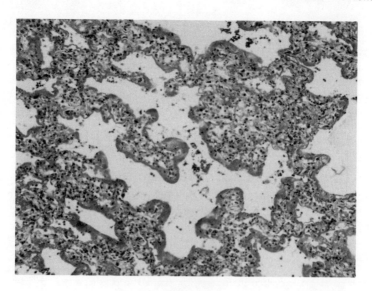

Figure 2

parameters after 6-12 hours of exposure to 100 percent oxygen
concentration. Their study involved measurement of $A-a_{O_2}$ gradient,
pulmonary shunting, pulmonary arterial pressure, total pulmonary
resistance, cardiac output and pulmonary extravascular water
volumes. The authors suggested that observed pathologic pulmonary
changes related to oxygen exposure were related to primary illness
and not to oxygen toxicity.

The weight of both experimental and clinical evidence favors
a role for oxygen as a pulmonary toxin when administered over an
appropriate time span and at an appropriate concentration. The
time and concentration necessary for toxicity seems to be species
dependent.

The pathologic picture is not pathopneumonic. Heavy lungs with
a beefy appearance and interspersed with pulmonary hemorrhages are
the notable gross findings. Light microscopy reveals hyperemia,
interstitial edema and hemorrhages, variable leucocytic infiltrates
and atelectasis (Fig. 2). The presence of a "hyaline membrane"
(Fig. 3) has not been uniformly present. If the patients live long
enough resolution by fibrosis is apparent. Examination of the tissue by electron microscopy has suggested early subcellar damage of
alveolar capillary endothelium and the type one pneumocyte. Later
disruption of the plasma membranes occurred and changes characterized

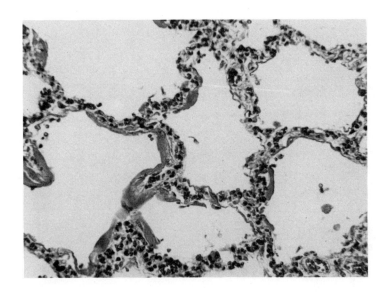

Figure 3

by light microscopy became apparent.

Physiologic parameters routinely observed to be abnormal after exposure to high concentrations of oxygen are: 1) Decreased $P_{a_{O_2}}$; 2) Increased right to left shunting; and 3) Decreased lung compliance.

I. LUNG DAMAGE RESULTING FROM BLOOD TRANSFUSION

Since the advent of cardiopulmonary bypass and techniques which enable massive blood transfusions, this entity has been postulated as a cause of pulmonary pathology. It is known that the pulmonary vascular bed serves as an effective filter in situations where substances 20 µ and larger are given intravenously. Etiologic factors considered important in transfusion embolization are the use of whole blood, the storage technique, age of the blood, volume of the transfusion, and the presence or absence of a adequate filtration system. The above factors, as well as the type of oxygenator and the use of hemodilution technique are of major importance in the production of micro-emboli within pump oxygenator systems.

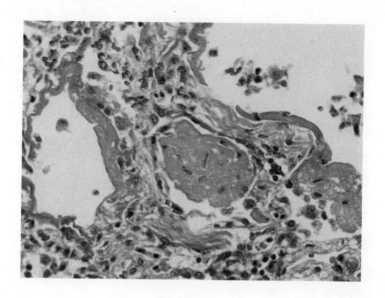

Figure 4

The pathologic lesion produced in the lung consists of emboli between 20 and 150 µ. It has been shown that the emboli consists primarily of fibrin platelet aggregates (Figure 4). Less frequently noted are embolic particles of bone marrow and fat. These can be virtually eliminated by techniques in the conduct of the open heart procedure which avoids the return of any blood from outside the cardiac chamber to the heart lung machine.

The physiologic result of micro-embolization in the lung is ventilation-perfusion imbalance with resultant hypoxemia. Pulmonary hypertension may result and may cause right sided heart failure if embolization is massive.

Swank, et al.[106] have described the use of screen filtration pressure (SFP) in an attempt to quantitate the influence of particulate emboli. This technique filters particulate matter and relates the amount of debris to a pressure necessary to force blood through the filter. This filter removes the same sized particles that are filtered by the pulmonary capillary bed. Regularly stored ACD blood begins developing significant micro-emboli within the first 48 hours and the number of emboli increases to a maximal number between days 7 to 10.[137,148,205] Heparinized blood begins to develop similar micro-emboli within 6 hours and a maximal number by 24 hours. A routine blood filter system is effective in removing these particles from only 1 or 2 units of blood.

Experimental work in dogs[137] has shown that transfusion of homologous blood with high SFP causes pulmonary hypertension. Further studies[138] using dogs have shown that these changes can be prevented by effective filtration. McNamara and co-workers[139] in a study of massively transfused combat casualties concluded that the lungs served as the final filter for this particulate debris. Statistical analyses by this group led to the conclusion that post injury hypoxia was related to the number of blood transfusions.

Similar considerations have been applied to those patients undergoing open heart surgery. Many of these patients undergo massive transfusion and virtually all are exposed to an added factor, the pump-oxygenator. The biophysical properties associated with oxygenation of blood is of major importance in the production of blood trauma. Of the three commonly used oxygenators the disc oxygenator and the bubble oxygenator produce the greatest amount of particulate debris while the membrane oxygenator produces the least as measured by SFP.[12] The trauma appears to occur at the blood-O_2 interface and this factor is virtually eliminated in membrane oxygenators. Another major factor which has reduced blood trauma and the formation of micro-emboli is the technique of hemodilution. This technique has minimized the total volume of blood transfused, but its major importance has been a reduction in blood viscosity and better micro-circulation.

Silicone which was extensively used as a debubbling agent in early oxygenators and which was frequently observed to embolize is no longer a major problem. The improvement in oxygenators and universal acceptance of hemodilution techniques have significantly decreased pump-oxygenator derived emboli as a problem in routine cardio-vascular surgery, although several recent studies,[12,153] comparing blood filtered to remove all particulate debris (20-150 μ) with unfiltered blood have failed to show any difference in the morbidity or mortality of randomized clinical open heart cases, almost all authors noted more micro-emboli through the use of filters and/or Dopper counters and several remarked that the patients with in line filters seemed better.[46]

As pumping and oxygenation systems are developed which permit prolonged perfusion, it seems certain that adequate filtration will be an integral part of any system.

Animal data has shown an impressive histologic and physiologic difference in favor of a filtration system which remove 20-150

particulate emboli from blood. Clinical studies on massively
transfused casualties are equally persuasive even though no controlled data is available.

REFERENCES

1. Achauer, B. M., Allyn, P. A., Furnas, D. W., Bartlett, R. H.:
        Ann. Surg. 177:311, 1973.

2. Adamson, I. Y., Bowden, D. H., Wyatt, J.P.: Arch. Pathol. 90:
        463, 1970.

3. Agar, J. M.: Med. J. Aust. 2:1182, 1966.

4. Alder, F., Peltier, L. F.: Surg. Forum 12:453, 1961.

5. Alfano, G. S., Hale, H. W.: J. Trauma 5:647, 1965.

6. Alican, F., Cayirli, M., Isin, E., Hardy, J. D.: JAMA 215:
        1301, 1971.

7. Alican, F., Cayirli, M., Isin, E. Hardy, J. D.: Ann. Surg.
        174:34, 1971.

8. Allardyce, D. B., Groves, A. C.: Surg. 66:71, 1969.

9. Allgower, M., Burr, C., Cuene, L., Engley, F., Fleish, H.,
        Gruber, U. F., Harder, F., Russell, R. G. G.: Ann. N. Y.
        Acad. Sci. 150:807, 1968.

10. Armin, J., Grant, R. T.: Clin. Sci. 10:441, 1951.

11. Ashbaugh, D. G., Petty, T. L.: Surg. Gyne. & Obstet. 123:493,
        1966.

12. Ashmore, P. C., Svitek, V., Ambrose, P.: Thor. & Cardiovasc.
        Surg. 55: 691, 1968.

13. Aub, J. C., Pittman, H., Brues, A. M.: Ann. Surg. 117:834, 1943.

14. Aviado, D. M.: Circ. Res. 7:1018, 1959.

15. Aviado, D. M., Schmidt, C. F.: Circ. 6:666, 1952.

16. Baker, P. L., Pazel, J. A., Peltier, L. F.: J. Trauma 11:
        1026, 1971.

17. Barber, R. E., Lee, J., Hamilton, W. K.: New Eng. J. Med.
        283:1478, 1970.

18. Beal, D. D., Conner, G. H.: Laryngoscope 80:25, 1970.

19. Beal, D. D., Lambeth, J. T., Conner, G. H.: Laryngoscope 78:396, 1967.

20. Beall, A. C., Bricker, D. L., Crowford, H. W., DeBakey, M. E.: Dis. Chest 49:568, 1966.

21. Bean, J. W., Beckman, D. L.: J. Appl. Physiol. 27:807, 1969.

22. Beckman, D. L., Bean, J. W.: J. Appl. Physiol. 29:631, 1970.

23. Beckman, D. L., Bean, J. W., Baslock, D. R.: J. Appl. Physiol. 30:394, 1971.

24. Beicher, H. K.: Ann. Surg. 117:825, 1943.

25. Benatar, S. R., Ferguson, A. D., Goldschmitt, R. B.: Quarterly J. Med. 41:85, 1972.

26. Benfield, J. R., Coon, B. S.: J. Thorac. Cardiov. Surg. 53:676, 1965.

27. Bergentz, S. E.: Chir. Scand. Suppl. 282:1, 1961.

28. Berger, J. C.: From the Office of the Surgeon General. Dept. of Army, 1962.

29. Bergmann, E. S.: Berl. Klin. Wschr. 10:385, 1873.

30. Berman, I. R., Ducker, T. B.: Submitted for Publication (personal communication).

31. Berman, R., Ducker, T. B., Simmons, R. L.: J. Neurosurg. 30: 532, 1969.

32. Birch, A. A.: J. Thorac. Cardiov. Surg. 55:196, 1968.

33. Blaisdell, F. W., Lim, R. C., Stallone, R. J.: Surg. Gynec. & Obstet. 130:15, 1970.

34. Blaisdell, F. W., Lim, R. C., Jr., Stallone, R. J.: Surg. Gynecol. Obstet. 130:15, 1970.

35. Bonbest, H. C.: Amer. Rev. Resp. Dis. 91:97, 1965.

36. Borja, A. R., Ransdell, H. T.: Amer. J. Surg. 122:81, 1971.

37. Brewer, L. A., III: Ann. Thor. Surg. 7:387, 1969.

38. Brewer, L. A., III: In Surgery in World War II. Thor. Surg.
         Editor F. B. Bery & E. M. McFetridge, U. S. Govern.
         Print. 2:462, 1963.

39. Brody, J. S., Fisher, A. B., Park, C. D., Hyde, R. W., Wald-
         hausen, J. A.: J. Appl. Physiol. 29:587, 1970.

40. Brown, F. K.: Amer. J. Physiol. 85:510, 1956.

41. Burford, T. H.: Surgery in World War II. Thor. Surg. Editor
         F. B. Bery and E. M. McFetridge. U. S. Government Print.
         2:16, 1963.

42. Burford, T. H., Burbank, R. B.: J. Thor. Surg. 14:415, 1945.

43. Cahill, J. M., Jouasset-Strieder, D., Byrnes, J. J.: Amer. J.
         Surg. 110:324, 1965.

44. Cameron, G. R.: Brit. Med. J. 1:965, 1948.

45. Campbell, G. S., Haddy, F. J., Adams, W. L., Visscher, M. B.:
         Amer. J. Physiol. 158:96, 1949.

46. Carlson, R. G., Lande, A. J., Baxter, J., Patterson, R. H., Jr.,
         Lillehei, C. W.: To be published.

47. Coalson, J. J., Hinshaw, L. B., Guenter, C. A.: Exp. & Mol.
         Path. 12:84, 1970.

48. Cohn, R.: Surg. Clin. North Amer. 52:585, 1972.

49. Conn, J. H., Hardy, J. D., Fain, W. R., Netterville, R. E.:
         J. Trauma 3:22, 1963.

50. Cooper, N., Matsuura, Y., Murner, E. S., Lee, W. H.: Amer.
         Surg. 33:882, 1967.

51. Cope, O.: Ann. Surg. 117:801, 1943.

52. Cushing, H.: Bull. Hopkins Hosp. 12:290, 1901.

53. Daly, I. DeB., Daly, M. DeB.: In Problems of the Pulmonary
         Circulation Ciba Foundation No. 8, Ed. A. V. S. de Reuck
         and M. O'Connor, Little, Brown, 1961.

54. Daniel, R. A., Cate, W. R.: Ann. Surg. 127:836, 1948.

55. Daniel, R. A., Jr., Cate, W. R., Jr.: Ann. Surg. 127:836, 1948.

56. DeMuth, W. E.:   J. Trauma 6:222, 1966.

57. DeMuth, W. E., Smith, J. M.:   Amer. J. Surg. 109:819, 1965.

58. DiVincenti, F. C., Pruitt, B. A., Jr., Reckler, J. M.:   J.
         Trauma 11:109, 1971.

59. Donnellan, W. L., Poticha, S. M., Holinger, P. M.:   JAMA 194:
         1323, 1965.

60. Ducker, T. B., Simmons, R. L., Martin, A. M.:   Amer. J. Dis.
         Child, 118:638, 1969.

61. Ducker, T. B., Simmons, R. L., Anderson, R. W.:   J. Neurosurg.
         29:475, 1968.

62. Ducker, T. B., Simmons, R. L.:   J. Neurosurg. 28:112, 1968.

63. Ducker, T. B., Simmons, R. C.:   Arch. Neurol. 18:123, 1968.

64. Ducker, T. B., Simmons, R. C.:   Arch. Neurol. 18:123, 1968.

65. Duvoisin, G. E., Fowler, W. S., Ellis, F. H., Jr., Payne, W. S.:
         Chest 58:102, 1970.

66. Durvoisin, G. E., Payne, W. W., Ellis, F. H., Jr., Fowler, W. S.:
         Chest 58:504, 1970.

67. Durvoisin, G. E., Payne, W. S., Ellis, F. H. Jr.:   Chest 58:
         96, 1970.

68. Edmunds, L. H., Jr., Graf, P. D., Nadel, J. A.:   J. Appl. Phy-
         siol. 31:722, 1971.

69. Eeles, G. H., Sevitt, S.:   J. Pathol. Bacteriol. 93:275, 1967.

70. Epstein, B. S., Hardy, D. L., Harrison, H. N., Teplitz, C.,
         Villarreal, Y., Mason, A. D., Jr.:   Ann. Surg. 158:924,
         1963.

71. Eraslan, S., Turner, M. D., Hardy, J. D.:   Surgery 56:970, 1964.

72. Erickson, D. R., Shinozaki, T., Beekman, E., Davis, J. G.:   J.
         Trauma 11:689, 1971.

73. Evarts, C. M.:   Surg. Clin. North. Amer. 50:493, 1970.

74. Farringotn, G. H., Saravis, C. A., Cossette, G. R., Miller,
         D. A., Clowes, G. H. A., Jr.:   Surg. 68:136, 1970.

75. Farringer, J. L, Carr, D.:   Amer. J. Surg. 85:474, 1959.

76. Faxon, N. W.:   Ann. Surg. 117:803, 1943.

77. Feller, I., Hendrix, R. C.:   Surg. Gynec. & Obstet. 119:1, 1964.

78. Fisher, A. B., Kollmeier, H., Brody, J. S., Hyde, R. W.,
         Hansell, J., Friedman, J. N., Waldhausen, J. A.:   J.
         Appl. Physiol. 29:839, 1970.

79. Fischer, J. F., Turner, R. H., Herndon, J. H., Risebarough,
         E. J.:   Surg. Gynec. & Obstet. 132:667, 1971.

80. Foley, F. D., Montcrief, J. A., Mason, A. D.:   Ann. Surg. 167:
         251, 1968.

81. Fuchsig, P., Bruckle, P., Blumel, G., Gottlab, R.:   N. E. J. M.
         276:1192, 1967.

82. Fulton, R. L., Peter, E. T.:   Surg. 67:499, 1970.

83. Fulton, R. L., Peter, E. T., Wilson, J. N.:   J. Trauma 10:719,
         1970.

84. Gardner, A. M. N., Harrison, M. H. M.:   J. Bone and Jt. Surg.
         39B:538, 1957.

85. Gauss, H.:   Arch. Surg. 9:593, 1924.

86. Gelin, L. E., Hallgran, B., Kerstell, J., Rudenstam, C. M.,
         Svanborg, A.:   Acta. Chir. Scand. 133:107, 1967.

87. Gerst, P. H., Rattenborg, C., Holaday, D. A.:   J. Clin. Inves.
         38:524, 1959.

88. Glas, W. W., Gerkin, T. D., Davis, H. L., Musselman, M. M.:
         Amer. J. Surg. 91:471, 1956.

89. Gray, A. R., Harrison, W. H., Couves, C. M., Howard, J. M.:
         Amer. J. Surg. 100:709, 1960.

90. Greenfield, L. J., Barkett, V. M., Coalson, J. J.:   J. Trauma
         8:735, 1968.

91. Gump, F. E., Mashima, Y., Jorgensen, S., Kinney: J. Surg. 70:
         262, 1971.

92. Haglin, J. J.:   Ph. D. Thesis, Univ. of Minn., 1964.

93. Haglin, J. J., Arnar, O.: Ann. N. Y. Acad. Sci. 162:404, 1969.

94. Hallgren, D., Kerstell, J., Kudenstam, C. M., Svanborg, A.:
    Acta. Chir. Scand. 132:613, 1966.

95. Hamilton, W. R., Hustrand, R. F., Peltier, L. F.: Surg. 56:
    53, 1964.

96. Hardman, J. M., Earle, K. M.: J. Neuropath. Exp. Neurol. 26:
    119, 1967.

97. Harrison, H. N.: Ann. N. Y. Acad. Sci. 150:627, 1968.

98. Harrison, W., Liebow, A. A.: Circulation 5:824, 1952.

99. Hausberger, F. X., Whitenack, S. H.: SGO 143:931, 1972.

100. Heaton, L. D., Hughes, C. W., Rosegay, H., Fisher, G. W.,
     Feighney, R. E.: J. Current Prob. Surg. November 1966.

101. Henzel, J. H., Smith, J. L., Pories, W. J., Burget, D. E.:
     Amer. J. Surg. 113:525, 1967.

102. Herndon, J. H., Riseborough, E. J., Fisher, J. E.: J. Trauma
     11:673, 1971.

103. Hewlett, T. H.: Sheft's Init. Mgmt. of Thor. & Thoracoabdom-
     inal Trauma. Charles C. Thomas, Springfield, 1968.

104. Hillman, J. W., LeQuire, V. S.: Surg. Forum 19:465, 1968.

105. Hinshaw, L. B., Kuida, H., Gilbert, R. P., Visscher, M. B.:
     Amer. J. Physiol. 191:293, 1957.

106. Hissen, W., Swank, R. L.: Amer. J. Physiol. 209:715, 1965.

107. Honer, R. J.: Anes. 29:702, 1968.

108. Hopkinson, B. R., Broder, J. R., Schenk, W. G.: J. Thor.
     Cardiovasc. Surg. 55:580, 1968.

109. Huggins, C. E.: Lancet 2:1059, 1959.

110. Hutchin, P., Feezor, M. C., Walker, E. L., Peters, R. M.: J.
     Thoracic Cardiov. Surg. 61:476, 1971.

111. Ingram, R. H., Szidon, J. P., Skalak, R., Fishman, A. P.:
     Circ. Res. 22:801, 1968.

112. Isawa, T., Benfield, J. R., Castagna, J., Johnson, D. E.,
     Taplin, G. V.: Amer. Rev. Resp. Dis. 103:76, 1971.

113. Jackson, T. M., Lee, W. H.: Arch. Surg. 87:937, 1963.

114. Jenkins, M. T., Jones, R. F., Wilson, B., Moyer, C. A.:
     Ann. Surg. 132:327, 1950.

115. Joseph, W. L., Morton, D. L.: Surg. Gynec. & Obstet. 133:
     821, 1971.

116. Joseph, W. L., Morton, D. L.: Ann. Thor. Surg. 11:442, 1971.

117. Juvenelle, A. A., Citret, C., Wiles, C. E., Jr., Stewart, J.
     D.: J. Thor. Surg. 21:111, 1951.

118. Keller, C. A., Schramel, R. J., Hyman, A. L, Creech, O., Jr.:
     J. Thor. Cardiov. Surg. 53:743, 1967.

119. Kerstell, J.: Amer. J. Surg. 121:712, 1971.

120. Koerner, S. K., Veith, F. J.: Chest 59:531, 1971.

121. Kuida, H., Hinshaw, L. B., Gilb ert, R. P., Visscher, M. B.:
     Amer. J. Physiol. 192:335, 1955.

122. Laurenzi, G. A., Yin, S., Guarneri, J. J.: New Eng. J. Med.
     279:333, 1968.

123. Lee, C., Lyons, J. H., Konisberg, S., Morgan, Fred, Moore,
     F. D.: J. Thor. & Cardiovas. Surg. 53:759, 1967.

124. Lee, C., Lyons, J. H., Moore, F. D.: J. Thor. & Cardiovasc.
     Surg. 53:770, 1967.

125. Lehman, E. P., Moore, R. M.: Arch. Surg. 14:621, 1927.

126. LeQuire, V. S., Shapiro, J. L., LeQuire, C. B., Cobb, C. A.,
     Jr., Fleet, W. F.: Amer. J. Path. 35:999, 1959.

127. Lewis, A., Pappas, A. M.: J. Trauma 9:49, 1969.

128. Lincoln, J. C. R., Lowenstein, E., Austen, W. G., Laver, M. B.:
     Ann. Surg. 172:877, 1970.

129. Lloyd, E. L, MacRae, W. R.: Brit. J. Anes. 43:365, 1971.

130. Long, J.: London Med. Gazzett 25:743, 1840.

131. Luisada, A. A.: Amer. J. Cardiol. 20:66, 1967.

132. Mackay, E. M.: Proc. Soc. Exp. Biol. Med. 74:695, 1950.

133. Mallory, T. B., Brickley, W. J.: Ann. Surg. 117:865, 1943.

134. Martin, A. M., Soloway, H. B., Simmons, R. L.: J. Trauma
          8:687, 1968.

135. Matsuura, Y., Najib, A., Lee, W. H.: Surg. Forum 17:86, 1966.

136. McNamara, J. J., Molot, M., Dunn, R., Burran, E. L, Stremple,
          J. F.: J. Thor. Cardiovasc. Surg. 63:968, 1972.

137. McNamara, J. J. : Surg. 71:594, 1972.

138. McNamara, J. J., Burran, E. L., Larson, E., Omeya, G., Suehiro,
          G., Yamase, H.: Ann. of Thor. Surg. 14, 2:133, 1972.

139. McNamara, J. J., Molot, M. D., Stremple, J.: Ann. Surg. 172:
          334, 1970.

140. Meek, R. N., Woodruff, B., Allardyce, D. B.: J. Trauma 12:
          432, 1972.

141. Mitchell, J. H., Linden, R. J., Sarnoff, J.: Circ. Res. 8:
          1100, 1960.

142. Mokkhavesha, S., Shim, S. S., Patterson, F. P.: J. Trauma
          9:39, 1969.

143. Montcrief, J. A.: New Engl. J. Med. 288:444, 1973.

144. Moon, V. A.: In: Shock: Its Dynamics, Occurrence, and
          Management: Lea & Febiger, 1942.

145. Moore, F. D.: Surg. Clin. North. Amer. 50:1249, 1970.

146. Moritz, E., Broder, J. R., Schueller, E. F., Schenk, W. F.:
          Arch. Surg. 105:275, 1972.

147. Moritz, A. R., Henriques, F. C., McLean, R.: Amer. J. Path.
          21:311, 1945.

148. Moseley, R. V., Doty, D. B.: Ann. Surg. 171:329, 1970.

149. Murphy, W., Nicoloff, D. M., Humphrey. E. W.: J. Surg. Res.
          April 1973, In Press.

150. Nash, G., Blennerhasset, J. B., Pontoppidan, H.: New Eng. J.
        Med. 276:368, 1967.

151. Neptune, W. B., Redondo, H., Balley, C. P.: Surg. Forum 3:
        379, 1952.

152. Nichols, R. T., Pearce, H. J., Greenfield, L. J.: Arch. Surg.
        96:723, 1968.

153. Osborn, J. J., Swank, R. L., Hill, J. D.: J. Thor. & Cardio-
        vas. Curg. 60:574, October 1970.

154. Paine, J. R., Lynn, D., Keyes, A.: J. Thor. Surg. 11:151,
        1941.

155. Pearson, F. G., Goldberg, M., Stone, R. M.: Can. J. Surg.
        13:241, 1970.

156. Peltier, L. F.: J. Trauma 8:812, 1968.

157. Peltier, L. F.: J. Trauma 11:661, 1971.

158. Peltier, L. F.: Clin. Ortho. & Related Res. 66:241, 1969.

159. Peltier, L. F.: Current. Prob. Surg. May 1967.

160. Peltier, L. F.: J. Bone & Ft. Surg. 38A:835, 1955.

161. Pennington, D. G., Hyman, A. L., Jaques, W. E.: Surg. 73:
        246, 1973.

162. Perez-Guerra, F., Walsh, R. E., Sagel, S. S.: JAMA 218:1568,
        1971.

163. Perry, J. F.: J. Minn. Med. 48:601, 1965.

164. Phillips, A. W., Cope, O.: Ann. Surg. 152:767, 1960.

165. Phillips, A. W., Cope, O.: Ann. Surg. 155:1, 1962.

166. Phillips, A. W., Cope, O.: Ann. Surg. 156:759, 1962.

167. Phillips, A. W., Tanner, J. W., Cope, O.: Ann. Surg. 158:
        799, 1963.

168. Pleukhahn, V. D.: Med. J. Aust. 2:1185, 1966.

169. Pruitt, B. A., Jr., DiVincenti, F. C., Mason, A. D., Foley,
        F. D., Flemma, R. F.: J. Trauma 10:519, 1970.

170. Raffer, P. K., Montemurno, R., Scudesc, Z., Sherr, S.: Surg. Forum 22:446, 1971.

171. Rapaport, F. T., Memirovsky, M. S., Bachvaroff, R., Ball, S. K.: Ann. Surg. 177:472, 1973.

172. Reed, G. G., Camp, H. L.: Ann. Otorhinolaryngo 78:741, 1969.

173. Reynolds, J., Davis, J. T.: Radiological Clin. North Amer. 4:343, 1966.

174. Ross, A. P.: Surg. 66:765, 1969.

175. Ross, A. P. J.: Surg. 65:271, 1969.

176. Sarnoff, S. J., Sarnoff, L. C.: Circulation 6:51, 1952.

177. Schatzki, R.: Ann. Surg. 117:841, 1943.

178. Schoenenberger, G. A., Allgower, M.: Brit. J. Surg. 56:704, 1969.

179. Schwartz, M. L., Northrup, W. F., Nicoloff, D. M., Humphrey, E. W.: In Press

179a. Schwartz, M. L., Murphy, W. R. C., Humphrey, E. W.: Fed. Proc. 32:439, 1973.

180. Scotti, T. M.: In Pathology. Edit. W. A. D. Anderson, C. V. Mosby Co., St. Louis 1971, Pg. 130.

181. Scully, R. E.: Amer. J. Path. 32:379, 1956.

182. Sealy, W. C., Ogino, S., Lesoge, A., Young, W., Jr.: Surg. Gynec. & Obstet. 122:754, 1966.

183. Secrist, W. L., Trummer, M. J.: Ann. Thor. Surg. 4:125, 1967.

184. Sevitt, S.: Lancet 2:826, 1960.

185. Sevitt, S., Gallagher, N.: Brit. J. Surg. 48:475, 1961.

186. Shook, C. D., MacMillan, B. G., Altemeier: Arch. Surg. 97: 215, 1968.

187. Siegelman, S. S., Sinha, S. B. P., Veith, F. J.: Chest 62: 475, 1972.

187a. Siegelman, S. S., Sinha, S. B. P., Veith, F. J.: Ann. Surg. 177:30, 1973.

188. Simmons, R. L., Ducker, T. B., Anderson, R. W.: J. Trauma
          8:800, 1968.

189. Simmons, R. L., Martin, A. M., Heisterkamp, C. A., Ducker,
          T. B.: Ann. Surg. 170:39, 1969.

190. Simmons, R. L., Ducker, T. B., Martin, A. M., Anderson, R. W.,
          Noyes, H. E.: Ann. Surg. 167:145, 1968.

191. Simons, R. L., Anderson, R. W., Ducker, T. B., Sleeman, H. K.,
          Collins, J. A., Boothman, K. P.: Ann. Surg. 167:158,
          1968.

192. Singer, M. M., Wright, F., Stanley, L. K., et al.: New Eng. J.
          Med. 283:1473, 1970.

193. Smith, C. W., Lehan, P. H., Monks, J. J.: J. Appl. Physiology.
          18:849, 1963.

194. Smith, J. L.: J. Physiol. 24:19, 1899.

195. Smyth, N. P. D., Hughes, R. K., Corndell, E. E.: Amer. Surg.
          27:770, 1961.

196. Sochor, F. M., Mallory, G. K.: Arch. Path. 75:303, 1963.

197. Soloway, H. B., Castillo, Y., Martin, A. M.: Annals Surg.
          168:937, 1968.

198. Sova, J., Jivasek, V.: Biochem, Clin. 4:263, 1964.

199. Spees, E. K., Strevey, T. E., Geiger, J. P., Aronstam, E. M.:
          Ann. Thor. Surg. 4:133, 1967.

200. Spencer, F. C.: Current Probl. Surg. January 1964.

201. Sproule, B. J., Brady, J. L., Gilbert, J. A. L.: Canadian Med.
          Assn. J. 90:1243, 1964.

202. Stone, H. H., Martin, J. D.: Surg. Gynec. & Obstet. 126:1242,
          1969.

203. Stone, H. H., Rhame, D. W., Corbitt, J. D., Given, K. S.,
          Martin, J. D., Jr.: Ann. Surg. 165:157, 1967.

204. Sugg, W. L., Webb, W. R., Ecker, R. R.: Surg. Gynec & Obstet.
          127:1005, 1968.

205. Swank, R. L.: New Eng. J. Med. 265:728, 1961.

206. Swank, R. L., Hissen, W., Bergentz, S. E.: Surg. Gynec &
        Obstet. 119:779, 1964.

207. Szabo, G., Serenyi, P., Kocsar, L.: Surg. 54:756, 1963.

208. Taylor, F. W., Gumbert, J. L.: Ann. Surg. 161:497, 1965.

209. Tedeschi, L. G., Castelli, W. P., Tedeschi, C. G.: Human
        Path. 2:165, 1971.

210. Thomas, D., Stein, M., Tanabe, G., Rege, V., Wessler, S.:
        Amer. J. Physiol. 206:1207, 1964.

211. Thompson, P. L., Williams K. E., Walters, M. N. I.: J. Path.
        97:23, 1969.

212. Ting, Y. M.: Amer. J. Roent. 98:343, 1966.

213. Tisi, G. M., Trummer, M. J., Cuomo, A. J., Ashburn, W. L.,
        Moser, K. M.: J. Appl. Physiol. 32:113, 1972.

214. Trimble, A. S., Kim, J. P., Bharadwaj, B., Bedard, P., Wells,
        C.: J. Thorac. Cardiov. Surg. 52:271, 1966.

215. VanDeWater, J. M., Kagey, K. S., Miller, I. T., Parker, D. A.,
        O'Connor, N. E., Sheh, J., MacArthur, J. D., Zollinger,
        R. M., Moore, F. D.: New Eng. J. Med. 283:621, 1970.

216. Veith, F. J., Richards, R. K.: Ann. Surg. 171:553, 1971.

217. Veith, F. J., Richards, R. K.: Science 163:699, 1969.

218. Veith, F. J., Siegelman, S. S., Dougherty, J. C.: Surg. Gynec.
        & Obstet. 133:425, 1971.

219. Veith, F. J., Sinha, S. B. P., Graves, J. S., Boley, S. J.,
        Dougherty, J. C.: J. Thor. Cardiovas. Surg. 61:804, 1971.

220. Wahrenbrock, E. A., Carrico, C. J., Amundsen, D. A., Trummer,
        M. J., Severinghaus, J. W.: J. Appl. Physiol. 29:615,
        1970.

221. Walhausen, J. A., Giammona, S. T., Kilman, J. W., Daly, W. J.:
        JAMA, 191:130, 1965.

222. Wanebo, H., VanDyke, J.: Journal Thor. Cardiovas. Surg. 64:
        537, 1972.

223. Weisman, S. J.: Surg. 6:722, 1939.

224. Welch, B. E., Morgan, T. E., Glamann, H. G.:   Fed. Proc.
          22:1053, 1963.

225. Wertzberger, J. J., Peltier, L. F.:   Surg. 64:143, 1968.

226. Whitenack, S. H., Hausberger, F. X.:   Amer. J. Path. 65:336,
          1971.

227. Widdcombe, J. G., Kent, D. C., Nade, J. A.:   J. Appl. Physiol.
          17:613, 1962.

228. Williams, J. W., Stembridge, V. A.:   Amer. J. Roent. 91:284,
          1964.

229. Willwerth, B. M., Crawford, F. A., Young, W. G., Jr., Sealy,
          W. C.:   J. Thor. Cardiovas. Surg. 54:658, 1967.

230. Wilson, J. W.:   Adv. in Microcirc. 4:197, 1972.

231. Wilson, J. W.:   Surg. Gynec. & Obstet. 134:675, 1972.

232. Wilson, R. D.:   Anes. & Analg. 49:716, 1970.

233. Worthen, M., Argano, B., Sinadlowski, W., Bruce, D. W.,
          MacCanon, D. M., Luisada, A. A.:   Dis. Chest. 55:45,
          1969.

234. Wright, W. B., Berg, P., Seki, Y.:   J. Thorac. Cardiovas.
          Surg. 62:473, 1971.

235. Wyche, M. Q., Jr., Marshall, B. E.:   Ann. Surg. 174:296, 1971.

236. Young, R. C., Nagano, H., Vaughan, T. R., Staub, N. C.:   J.
          Applied. Physiol. 18:264, 1963.

237. Zikria, B. A., Sturner, W. Q., Astarjian, N. K., Fox, C. L.,
          Ferrer, J. M., Jr.:   Ann. N. Y. Acad. Sci. 150:618, 1968.

PHYSICAL MALTREATMENT OF CHILDREN

Robert W. ten Bensel

University of Minnesota

A Clinical Lament
Poor forlorn babe, barely started in life
But victim already of cruel family strife:
Your parents' tongues locked in silence,
Hush untold tales of secret violence.
When we flush your flesh with radiant streams,
Sick bones shine clear in truthful gleams.
It's shake, shake and shake more than bash and batter,
That bruise brain, bone and dura mater.
Remember your mother is not a fiend partaken,
Just your mom in distress, by the world forsaken.

John Caffey, M. D.[1]

In order to understand physical maltreatment of children one must look first at the historical perspective. Violent abuse and physical neglect has been performed upon children since the beginning of recorded time.

Infanticide has been practiced in almost every nation, both civilized and uncivilized, so much so that it can almost be considered a universal phenomenon.[2] Solomon states that infanticide has "been responsible for more child deaths than any other single cause in history, other than possibly bubonic plague."[3] Infanticide usually refers to the willful killing of a newborn with the consent of the parents, family, or community. The removal of children by this method has been justified on several bases. Bakan[2] feels that infanticide has been practiced in many cultures both primitive and civilized for the purpose of controlling population or elimina-

249

ting weak or deformed infants. Other rationales for this practice
have been ascribed to religious appeasement or reactions to proph-
ecies of doom. From a psychosociological point of view the individual responsiblity for the killing of an infant is transferred to a
higher institution thus leaving the individual's psyche undamaged
in the process.

The ritual killing or maiming of older children in attempts to
educate them, exploit them, or to sometimes rid them of evil spirits
has also been part of man's culture since early biblical times.
With the advent of urbanization and the resulting technological
changes, more value was placed upon the child by society; the phenomenon became defined as one of maltreatment of children. Physical
maltreatment of children implies a spectrum of both physical neglect
and abuse. There is a continuum of maltreatment from inappropriate
discipline such as a facial slap to repeated beatings to severe
malnutrition. Forms of mutilation of children have been recorded
over the centuries as part of religious and ethnic traditions.[4]
This has been referred to as ritual surgery and has included castra-
tion to produce eunuchs or singers, footbinding of female children
in China, wrapping the neck with wiring, splinting the lips and
nose by certain cultures in Africa, tatooing and cranial bindings
as seen in North American Indians. Uvulectomy was performed among
the Berbers of Nothern Morocco and in Roman times children were
used as professional beggars after they were hamstrung, deformed,
had gouged eyes or been blinded.

Charles Dickens wrote about the problems of children growing up
in an industrialized society.[3] Receiving "the Dickens" remains a
euphemism for the beating of children.

The first reported case of physical maltreatment of children in
the United States was the infamous Mary Ellen case in 1874.[5] Mary
Ellen had been beaten by her foster parents and in an attempt to
get help for the child the social worker finally took Mary Ellen
to the Society for the Prevention of Cruelty to Animals. It was
out of this experience that the Society for the Prevention of
Cruelty to Children was organized.

In the medical literature, child maltreatment was not appreciated
even in 1946 when Dr. Caffey[6] reported six children with multiple
fractures in the long bones of infants suffering from chronic subdural hematomas. Caffey concluded that the "fractures appear to be
of traumatic origin but the traumatic episodes and the causal
mechanism remain obscure."[6] Silverman in 1953[7] reached similar
conclusions as Caffey and noted the "excellent prognosis of traumatic
lesions compared with that of other diseases considered in the differential diagnosis." In 1954 Drs. Woolley and Evans[8] wrote an
article suggesting the possibility that parents were indeed respon-

sible for the fractures and subdural hematomas and concluded that
"it is difficult to avoid the overall conclusion that skeletal
lesions having the appearance of fractures, regardless of history
or injury or the presence or absence of intracranial bleeding, are
due to the undesirable vectors of force."

It was Dr. C. Henry Kempe's monumental work published in 1962[9]
which brought the full impact of physical maltreatment before the
medical and general public's attention. His article was titled
"The Battered Child Syndrome," and this term has been very helpful
in attracting attention to this still neglected medical and social
problem. However, it is important to realize that the Battered
Child Syndrome is only one part of the spectrum of physical maltreatment of children.[10]

At the present time the terminology used to best describe the
spectrum of physical maltreatment against children includes both
physical abuse and neglect. (Table 1.) Physical abuse to active
interaction with the child by another person irrespective of age
of the abuser. Physical abuse can be further subdivided into
mild, moderate, or severe abuse, ranging from bruises from spanking
to a single blow fracturing an arm or striking the head causing
death. Physical abuse usually refers to either a single episode
of abuse or repeated milder forms of abuse as seen in inappropriate
disciplinary practices. It is important to remember that death
can occur from a single episode of physical abuse.

Physical neglect refers to the failure to provide basic physiological needs for the child's existence. This would include failure to provide for adequate nutrition and clothing for the child;
it would also encompass proper medical care and provision for a
safe environment for the child to live. Neglect is often manifested by poor skin hygiene, lack of adequate nutrition, and is
a continuum with "failure to thrive." Moreover, there is a close
correlation between children with "failure to thrive," emotional
neglect, and concomitant mental retardation.[11]

The Battered Child Syndrome applies to the child who has had
repetitive, severe injuries which usually involve the fractures of
bones, internal injuries, severe skin injuries without fractures,
or central nervous system damage. The hallmark of this type of
maltreatment is one where the injuries described by the parents do
not adequately explain the types of injuries sustained by the child.
The legal interpretation of abuse implies that the injuries were
of a "nonaccidental" nature.[12]

Sexual abuse: Sexual attacks against children run a wide spectrum in this society from simple exposure or molestation to incest,
sodomy, and rape.

TABLE 1

MALTREATMENT OF CHILDREN:

NEGLECT	ABUSE
Emotional	Emotional
Physical	Physical
Medical	Mild
Nutritional	Moderate
	Severe
	Battered Child Syndrome
	Infanticide (Homicide)

THE DIMENSIONS OF THE PROBLEM

The maltreatment of children can be considered a disease affect-
ing the child and the parents. Some hard statistics are available,
but due to the nature of the problem there are probably inaccuracies
in reporting. Reported statistics include both cases of abuse and
neglect and there is no nation-wide consistency. There appears to
be a rapidly increasing rate of maltreatment of children.[10] This
is due not only to better identification and reporting but also
to a probably absolute increase in the number of children who are
maltreated. In New York City in 1968 there were 956 cases of
abuse reported; in 1969, 1,600; in 1970, 3,000; and in 1972, 5,200
cases. Data from California and Colorado when extrapolated yield
a conservative estimate of 200,000 to 250,000 children in the
United States who are maltreated annually.[3] The incidence is given
at approximately 350 per 100,000 children at risk per year.[13] Some
workers[13] state that for every child reported a hundred children
go unreported. Recent Congressional testimony[14] states that an
estimated 6,000 die annually as the result of child abuse. This
represents about 10 percent of children who suffer from physical
maltreatment.[14]

THE ABUSING PERSON

It is now well known that not only parents but also other
adults and even children may maltreat children. Dr. Adelson[15]
has described the "Battering Child" where preschool children can
experience homicidal rage and kill other children.

In order for abuse to take place the potential for abuse needs to be present. As the type and degree of physical attack varies greatly so does the psychology of the individuals involved. At the one extreme there is direct murder of children which is usually the result of a frank psychosis. At the other extreme are those cases where only mild physical abuse may have occurred. In these latter cases, the abusers may not differ greatly psychologically from an ordinary sample of people.

From the studies published there seems to be at least four categories relating to potential for abuse. These categories are taken from Dr. Helfer's recent article.[16]

1. How the parents themselves were reared. Thirty to sixty percent of abusing parents claim that they have been abused as children themselves.[3] This life style imprinting has an effect not only on abuse when they become parents but also is an etiological factor in juvenile deliquency and adult homicide.[17,18,19]

2. The inability of families and of mothers particularly, to use people to help them when they are "up tight" with their children. Parents who abuse children have not developed skills or abilities to use other people to help them when they are in crisis with their children. They are often isolated from their families and often have a very low self-image.

3. The marriage relationship between the spouses is often manifested in an unstable relationship. There is a high proportion of pre-marital conception, youthful marriage, unwanted pregnancies, forced marriages, and emotional or financial difficulties in the marriages of abusing individuals.

4. How parents see the child. The parents who have the potential for abuse often expect the child to do things for them at a very early period of life. They have an inappropriate expectation of the child's level of functioning and capacity to give at any given developmental age.

THE ABUSED CHILD

From national studies[3] we see the profile of a young child who is usually under age four (fifty percent under age two), more males than females abused, and a child who is somewhat "different." Hyperactive children, mentally defective children or children with other physical handicaps have a higher frequency of being abused children. Children may be unwanted from the financial or psychological perspective of the parents. Older children often provoke adults into assaulting them.

THE CRISIS

It is generally conceded that most cases of physical abuse or repetitive abuse of children arise out of a single or multiple series of crises within a family. Most abuse occurs in the lower socioeconomic groups; however, it has been reported in all socio-economic and educational groups. Being poor does create extra stresses or crisis situations within families and having financial resources may lead to under-reporting. The crisis is not a basic etiological factor but is rather a precipitating one.[16]

THE RESULTS OF PHYSICAL ABUSE OR NEGLECT UPON THE CHILD

In the general physical examination of abused or neglected children one will find a wide range of manifestations. There may be general signs of neglect manifested by poor skin hygiene, malnutrition, withdrawal, irritability, or repression of personality. Bruises, burns, abrasions, or old healed lesions may be present on examination of the skin. There may be x-ray evidence of old fractures or dislocation. Many children may present in coma, convulsions, or dead as a result of subdural hematomas. Intentional administration of drugs to children is a subtle form of abuse which is often very difficult to detect unless one has taken a thorough history for the potential for abuse.[20]

Radiological-Skeletal Manifestations

When one suspects physical abuse or the Battered Child Syndrome, x-rays of the long bones, the ribs, the spine, and skull are mandatory.[21,22,23] When there is trauma the growing bones of the child often will record the past as well as current episodes. The findings of new and old skeletal lesions is the hallmark of repeated child abuse. In addition to the more classical fractures and dislocations, growing bones manifest changes in the cartilage plate, cupping of the metaphysis, and external cortical thickenings. Dr. Caffey[1,24] feels that external cortical thickenings in association with small peripheral cortical fractures at the metaphysis level are the principal findings in child abuse. Cortical metaphyseal fragments are present immediately following injury and permit the early radiographic diagnosis of new trauma. Periosteal elevation in contrast usually does not become visible radiographically until seven to fourteen days after the injury although subperiosteal soft tissue swelling may be present. Traumatic metaphyseal cuppings develop slowly and usually are not diagnostic until several weeks or months following the injury. It is important to realize that there may be marked changes present in bones with little evidence of injury to overlying tissues.

Dr. Caffey has pointed out[24] that often both the proximal and distal ends of the single joint are affected especially at the knee. His conclusion is that "the metaphyseal avulsion fragments appear to result from indirect traction, stretching, and shearing, acceleration, and deceleration stress on the periosteal and boney structures rather than direct impact stresses such as smashing blows on the bones itself."[24] Traumatic involucrum commonly develop in the same terminal segment of the boney shaft where metaphyseal avulsions are located. These are usually the largest and most conspicuous of all the traumatic lesions in growing bones. Dr. Caffey[1] states these develop secondary to traumatic eruption of the abundant normal peripherating blood vessels which course through between the cortical walls of the periosteum and medullary cavity and which are severed at the internal edge of the periosteal with the external edge of the cortex. Also there may be associated, diffuse sclerosis of shafts of some of the affected bones as revealed by brittle, chalk-like appearing bones.[25]

Dr. Caffey has made a good case that shaking may be a common and often disregarded form of violence in child abuse.[24] Radiographic bone changes seem to support traumatic factors in many cases of child abuse and are due to rough handling of the infant's arms and legs such as by grabbing, wringing and jerking motions. This also predisposes the child to whiplash-shaking injuries of the head. It is important to realize that all of these injuries can occur in the absence of bruises of the arms and legs.

It is also important to remember that no bones have been spared, although larger bones are more commonly affected than the smaller ones. Overt rib fractures in different stages of repair are seen in more severe cases and in other cases there may be only local thickening of ribs indicating the sites of contusion with subperiosteal hemorrhage. The vertebrae bodies may show an irregular hollowed-out appearance or notching. The cranium may show signs of increased pressure which is generally nonspecific in nature and not diagnostic of trauma.

The differential diagnosis in small children must include the trauma induced by obstetrical injuries during breech deliveries or the normal birth process. These lesions are usually symmetrical in appearance. The x-ray findings can be identical in appearance and often confused with child abuse. Congenital lues, Gaucher's disease, scurvy, leukemia, and metastatic neuroblastomas must also be considered when there is periosteal elevation present.[26,27] With systemic disease there tends to be more symmetrical x-ray findings.

Cutaneous and Mucosal Lesions

Sussman[28] has characterized the skin lesions in child abuse
so that one may distinguish them from other skin conditions.
The ecchymosis or hematomas are more often concentrated in clusters
on the trunk and buttocks and to a lesser extent on the head and
proximal segments of the extremities. The lesions are morphologi-
cally similar to the implements used to inflict trauma; e. g. the
hand, belt buckle, strap, coat hanger, etc. The bleeding into the
skin is purpuric and almost never petechial. Hamlin has described
a case of subgaleal hemorrhage from a hair-pull as well as widespread and localized alopecia.[29]

Burns are fairly common and may be due to cigarette burns to
the skin used in physical discipline or from immersing a child's
extremities or buttocks into boiling water.

Facial and mucosal lesions are quite common in some series.[30,31]
In Cameron, et al, study of twenty-nine fatal cases, half of the
children had laceration of the inner aspect of the upper lid near
the frenulum and a tearing of the lip from the gingiva.[31]

Neurological Manifestations

The early studies related to child abuse showed that one-third
of the children who survived child battering would be left with
mental retardation. The major cause of death in all series of
fatalities reported are due to acute subdural, epidural, or suban-
achnoid hemorrhage. It is apparent from the vectors of force
applied to small children's heads that they can be left with a
variety of residual neurological signs.[32] Vomiting, irritability,
and an enlarging head may be signs of a chronic subdural hematoma.
Retinal hemorrhages as well as complete retinal detachment have
been reported.[33,34] Swischuk has recently reported seven cases
of child abuse with spinal cord injury.[35] Six of their cases had
hooked or notched vertebrae and was the most prevelant lesion in
their series. One case had a fracture and dislocation in both the
cervical and lumbar areas of the spine which was attributed to
violent shaking.

Lindenberg and Freytag[36] published one of the few reports relating to the morphology of blunt trauma in early infancy. They
point out that the cerebral lesions in young children represent a
very characteristic entity in contrast to contusions to the brain.
The infantile lesions consist of grossly visible tears in the cerebral white matter and microscopic tears in the outer most layer of
the cortex. The white matter tears were more frequent in the
occipital and temporal lobes. In their study the tears occurred
only in infants five months old or younger and they hypothesized

that tears and not contusions occur due to the soft consistency of
the not yet myelinated brain, the pliancy of the skull, the smooth-
ness of the inner cranial fossae, and the shallowness of the subarachnoid space.

There is a paucity of material in the pathology literature to
more fully substantiate the neurological findings present in the
maltreatment of children.

Gastrointestinal Manifestations

Abdominal complications have been described in children who
have presented with the clinical findings and symptoms of an acute
abdomen. Laceration and contusion of solid organs such as the
liver or spleen are frequently reported.[37,38,39] Gornall, et al,[40]
describes six cases of intra-abdominal visceral injuries over a
nine year period with two of the children dying. Camps' study in
1969 (England) showed that liver injuries accounted for 19 of 100
fatal cases of child abuse deaths.[41] Rupture of the small or
large intestine is the most commonly reported lesion followed by
intraluminal hematomas of the duodenum and jejunum. Avulsion of
the common bile duct, laceration of the pancreas, (with pancreatitis
and pseudocyst formation) avulsion of the root of the mesentery, and
rupture of the inferior vena cava have all been reported.

The location of the majority of abdominal injuries center in
the midepigastrium and has been explained as a consequence of a
decelerating force (kick or punch) to a relative fixed point in the
midabdomen. The crushing force against the tissues in this region
between the external objective and the vertebral bodies is thought
to result in small intestinal ruptures or intraluminal hematomas.
If the midepigastrium force is displayed laterally one is then likely
to see rupture of the liver or spleen.

Pathological Manifestations of Neglected Children

Dr. Weston[42] has published one of the few reports regarding the
pathology of child abuse. His description of these infants who
were neglected was "associated with gross dereliction of all the
amenities of food, clothing, and sanitation which the helpless infant
is unable to provide." The children have extremely soiled clothing,
frequently caked or matted with vermin or feces. Virtually no turgor
is present nor is there palpable subcutaneous fat. Their skin is
often infested by insects including ants, bees, and roaches.

It is now apparent that not only can chronic neglect and malnutrition be present but also children may die as a result of acute[43,44]
starvation and either hypertonic or acute hypotonic dehydration.
Dr. John Coe's work in the use of the vitreous humor in determining

TABLE 2

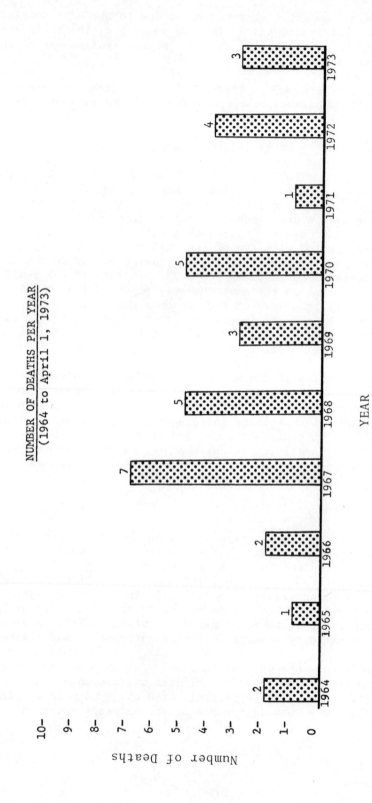

NUMBER OF DEATHS PER YEAR
(1964 to April 1, 1973)

TABLE 3

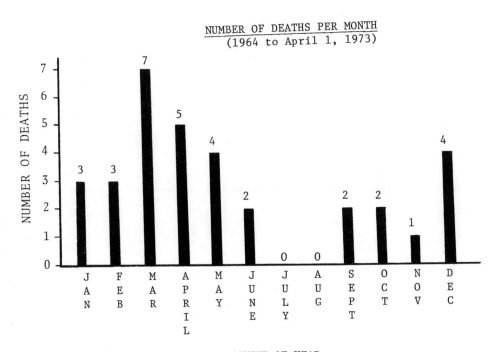

NUMBER OF DEATHS PER MONTH
(1964 to April 1, 1973)

MONTH OF YEAR

the post-mortem electrolytes has been extremely helpful in document-ing cases of dehydration after the child has died.[45]

HENNEPIN COUNTY MEDICAL EXAMINER'S DATA

The purpose of this part of the discussion will relate to a survey of the deaths due to physical maltreatment of children in Hennepin County of Minnesota from January 1964 to April 1973. All autopsy records of children were reviewed and there were thirty-three (33) cases which fit the definition of physical maltreatment of children. This period of time was chosen as there had been a con-sistent medical examiner, a rather uniform protocol including total body x-rays of all children, and represents the period of time since the passage of the State law requiring mandatory reporting by physicians of child abuse and neglect.

TABLE 4

AGE–RACE–SEX of 33 VICTIMS OF PHYSICAL ABUSE
(1964 to April 1, 1973)

AGE:	WHITE		NON–WHITE		TOTALS
	Male	Female	Male	Female	
Months:					
0–3	2	3	2	1	8
3–6	1	–	–	–	1
6–12	4	3	3	2	12
Years:					
1–2	3	–	2	1	6
2–3	–	1	1	–	2
3–4	–	1	–	–	1
4–5	–	–	–	1	1
5	1	1	–	–	2
TOTAL	11	9	8	5	33

Demographic Studies

The number of cases that were seen in a given year range from
one case to seven cases (Table 2). More striking than the number
of cases per year is the number of deaths in any given month.
Table 3 shows dramatically the peak incidences from December through
May with no deaths occurring in July or August. These findings are
interesting in that Weston's[42] findings showed that there was no
seasonal variation. This may represent a unique feature of the
environmental factors operating in Minnesota. The peak incidence
in the winter and spring also parallels the general incidence of
maltreatment seen in our population. It is our hypothesis that
during the summer time the children are more distributed outside

the home and adults have more alternatives in handling stressful
situations with their children.

Twenty-two children have died in the hospital, nine in their
parents' home, one in a foster home, and one at a babysitter's
home. In attempting to look at the relationship between the time
of death and possible association with Sudden Death In Infancy
Syndrome, five children were found dead at 6:00AM, four children
died between noon and 6:00PM, and two children died from 6:00PM
to midnight. This represents too small a sample to draw any conclusions relating to time of death and thus an epidemiological
association with Sudden Infant Death. In two of our cases they
were initially suspected as Sudden Death Syndrome but the x-rays
revealed the existence of old fractures. In three cases the children were found abandoned with no adult in the care of the child.

The age, sex, and race distribution of the thirty-three victims
are presented in Table 4. Our general child abuse statistics show
a 58:42 male-female distribution. Also there is noted an increase
in the non-white death rate as compared to the general incidence
(white-89%; black-5%; and American Indian-5%) in our community.
Six of the children who died were black, six American Indian and
one Chinese. Perhaps the increase death rate in non-whites may
be attributed to their failure to seek early medical attention or
to more severe forms of physical maltreatment.

The classification of death based upon x-ray findings, past
history of the patient, and the potential of abuse in the parents
and/or others persons, show that nineteen children represented
the Battered Child Syndrome, ten children died as a result of a
single episode of physical abuse, and six children died from physical neglect. Two children had physical abuse combined with
sexual abuse.

The primary cause of death in our series was as follows:
Twenty-five due to central nervous system injury; three from acute
and chronic malnutrition; three from hypertonic dehydration; one
from laceration of the liver; and one from laceration of the jejunum
with peritonitis and a ruptured spleen.

Skeletal Trauma

In our study twenty-nine cases had post-mortem x-rays taken
and twenty of these were positive for skeletal trauma. In eleven
cases there were skull fractures, nine had multiple rib fractures,
six had fractures of the femur, three had humeral fractures, two
elbow, one clavicular fracture, one multiple finger fractures, and
one soft tissue calcification. In eleven cases there was evidence
of old and new fractures.

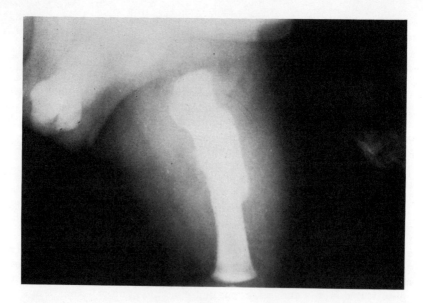

Fig. 1. Radiograph of the left femur showing a healing mid-shaft
fracture with surrounding periosteal elevation. Also note the
deformed and dislocated femoral head which is probably the result
of twisting of the hip joint.

    Figures 1 and 2 show the type of periosteal reaction to old
trauma of the femur.

Case Presentation (CASE #1)   (J. R., 2½-month-old, American Indian,
                                male)

    This child was the product of a full term, normal delivery.
The child had not been seen by a physician since birth.

    On the date of death, the deceased's father was at home when
he noticed that the child was crying when he picked him up, squeezed
him hard; and put him back in the buggy so that the child's head
hit the metal frame. At about this time he noticed that the
child stopped breathing. He waited three hours before calling his
wife who notified the police. The child was brought to the
Meical Examiner's office D. O. A.

    At autopsy there was no evidence of bruising noted over the
extremities and there was only mild inflammation of the glands of
the penis. Otherwise the external examination appeared normal.

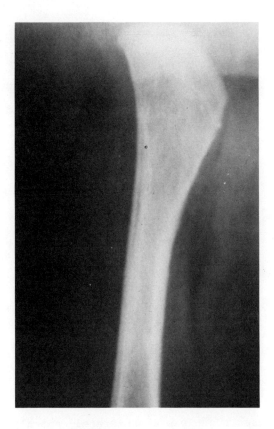

Fig. 2. Periosteal elevation of the shaft of the femur in Case 1.

On the internal examination it was noted that a large fracture of
the left parietal bone (Figure 3) with bilateral acute subdural
hematomas and massive acute subarachnoid hemorrhage. On examination
of the rib cage there were multiple old rib fractures.

    X-rays were taken which revealed fractures of the right third,
fourth, fifth, sixth, seventh, and eighth and questionably ninth
ribs and the left fourth, fifth, sixth, seventh, eighth, ninth
and tenth ribs. These fractures all appeared to be in the healing
stage. There was healing fractures of the left second and third
metacarpals and the right first, second, third, fourth, and fifth
metacarpals. There was a healed distal fracture of the right femur,
as well as a healing fracture of the proximal left tibia. Periosteal
new bone formation was noted about the left humerus.

    This case represents the severest form of maltreatment – the
Battered Child Syndrome.

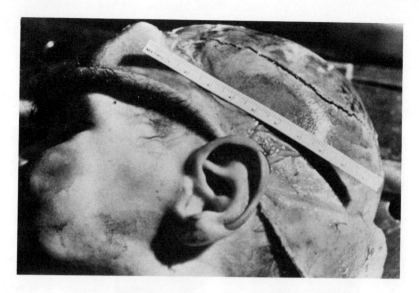

Fig. 3. Photograph of the large parietal skull fracture in Case 1.
Also note the clustering of bruises on the left cheek.

Central Nervous System Pathology

In twenty-six cases there was demonstrated central nervous system
pathology. It was judged that in twenty-three patients their central
nervous system injury was the cause of death. The causes of death
were due to acute subdural hematoma in eighteen cases, two cases of
acute epidural, two cases of acute subarachnoid hemorrhage, and one
patient with cerebral contusions and cerebral edema. In addition,
there were three cases with chronic subdural hematomas, one case of
chronic subdural hematoma of the spinal cord and one patient with
mild cerebral contusions. Of the three patients under age five
months who died from central nervous system causes we did not find
the type of blunt pathology described by Lindenburg and Freytag.[35]
In only two of our cases was the spinal cord examined at the time
of autopsy.

Skin Lesions

In sixteen cases there were skin lesions found at the time of
autopsy. In eleven cases there were multiple clustered bruises
of the type described by Sussman.[28] Eight of these were character-

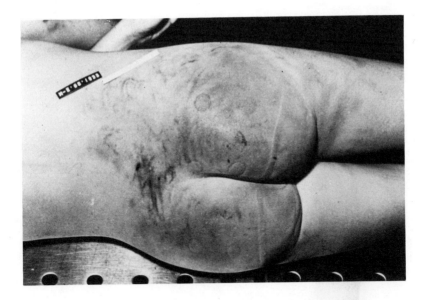

Fig. 4. Skin lesions of the buttocks of Case 2. The lesions are
those caused by the looping of an electrical cord.

ized as severe and grouped around the head, back, or buttocks. In
three cases the skin bruises were considered mild and localized
to the extremities.

Sixteen cases manifested poor skin care as showed by severe
diaper rash, abraded skin, and untreated skin conditions such as
eczema and impetigo.

Three patients had burns of the hands and feet presumably by
hot water and in one patient multiple cigarette burns were identified.

Case Presentation (CASE #2) (S. W., five-year-old, black, female)

This five-year-old child had a history of being beaten repeatedly
by her mother for disciplinary reasons. The history revealed that
she had been beaten on the buttocks with a shoe. The child had gone
to school on the day of death and had apparently been normal. After
school the mother beat her again and after the beating her breathing
was noted somewhat labored. The child was then in bed and several
hours later was found dead; lying nude in bed.

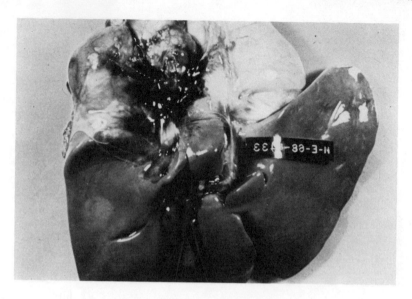

Fig. 5.   Laceration of the liver in Case 2.

Examination at the scene showed multiple bruises of the buttocks,
face, and back resembling the loop of an electrical cord (Figure 4).
The autopsy showed extensive fresh bruises of the buttocks, head,
abdomen, and extremities. A lacerated liver was found with acute
intraabdominal hemorrhage (Figure 5). Examination of the brain
revealed marked cerebral edema with acute subarachnoid and subdural
bleeding. X-rays were taken and showed no evidence of any fractures.

This case represents the severity of a single episode of physical
abuse without skeletal trauma. Figure 6 shows the imprint of the
bristles of a hairbrush in association with an acute subdural.
Figure 7 is the incised area of the buttocks showing the depth of
the subcutaneous hematomas.

Nutritional Status

Physical neglect accounted for six deaths in our series.

Case Presentation (CASE #3) (L. J., 13-month-old, black, male)

This child and his two-year-old sibling were found by friends

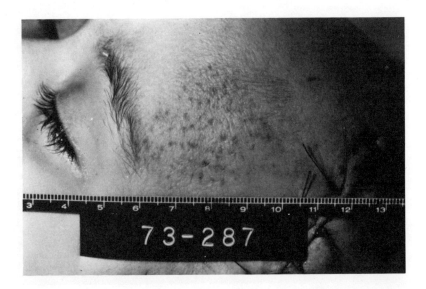

Fig. 6. Hairbrush bristle marks on left forehead region of a five-year-old child who presented with coma and convulsions. The underlying pathology was an acute subdural hematoma associated with a skull fracture.

alone in the home when the parents had not been seen for some time. The deceased was found dead in the crib and his two-year-old brother was taken to the hospital severely dehydrated. When seen by the Deputy Medical Examiner, the deceased was lying supine in the crib in slacks, diapers, and a shirt. There was full rigor of extremities and the body was cold. There was marked dehydration, sunken eyes, and dry mucosal membranes. A small bruise was noted on the left temporal area. The autopsy showed severe dehydration, acute starvation with fatty metamorphosis of the liver. The vitreous electrolytes were: sodium 170 mg% and urea nitrogen 58 mg%.

This case presents the value of post-mortem chemistries in assessing the state of hydration.[45]

In three children chronic malnutrition was present and one child had hyponatremia on the basis of vitreous electrolytes, four children had hypertonic dehydration both clinically and by vitreous electrolytes, and one child had Vitamin D deficient rickets by x-ray. In nine cases there was mild malnutrition as judged by the pathologist.

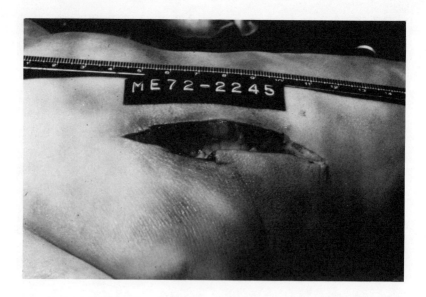

Fig. 7. Incision into the buttocks region of a patient with multiple bruises and acute subdural hematomas. The photograph shows the
depth of the hematomas in the subcutaneous tissue.

Gastrointestinal Manifestations

Two children died as a result of gastrointestinal trauma. In
one case a laceration of the liver was present and another child
died with a ruptured spleen, laceration of the jejenum, and secondary
peritonitis. In addition there was one case with a perforated duo-
denal ulcer and secondary peritonitis. Another child had ecchymosis
of the colon. In none of our cases were intramural hematomas found
in the duodenum or jejenum. There were no cases of vascular injuries
or injuries to retroperitoneal structures.

SUMMARY

We have attempted to define the spectrum maltreatment of children
as seen in our current society. The potential for abuse, a child
who is somewhat "different," and a stressful situation is the interaction which produces maltreatment. The physical abuse rendered to
the child is the result of a variety of mechanisms from direct blows
from a variety of objects to violently shaking the child. The
spectrum of the pathological findings are as varied as the means to
inflict the trauma and involve every organ system. The predominate

pathology is located in the central nervous system, bones, and cutaneous tissue.

ACKNOWLEDGMENT

The author appreciates the valuable help of Ms. Kaye Morgan in preparing this manuscript.

REFERENCES

1. Caffey, John: Pediatric X-ray Diagnosis- 6th Edition, pp. 1132-1147. Publishers: Year Book Medical Publishers Inc. Chicago, 1972.

2. Bakan, David: Slaughter of the Innocents: A Study of the Battered Child Phenomenon. Publishers: Beacon Press, Boston, 1972.

3. Solomon, Theo: History and Demography of Child Abuse. Pediatrics 51(4): Part II, 773-776, April 1973.

4. Radbill, Samuel X.: A History of Child Abuse and Infanticide. The Battered Child, Edited by Ray E. Helfer and C. Henry Kempe: Chicago and London: The University of Chicago Press, 1968.

5. Fontana, Vincent J.: The Maltreated Child: The Maltreatment Syndrome in Children. Second Edition, Springfield, (Ill.): Thomas, 1971.

6. Caffey, John: Multiple Fractures in the Long Bones of Infants Suffering from Chronic Subdural Hematoma. Am. J. Roentgenol. 56(2):163-167, 1946.

7. Silverman, F. N.: The Roentgen Manifestations of Unrecognized Skeletal Trauma in Infants. Am. Jour. Roentgenol. Radium Ther. Nucl. Med. 69(3):413-427, March 1953.

8. Woolley, P. V., Jr., Evans, W. A., Jr.: Significance of Skeletal Lesions in Infants Resembling Those of Traumatic Origin. JAMA 158(7):539-543, June 18, 1955.

9. Kempe, C. Henry: The Battered Child Syndrome. JAMA 181(1):105-112, July 7, 1972.

10. Fontana, Vincent J.: The Diagnosis of the Maltreatment Syndrome in Children. Pediatrics (Supplement): Symposium on Child Abuse 51(4), Part II: 781-792, April 1973.

11. Koel, B. S.: Failure to Thrive and Fatal Injury as a Continuum.
        Amer. J. Dis. Child 118:565-567, October 1969.

12. Gregg, Grace S., Elmer, Elizabeth: Infant Injuries: Accident
        or Abuse? Pediatrics 44(3):434-439, September 1969.

13. Gil, David G.: Violence Against Children. Cambridge, Mass.:
        Harvard University Press, 1970.

14. Congressional Record-Proceedings and Debates of the 93rd Con-
        gress, First Session (Senate), United States of America.
        Volume 119(39):S4444, pp. 1-9, Washington (Tuesday-March
        13, 1973).

15. Adelson, L.: The Battering Child. JAMA 222(2):159-161, Oct-
        ober 9, 1972.

16. Helfer, Ray: The Etiology of Child Abuse. Pediatrics (Supple-
        ment): Symposium on Child Abuse 51(4), Part II: 777-779,
        April 1973.

17. Duncan, Glen M., et al: Etiological Factors in First-Degree
        Murder. JAMA 168(13):1755-1758, 1958.

18. Steele, Brandt F.: Violence in Our Society. The Pharos, pp
        42-48, April 1970.

19. Silver, L. B. et al: Does Violence Breed Violence? Contribu-
        tions from A Study of the Child Abuse Syndrome. Am. J.
        Psychiat. 126(3):152-155, September 1969.

20. Dine, Mark S.: Tranquilizer Poisoning: An Example of Child
        Abuse. Pediatrics 36:782-785, 1965.

21. Caffey, John: Parent-Infant Traumatic Stress Syndrome. Am.
        Jour. Roentgenol. Radium Ther. Nucl. 114:218-229, Feb-
        ruary 1972.

22. Kempe, C. Henry: The Battered Child and the Hospital. Hospi-
        tal Practice, pp 44-57, October 1969.

23. Lloyd, Robert G.: The Diagnosis of Injury to Bones and Joints
        in Young Babies. Proc. Roy. Soc. Med. 61:1299-1300, Dec-
        ember 12, 1968.

24. Caffey, John: On The Theory and Practice of Shaking Infants.
        Amer. J. Dis. Child 124(2):161-169, 1972.

25. Hiller, H. G.: Battered or Not: A Reappraisal of Metaphyseal

Fragility. Am. Jour. Roentgenol. Radium Ther. Nucl.
Med. 114:241-246, February 1972.

26. Baker, D. H., Berdon, W. E.: Special Trauma Problems in Child-
    ren. Radiol. Clin. North Amer. 4:289-305, 1966.

27. Silverman, F. N.: Unrecognized Trauma in Infants, The Battered
    Child Syndrome, and the Syndrome of Ambroise Tardieu.
    Radiology 104:337-353, August 1972.

28. Sussman, Sidney J.: Skin Manifestations of the Battered Child
    Syndrome. J. Peds. 72(1):99-101, January 1968.

29. Hamlin, H.: Subgaleal Hematoma Caused by Hair-Pull. JAMA 204:
    339, April 22, 1968.

30. Tate, R. J.: Facial Injuries Associated with the Battered Child
    Syndrome. Br. J. Oral Surg. 9:41-45, July 1971.

31. Cameron, J. M., Johnson, H. R., Camps, F. E.: The Battered
    Child Syndrome. Med. Sci. Law 6:2-21, January 1966.

32. Baron, M. A.: Neurological Manifestations of the Battered Child
    Syndrome. Pediatrics 45:1003-1007, June 1970.

33. Harcourt, B., et al: Opthalmic Manifestations in the Battered
    Baby Syndrome. Br. Med. J. 3:398-401, August 14, 1971.

34. Kiffney, G. T., Jr.: The Eye of the Battered Child. Arch.
    Opth. 72:231-233, August 1964.

35. Swischuk, L. E.: Spine and Spinal Cord Trauma in the Battered
    Child Syndrome. Radiology 82:733-738, March 1969.

36. Lindenberg, Richard, Freytag, Ella: Morphology of Brain Lesions
    from Blunt Trauma in Early Infancy. Arch. Path. 87:298-
    305, March 1969.

37. McCort, J., et al: Viseral Injuries in Battered Children.
    Radiology 82:424-428, March 1964.

38. Bratu, M. et al: Jejunal Hematomas, Child Abuse, and Felson's
    Sign. Conn. Med. J. 34:261-264, April 1970.

39. Toulonkian, R. J.: Battered Children with Abdominal Trauma.
    G. P. 40:106-109, December 1969.

40. Gornall, P., Ahmed, S., Jolleys, A., Cohen, S. J.: Intra-
    Abdominal Injuries in the Battered Baby Syndrome. Arch.

Dis. Childhood 47:211-214, 1972.

41. Camps, F. E.: Injuries Sustained by Children Violence. Re-
     cent Advances in Forensic Pathology, Edited by F. E.
     Camps: London: Churchill, 1969.

42. Weston, James Tuthill: The Pathology of Child Abuse. The
     Battered Child, Edited by Ray E. Helfer and C. Henry
     Kempe: Chicago and London: The University of Chicago
     Press, 1968.

43. Pickel, S., et al: Thirsting and Hypernatremic Dehydration -
     A Form of Child Abuse. Pediatrics 45:54-59, January,
     1970.

44. Adelson, Lester: Homicide by Starvation: The Nutritional
     Variant of the 'Battered Child.' JAMA 196(5):104-106,
     November 2, 1963.

45. Coe, John: Use of Chemical Determination in Vitreous Humor in
     Forensic Pathology. J. Forensic Sciences 17(4):541-546,
     1972.

ANESTHETIC PROBLEMS IN TRAUMA MANAGEMENT

John R. Gordon

University of Minnesota

The discussion of anesthetic problems in trauma management is restricted to the role of recently developed drugs in general anesthesia. It deals with two new inhalation anesthetics enflurane and isoflurane; with the intravenous agents ketamine hydrochloride, a dissociative anesthetic, and droperidol-fentanyl as employed in neuroleptanalgesia; and with one new, and one old, neuromuscular blocking drug. In addition the perennial problem of the trauma victim with a full stomach who requires general anesthesia is presented in detail.

NEW INHALATION AGENTS IN TRAUMA

In a brief history of isoflurane Vitcha[131] described how the purification of uranium isotopes for atomic research developed a fluorine chemical technology for the synthesis of nonflammable, halogenated anesthetics. The most recent were two methyl ethyl ethers, enflurane and isoflurane discovered by R. C. Terrell in 1963 and 1965. Enflurane was approved for clinical use in the United States in 1972 while isoflurane was undergoing clinical investigation.

Early studies of enflurane[130,15,44] and isoflurane[40,21,22] indicated they possessed many qualities of the so-called ideal anesthetic agent. They were nonflammable, potent and stable and there was a wide margin of safety between anesthetic concentrations and those producing circulatory arrest. Enflurane however, exhibited a disturbing property not seen with any other clinically useful inhalation agent. It produced electroencephalographic patterns,

273

and occasional motor movements [9] interpreted as representing cerebral cortical irritability which was enhanced by hyperventilation,
by decreased arterial carbon dioxide tension and by increased depth
of anesthesia.[90] The cortical irritability was not considered a
major problem because there was no evidence of cerebral hypoxia
from measurements of cerebral blood flow, metabolic rate and pattern
of brain carbohydrate metabolism during and after deep ethrane anesthesia in man; behavorial patterns postoperatively were like those
seen with other inhalational agents; the incidence of abnormal motor
movements was low; and the electroencephalographic pattern indicative of cerebral irritability could be readily and rapidly reversed
by reducing the depth of anesthesia. Isoflurane, on the other hand,
did not give rise to cerebral hyperactivity and unlike enflurane
and other methyl ethyl ethers in current use, did not sensitize the
myocardium to epinephrine.[67] In common with other inhalation anesthetics both enflurane and isoflurane were respiratory depressants,
the latter being especially potent in this respect[58] and unique
among agents studied[72] in that respiratory frequency did not increase as anesthesia deepened. The respiratory depression by isoflurane was antagonized by surgical stimulation.[50]

Both agents underwent minimal metabolism in the body with only
2.4 percent of enflurane recovered as metabolite in man[25] and no
hepatic extraction of isoflurane detected in miniature swine.[62]
This was considered an important advantage of the agents as compared
to halothane and methozyflurane which underwent significant metabolism, because it was suggested that toxicity was related to metabolic products. The lack of in-vivo biotransformation was especially
important in severely traumatized patients where hepatic circulation
was impaired by shock or hypovolemia rendering the liver more susceptible to damage by toxic metabolites. Another factor was that
such patients tolerated only low concentrations and Sawyer and others[103] reported that the liver metabolized more anesthetic at low,
than at high, concentrations. An additional property of the new
agents was that, in common with other halogenated anesthetics, they
produced few significant metabolic alterations in patients apart
from a rise in blood sugar and plasma cortisol levels[39] which perhaps reflected an unimpaired response to stress.

The role of the new agents in the anesthetic management of
trauma however depended largely on their actions on the cardiovascular system. Skovsted and Price have investigated the effect of
many anesthetics on sympathetic nervous activity in cats before and
after complete baroreceptor denervation.[111,107,108,109] In each
instance, they measured mean arterial blood pressure, sympathetic
nervous activity and end-expiratory anesthetic concentrations. They
recorded sympathetic activity from strands of the left cervical sympathetic trunk and assessed barostatic reflexes by electrical stimulation of the central end of the cut left aortic depressor nerve.

They were able to divide all anesthetics studied into two major classes on the basis of changes in sympathetic nerve impulse traffic, arterial blood pressure and baroreceptor function in response to increasing concentrations. Group one agents which included diethyl ether, cyclopropane, chloroform, divinyl ether, trichloroethylene and fl023, increased sympathetic nervous activity by inhibiting barostatic reflexes. Group two anesthetics were the barbiturates, halothane and methoxyflurane, all of which decreased sympathetic nervous activity and had no significant effect on barostatic reflexes. In interpreting their date they adopted the concept of Alexander[1] who, in 1946, stimulated pressor and depressor regions in the medulla of the cat and postulated that impulses from the depressor center descended to spinal cardiovascular centers and tonically inhibited them. The depressor center was excited by impulses from peripherally located baroreceptors while the pressor region enjoyed tonic activity.

The cardiovascular effects of the two agents have been studied recently in man. Stevens et al.[116] administered isoflurane to seven volunteers at constant carbon dioxide tension and body temperature. In contrast to halothane, the most commonly used potent inhalation agent, cardiac output was maintained by an increase in heart rate which offset the decrease in stroke volume as anesthesia was deepened with isoflurane. Myocardial contractility was sustained even at deep levels of anesthesia and there was only minimal elevation in right atrial pressure. Arterial pressure and peripheral resistance declined with the vasodilation apparently occurring mainly in muscle and skin. Spontaneous ventilation altered the cardiovascular response[33] because the resulting increase in arterial carbon dioxide tension elevated heart rate and cardiac output. The cardiovascular actions of isoflurane, together with its minimal degree of metabolism and its compatibility with epinephrine suggested a useful place for it in the anesthetic management of trauma.

NEW INTRAVENOUS AGENTS IN TRAUMA

Wyant has described three major epochs since the discovery of general anesthesia in the 1840's.[137] In the first period most anesthesias involved a single potent agent administered in concentrations sufficient to provide adequate analgesia and muscle relaxation. The succeeding era began when the introduction of tubocurarine made possible the combination of light general anesthesia and a peripherally acting muscle relaxant. The third era promised drugs which altered the central nervous system in a manner unlike any past agent. The new anesthetics apparently interrupted the individual's contact with the surroundings and produced profound general analgesia. These interesting compounds caused the introduction of two new terms in anesthesiology: dissociative anesthesia and neuroleptanalgesia.

Dissociative Anesthesia

The term "dissociative anesthesia" was suggested by Domino, Chodoff and Corssen[45] to describe the state produced by phencyclidine and its derivative ketamine hydrochloride which, they believed, specifically depressed some asociation areas so that sensory input which reached the cortex was not perceived. The interference with proper association of different impulses resulted in dissociation or dissociative anesthesia.[30] Subsequent workers disagreed with the term and placed more emphasis on the ability of the phencyclidines to induce a state of catalepsy. Mori and his colleagues[88] who studied cats with chronically implanted electrodes described three distinct patterns of ketamine-induced central nervous system electrical activity. An initial excitation with catatonic behavior was followed by a catatonic-anesthetic state and finally by electrographic seizures not manifested clinically. Ferrer-Allado and others[52] administered ketamine in various doses to nine patients being investigated for seizure disorders with cortical, limbic and thalamic electrode implants, and correlated electrical activity with gross behavior. Six patients received two to four mg per kg intravenously and exhibited electrical seizure activity in the limbic and thalamic areas with behaviors ranging from unconsciousness and immobility to actual tonic and clonic motor activity. Surface electrographic tracings did not always indicate limbic electrical activity. One patient who received a dose of only 0.5 mg per kg showed seizure activity via a depth electrode while still conscious. The authors agreed with Mori's group that the action of ketamine was more correctly labeled catatonic anesthesia. On the other hand Chen[26] discussed the possible mode of action of the phencyclidine derivatives and its resemblance to the central nervous depressant effect of the sympathomimetic amines; he recalled that many sympathetic amines given intraventricularly caused marked depression in most species including man, and postulated that ketamine's anesthetic action might be related to its sympathetic activity. He preferred the term sympathomimetic anesthesia for the state produced by phencyclidine and ketamine.

In spite of disagreement over the classification of ketamine there was unanimity of opinion on the unique properties of the drug. In a progress report several years after its introduction Corssen[32] listed among its disadvantages, occasional extrapyramidal activity and hallucinatory reactions. The agent's advantages included profound analgesia without significant impairment of respiratory function, stimulation of cardiovascular activity, preservation of protective reflexes and maintenance of an unobstructed airway regardless of position. It was the respiratory and cardiovascular effects of ketamine which made it useful in the anesthetic management of victims of severe trauma.

The phencyclidine group of drugs unlike previous anesthetics rarely depressed ventilation significantly, and maintained intact the tone of the muscles of the tongue and upper airway while preserving pharyngeal and laryngeal reflexes. This made the drugs particularly valuable in acute burns and other trauma involving the head and neck by permitting spontaneous breathing through an open airway and providing profound analgesia for initial surgery without endotracheal intubation and deep anesthesia. It also proved convenient and safe for procedures such as debridements of burns and dressing changes often requiring position alterations or the prone position on repeated occasions.

The stimulant effect of ketamine on the cardiovascular system suggested its use as an induction agent in severely injured patients although the mechanism of its circulatory support was not elucidated. Most workers agreed that the response consisted of an initial hypotension probably caused by direct myocardial depression usually too brief to be detected clinically, followed by a more prolonged increase in arterial blood pressure and heart rate. Dowdy and Kaya[47] who investigated the biphasic response in dogs attributed the stimulation to a decreased responsiveness of baroreceptors which normally inhibited sympathetic vasomotor centers. Additional studies of the pressor response to ketamine in dogs by Traber and his group demonstrated that ketamine failed to produce a pressor response during epidural anesthesia[127], that sympathetic ganglion blockade prevented the response and that it was diminished by pretreatment with phentolamine and abolished by pretreatment with phentolamine and atropine.[129] They concluded that ketamine raised blood pressure and heart rate in the dog by augmenting alpha-adrenergic activity and blocking the vagus.

Several groups have reported that ketamine increased plasma catecholamines in man. Dundee and others[48] detected an increase in 14 of 17 patients after ketamine and cited this as the most likely cause of the hypertensive response to the drug. Takki and his colleagues at the University of Helsinki[120] studied changes in blood pressure, pulse rate and plasma catechols in adults and children during induction of anesthesia with ketamine alone and with ketamine followed by succinylcholine. Both kinds of induction caused a rise in blood pressure but the only significant increase in noradrenaline and adrenaline followed ketamine-succinylcholine in adults. Baraka's group[8] measured arterial plasma levels of norepinephrine and epinephrine in three healthy adults 5 and 10 minutes after the intravenous injection of ketamine in a dose of 2 mg per kg. Arterial blood pressure and heart rate increased in each case and were associated with increased levels of the catechols. Oyama and others[95] noted significant increases in plasma free-cortisol levels in 21 patients anesthetized with ketamine before operation and suggested the hypertensive response was caused by ketamine-induced stimulation of the adrenal cortex.

The beneficent circulatory effects of ketamine persuaded Chasapakis and his colleagues[24] to use it for anesthesia for patients in hemorrhagic shock. They reasoned that the drug's ability to raise blood pressure without increasing peripheral resistance made it preferable to other intravenous anesthetics in common use, all of which caused a fall in pressure. Their technique consisted of premedication with atropine intravenously just before induction of anesthesia with 2 mg per kg of ketamine intravenously and intubation with the aid of succinylcholine. Blood pressure rose in all patients from 15 to 45 torr with an average of 30 torr. The authors apparently relied on active pharyngolaryngeal reflexes during ketamine to prevent pulmonary aspiration in those patients who may have had full stomachs but they were aware of a report by Penrose[98] of aspiration pneumonitis following ketamine induction for general anesthesia. Nettles, Herrin and Mullen[91] have emphasized the value of ketamine as an induction agent in acutely traumatized patients. They induced anesthesia with ketamine in 30 patients aged 16 months to 87 years all classified as poor risks and most of whom had suffered acute, severe gunshot wounds. No untoward cardiovascular or respiratory effects resulted in 29 patients but hypotension and bradycardia occurred in one, a 49-year-old woman who responded to methylprednisolone. There were no postanesthetic phychic disturbances. The authors were of the opinion that ketamine satisfied many of the requirements for an ideal induction agent in poor-risk, traumatized patients.

Caution in the use of dissociative anesthetic agents in patients with intracranial injury has been advised because of reports of increases in cerebrospinal-fluid pressure associated with their use. Gardner and colleagues[59] administered ketamine intravenously in a dose of 2 mg per kg to eleven healthy unmedicated male volunteers and measured a mean increase of 253 mm water in cerebrospinal-fluid pressure. The marked change was related to a mean increase of 28 torr in intra-arterial pressure and could not be attributed to changes in arterial carbon dioxide tension. Similar advice has been given with regard to patients with seizure disorders by Ferrer-Allado.[52]

Neuroleptanalgesia

In the years immediately following 1950 Laborit and his colleagues used the so-called lytic cocktail, a mixture of sedatives, phenothiazines and autonomic blocking drugs to produce a state of artificial hibernation.[71] The technique never became popular because of instances of protracted, severe circulatory instability and respiratory depression.[93] Some ten years later certain butyrophenone derivatives which possessed phenothiazine-like actions were advocated for use in anesthesia to produce a state of apathy and mental detachment. The combination of one of these neuroleptic agents and a narcotic analgesic was recommended by Nilsson and Janssen[94] as an alternative to general anesthesia. Earlier experiences with such combinations had resulted in hypotension, extrapyramidal disturbances and psychic changes but by 1963 Janssen had synthesized the butyrophenone, dehydrobenzperidol or droperidol, and the potent, short-acting analgesic fentanyl, and favorable reports of neuroleptanalgesia appeared.

Early clinical studies[63,31] indicated that the combination of droperidol and fentanyl induced a state of altered consciousness or "mineralization" in which patients lay resting and quiet during surgery. When nitrous oxide was added to cause sleep and amnesia, and muscle relaxants to decrease muscular tone all types of surgery could be performed. Initial reports which listed some of the advantages of the new technique as profound analgesia and minimal hypotension with cardiovascular stability, suggested that droperidol possessed alpha-adrenergic blocking action of value in cases complicated by massive blood loss or shock. One of the disturbing properties apparent by that time was the occasional production of muscle rigidity involving the thorax with severe resistance to inflation of the chest. This was curable by succinylcholine.

The term neurolepsis referred to the suppression of neuronal activities in the nervous system; and neuroleptanalgesia meant the state of tranquil indifference and intense analgesia with little hypnosis. Cerebral activity was relatively unaffected so the patient could communicate and understand but when not disturbed, exhibited psychic detachment and motor sedation, or mineralization, because of reduced brain stem activity. When nitrous oxide was added to inhibit awareness and neuromuscular junction blockers to provide relaxation, neuroleptanesthesia ensued. As usually practiced neuroleptanalgesia was induced by the intravenous administration of a mixture of the neuroleptic agent, droperidol (2.5 mg per ml) and the analgesic fentanyl (0.05 mg per ml) in a 50:1 ratio.

The use of neuroleptanesthesia for patients suffering severe trauma was encouraged by experimental studies which indicated that droperidol protected rats from traumatic shock[66] and had a prophylactic effect in hemorrhagic shock in dogs.[36] Dixon and his group who also studied dogs[38] reported that droperidol-fentanyl did not depress myocardial contractility but its administration was followed by a marked, immediate fall in peripheral vascular resistance and a more slowly onsetting increase in peripheral vascular capacitance. Moran and others[87] measured a decrease of 38 percent in peripheral vascular resistance in dogs but this was offset by a rise in cardiac output of 24 percent above the control value. Most workers agreed that the circulatory effects of droperidol resulted from alpha-adrenergic blockade but Puddy,[99] who demonstrated a consistent blocking of vasoconstriction produced by noradrenaline, histamine, and potassium and sympathetic nerve stimulation in the isolated auricular artery of the rabbit, postulated a non-specific inhibition of vaso-constriction rather than alpha blockade. Greene[61] utilized a guinea pig ileum preparation to show that droperidol's antagonism to acetylcholine and histamine was non-competitive.

Clinical evaluations confirmed the finding in animals that droperidol-fentanyl maintained cardiovascular stability except for mild hypotension. Zauder and others[139] measured cardiac output, mean arterial blood pressure and arterial and venous blood gases before and after the rapid intravenous administration of droperidol-fentanyl to nine geriatric patients prior to major surgery. They found only one significant change, an average decrease in mean blood pressure from 97.6 to 87.4 torr and concluded that neuroleptanalgesia was without significant effects on the cardiovascular system of unstressed surgical patients. Tarhan and his group[122] studied the hemodynamic and blood-gas changes when the mixture was administered to 18 adult patients with advanced cardiac disease and reported significant decreases in systemic systolic, diastolic and mean arterial blood pressures due solely to decreased systemic resistance with no alteration in cardiac indices. Whitwam and Russell[136] noted the effects of droperidol on the arterial pressures of patients undergoing translumbar aortography, heart surgery with cardiopulmonary bypass or vascular surgery. Direct injection into the aorta prompted an immediate fall in pressure caused by a direct action on peripheral vessels. Injection of droperidol into the oxygenator during cardiopulmonary bypass decreased arterial blood pressure transiently and, for up to ten minutes, completely suppressed or severely altered the peripheral vasoconstriction of adrenaline and noradrenaline. They decided droperidol was an adrenergic blocker and stated that their evidence did not contradict the hypothesis that its action was mediated by blockade of alpha-adrenergic receptors.

The role of neuroleptanesthesia in major trauma has not been completely settled. Small amounts of droperidol-fentanyl have been employed successfully during the technique of awake intubation. The drugs were given in increments to induce a state of tranquil cooperation in which the patient tolerated topical anesthetization of the oropharynx and glottis while he maintained his protective airway reflexes[27] during endotracheal intubation. Judicious administration of neuroleptanesthesia has been combined with nitrous oxide-oxygen and the muscle relaxant pancuronium for surgery in severely injured patients. If significant hypovolemia was present it was corrected by blood and fluid replacement to prevent undue hypotension as droperidol decreased systemic vascular resistance. Once again the agents were used in small increments to avoid relative overdose in these extremely sensitive patients.

MUSCLE RELAXANTS IN TRAUMA

Succinylcholine Chloride

In 1959 Forrest[57] reported a cardiac arrest in a 14-year-old
girl who had sustained third degree burns over forty percent of
her body. She had been anesthetized uneventfully several times
prior to this for skin grafting and debridement using thiopental
and nitrous oxide-oxygen, with tubocurarine as the muscle relaxant
for endotracheal intubation. On the occasion reported the anesthetic technique was altered in that succinylcholine replaced tubocurarine as the relaxant. The circulatory arrest occurred several
minutes after this; the patient was successfully resuscitated and
underwent two subsequent anesthesias, without succinylcholine, uneventfully. In his discussion of the etiology the author incrimi-
nated a combination of factors including diminished blood volume,
thiopental and positive pressure ventilation. He did not mention
the role of succinylcholine which at that time was considered free
of ill effects in burned patients although several people had
pointed out that the thermally injured were poor anesthetic risks.

Moncrief[86] presented the complications encountered in the
treatment of 1000 burns at Brooke Army Medical Center, Texas, in
the years 1950 through 1956. There were five cardiac arrest, four
of which occurred during the induction of anesthesia. The author
listed few details of anesthetic methods but, as all arrests were
in 1953, succinylcholine may have been involved as it was introduced in the United States the previous year.[56] The increased hazard of anesthesia for the burned patient also was emphasized by
Masters and his associates[79] who described the complications encountered in 3008 general anesthesias for plastic surgery. The
highest rate of complications (19 percent of all complications)
was in burns; of a total of seven cardiac arrests three were in
burned patients. Anesthetic management was not discussed in detail
because of the broad scope of the report. Finer and Nylen reported a double cardiac arrest with survival in a thermally injured
patient[53] and discussed the incidence of cardiac arrest during
anesthesia for the treatment of burns at their institution, the
University of Uppsala Hospital, Sweden.[54] During a period of four
years and four months, 3372 plastic surgical operations were performed utilizing general anesthesia and of these 628 involved
burned patients. The incidence of cardiac arrest was 1:209 anesthesias in burn management and 1:2744 for other plastic procedures.
In a review of the literature up to 1961, Finer and Nylen noted
15 additional cases of arrest in burns, all of which were associated
with endotracheal intubation. The following year Bush and others[20]
suggested that the combination of succinylcholine and endotracheal
intubation resulted in cardiac arrest by further enhancing the already increased vagal activity in burned patients and, on this

basis, explained circulatory collapses in two burned patients
immediately following intubation facilitated by intravenous succinylcholine. Fleming and his colleagues[55] blamed a similar incident
occurring in the presence of succinylcholine on vagal stimulation
provoked by intubation and added that the patient with severe
toxemia from infected burns was at increased risk from endotracheal
intubation. They did not exonerate succinylcholine but noted that
the drug had been implicated by Bullough[19] and by Leigh and his
group[73] in the production of severe bradycardia and brief periods
of asystole in clinical anesthesia for non-burned cases.

A more explicit etiology was proposed in 1961 by C. M. Allan and
his colleagues[2] who reported a case of ventricular fibrillation in
a burned boy. They suggested that the lethal arrhythmia may have
resulted from succinylcholine-induced changes in intracellular and
extracellular potassium concentrations. However, as McCaughey[80]
pointed out the next year, they did not advance any evidence to
support the theory. It had been known for some time that depolarizing relaxants such as succinylcholine increased serum potassium
values. In 1954 Klupp and Kraupp[70] reported that succinylcholine
produced hyperkalemia in dogs and two years later Paton[96] pin-
pointed one source of the potassium efflux as skeletal muscle when
he demonstrated the effect in hepatectomized dogs in which the liver
could not have been the source. He questioned if a similar process
might be important in clinical work when he measured an average rise
in potassium of 0.5 mEq per liter in patients receiving succinylcholine.[97] The nondepolarizing relaxant gallamine caused no change.
Paton warned that a much greater hyperkalemia may follow depolarizing
agents in patients with disturbed electrolyte balance.

Presumably patients with abnormal striated muscles behaved
similarly to those with ionic imbalance. It was well known that
organs chronically deprived of their motor nerves developed "denervation hypersensitivity" and were hyper-responsive to their neuro-
humoral transmitter and other agents. Axelsson and Thesleff[3,4]
studied the denervated tenuissimus muscle of the cat to learn that,
as time elapsed after neural deprivation, acetylcholine which produced depolarization only when applied to the end plate region in
the innervated cell, now depolarized an ever increasing area; after
two weeks the whole membrane was as sensitive as the end plate. And
just as it did at the end plate in normal muscle, acetylcholine
raised the permeability to potassium of the entire membrane of the
denervated cell.

Depolarizing muscle relaxants acted in a manner similar to acetylcholine to produce their neuromuscular block by prolonged depolarization. It was not unreasonable to conjecture that succinylcholine-
induced depolarization produced dangerous and prolonged potassium
efflux from denervated muscle; and although the analogy was not per-

fect for the muscles of burned patients where no obvious denervation ensued, it was tempting to speculate that in burned patients
potassium leaked from their depolarized striated muscle cells which
if not denervated were somehow damaged. And the effects of hyperkalemia on cardiac action were well recognized as capable of producing cardiac arrest.[113]

   Reports of circulatory collapse following the administration of
succinylcholine to burned patients continued to accumulate[10] but,
as late as 1967, opinions differed as to the significance[81] and
degree[75] of the potassium rise. It was in that year, however, that
Tolmie, Joyce and Mitchell[126] correlated serial serum potassium
values after succinylcholine with electrocardiographic changes in
a young marine with massive soft tissue destruction and extensive
third degree burns. The patient had undergone eleven uneventful
dressing changes or debridements with thiopental-halothane-nitrous
oxide-oxygen anesthesia and succinylcholine intravenously to facilitate endotracheal intubation. The twelfth anesthetic, which was
administered thirty-one days following injury, was complicated by
cardiac arrest within four minutes of succinylcholine administration and intubation. Resuscitation was accomplished and the re-
dressing completed satisfactorily. During several subsequent anesthetics, blood drawn for serum electrolytes yielded information
that potassium values increased steadily and rapidly to over 8 mEq
per liter after succinylcholine. The electrocardiogram revealed
progressively widening QRS complexes, disappearance of the P-wave
and finally, ventricular fibrillation.

   Tolmie's group believed the circulatory arrest was most likely
due to hyperkalemic arrhythmias and cardiac atony. They noted that
electrocardiographic changes were less severe when smaller amounts
of succinylcholine were used, and were absent when a nondepolarizing
relaxant, gallamine, was injected to produce satisfactory conditions
for intubation. And they were unable to produce significant abnormalities by vagal stimulation such as carotid massage, eyeball pressure and endotracheal tube movement following resuscitation.

   The report of Tolmie's group left little question that hyperkalemia secondary to succinylcholine was the major cause of cardiac
arrhythmias and arrests when the relaxant was administered to severely burned patients. Still not clear were the onset and duration of
the period of sensitivity, and the mechanism by which succinylcholine promoted potassium efflux. Most reports indicated that the
danger period began about two weeks after the burn and lasted four
to six weeks.[104] Some clinicians[112] used no depolarizing agents
in burned patients while others avoided them only after the first
24 to 48 hours.[124] The mechanism of the potassium release was explained in several ways. Waters and Mapleson[133] put forward a
hypothesis to explain the cause of the muscle pains that often

followed the administration of succinylcholine to normal patients. They attributed the pain to damage produced in muscle by the unsynchronized contraction of adjacent muscle fibers just before the onset of paralysis. The incoordination was related to slight differences in the concentrations of depolarizer required to cause parallel, adjacent fibers to contract in the initial stage of fasciculation prior to relaxation. The unsynchronized contractions applied a shearing force to the connective tissue between fibers producing damage and later, pain. Such minor injury may have been responsible for the small increase in serum potassium which followed succinylcholine in normal individuals. The effect was magnified in burned patients due to thermal injury to muscle or due to a generalized metabolic derangement peculiar to burns.

Additional evidence that depolarizing relaxants altered the integrity of muscle membranes was provided by the demonstration of increased serum concentrations of creatine phosphokinase by Tammisto and Airaksinen[121] at the University of Helsinki, Finland, who measured creatine phosphokinase values before, and 24 to 48 hours after, anesthesia involving succinylcholine, nitrous oxide and halothane. Succinylcholine increased the serum concentrations of the enzyme especially when administered in divided doses with the most marked increases resulting from intermittent doses in the presence of halothane. Innes and Stromme[65] measured creatine phosphokinase activity in serum at induction of anesthesia and after 24 hours in children undergoing eye operations involving minimal surgical trauma. Five anesthetic sequences were used: halothane-succinylcholine-halothane; thiopental-succinylcholine-halothane; diethyl ether-succinylcholine-halothane; halothane alone and diethyl ether alone. Only those groups receiving succinylcholine showed significant rises in serum creatine phosphokinase. Of the induction agents halothane increased the rise most. In addition to muscle enzymes, myoglobin also increased in serum after depolarizing agents. Bennike and Jarnum[12] presented a case of myoglobinuria with acute renal failure following a minor operation under halothane anesthesia during which succinylcholine was used. The patient exhibited particularly vigorous fasciculations which the authors believed precipitated the rhabdomyolysis. Creatine phosphokinase increased markedly in serum in the period following myoglobinuria. Further confirmation was presented by Ryan and his associates[101] who administered succinylcholine intravenously or intramuscularly to children and adults for minor surgical procedures. Myoglobinemia was detected in 40 percent of the children given the agent intravenously but in none of the children receiving the drug intramuscularly. Only 1 of 30 adults had myoglobinemia.

Subsequent to the demonstration that depolarizing relaxants caused lethal hypokalemia in burned patients similar problems were noted in other types of trauma. Birch and his colleagues[13] investigated the response to an intravenous dose of 100 mg of succinylcholine in patients with severe trauma including multiple fractures, bilateral amputations, extensive missile wounds as well as burns, by measuring serum potassium before and 3 minutes after succinylcholine. No significant rise was detected until one week after injury when a progressive hyperkalemia began and increased until approximately the third week when three patients with severe injuries showed an elevated level capable of causing cardiac arrest. Peak concentrations appeared within three to six weeks and returned to normal within two months. Simultaneously-drawn arterial and venous samples from the non-traumatized leg of an injured patient showed consistently higher values on the venous side. In addition when tourniquets were placed proximal to the areas of trauma before succinylcholine administration significant hyperkalemia was produced. It seemed apparent that potassium effluxed from non-injured as well as from injured, sites. There was perhaps, a general metabolic defect and also some role attributable to denervation hypersensitivity.

Other neuromuscular afflictions led to dangerous hyperkalemia following succinylcholine administration.[123,100] Cooperman[29] studied 37 patients, 16 of whom had paraplegia from spinal cord trauma, 6 were paraplegic from tumor, 6 hemiplegic from cerebrovascular accidents and the remainder of whom suffered from multiple sclerosis, parkinsonism or muscular dystrophy. Following succinylcholine (1 mg per kg body weight) fifteen of the patients sustained increases in serum potassium concentration from 1 to 6 mEq per liter. Cooperman stated that most increases over 1 mEq per liter occurred in those who had been ill less than six months or, if longer, who had disease of a progressive nature; and the degree and extent of pathologic muscle paralysis seemed directly related with relaxant-induced hyperkalemia.

The knowledge that succinylcholine fasciculations and muscle pain were alleviated or prevented by pretreatment with small doses of non-depolarizing relaxants encouraged several groups to investigate this means of preventing hyperkalemia in trauma patients. Weintraub, Heisterkamp and Cooperman[134] studied 20 patients who had sustained trauma such as burns, open fractures and gunshot wounds 4 to 55 days before and who were to have corrective procedures. They were given intravenously either succinylcholine 100 mg or tubocurarine 0.1 mg per kg body weight five minutes before 100 mg of succinylcholine. Those receiving tubocurarine had increases in potassium one half of those receiving succinylcholine alone. Takki and his associates[119] did not measure potassium values but found that fasciculations could be prevented while allowing good conditions for intubation by intravenously administering 1.5 mg succinylcholine per kg of body weight three mintues after a dose of tubocurarine of 0.05 mg per kg. Cullen[34] used the same dose of succinylcholine but preferred another nondepolarizer, gallamine in a dose of 10 to 20 mg

to tubocurarine which he found delayed the onset of paralysis.
Tobey and his group[125] determined potassium changes in patients
with neural injuries following succinylcholine, gallamine, tubocurarine, and succinylcholine preceded by 6 mg tubocurarine. Gallamine and tubocurarine provoked no significant change in serum potassium. Tubocurarine prior to succinylcholine reduced the potassium rise but the authors suggested that the increase was enough
to produce serious arrhythmias and advised that succinylcholine be
replaced with nondepolarizing relaxants for patients with lower
motor neuron disease.

The effect of other anesthetic drugs on serum potassium was not
as marked as that of the depolarizing relaxants. All inhalation and
intravenous general anesthetics decreased serum potassium.[43,41]
List and Stovner et al.[118] measured the effect of different induction agents on succinylcholine hyperkalemia. Halothane augmented
the effect sufficiently to indicate that the combination be used
with caution in patients with major trauma, burns, tetanus, uremia
and certain neuromuscular diseases.

Pancuronium Bromide

The nondepolarizing muscle relaxants tubocurarine and gallamine
did not produce significant hyperkalemia in injured patients but
they had other side effects which were unwelcome in trauma where
cardiovascular and respiratory deteriorations were common. Tubocurarine, the first peripheral muscle relaxant used in anesthesiology, had several disadvantages. It increased blood histamine
levels in four of eleven patients studied by Westgate and Van Bergen[135] who injected 0.44 mg per kg tubocurarine intravenously and
produced hypotension and increased airway resistance. Other investigators reported clinical evidence of histamine release such as
skin rashes, erythema and hypotension.[82,11,92,16] Gallamine, the
first synthetic relaxant, caused little histamine release but had
marked vagolytic activity[46] which resulted in tachycardias often
detrimental to patients in shock.

The recently introduced relaxant, pancuronium bromide, seemed
relatively free of serious side effects. It was synthesized in 1964
by Hewett and Savage who were investigating a series of amino steroids. While it lacked hormonal action it appeared to have potent
neuromuscular blocking properties. In 1966 Buckett and Bonta[18]
studied it in various animal species and reported the potency to
be greater than that of tubocurarine and of equal duration. It was
devoid of histamine-releasing properties, it was readily reversed
by neostigmine and possessed only one-tenth the ganglionic blocking
effect of hexamethonium. The following year Baird and Reid of the
Department of Anesthetics, Royal Infirmary in Glasgow, Scotland,
reported the first use of the compound in humans.[5] They administered

it to six healthy females aged 23 to 51 years, undergoing minor
gynecological procedures. Their electromyographic data showed the
block to be of the nondepolarizing type reversible by anticholine-
sterases and its potency was approximately five times that of tubocurarine. In a study[117] comparing the hemodynamic effects of pancuronium and tubocurarine Stoelting reported premature ventricular
contractions in two of four patients receiving pancuronium alone and
none in others who were given both relaxants. The arrhythmias which
developed concomitantly with increases in heart rate and blood
pressure and when arterial blood gases and serum potassium values
were normal, disappeared spontaneously in ten minutes. The author
did not believe thephenomenon implied that pancuronium increased
ventricular irritability. Coleman and his colleagues[28] measured
heart rate, arterial blood pressure and cardiac output in 28 anesthetized patients after the injection of pancuronium. There were
significant increases in all three measurements which could be
prevented by atropine. Most investigations suggested that the
absence of hypotension made pancuronium a promising agent for use
in poor risk patients in whom hypotension should be avoided.[35,68,]
[69,42] As mentioned previously Chasapakis and his co-workers[24]
selected it as the relaxant to be used with ketamine anesthesia for
patients in hemorrhagic shock.

TRAUMA AND THE FULL STOMACH

Reports of the aspiration of stomach contents into the airways
during general anesthesia for victims of trauma appeared soon after
the introduction of anesthesia. In August of 1862, Balfour[6] described a new cause of death under chloroform which had occurred
during the Burmese war in 1853. The case involved a soldier who
suffered a gunshot wound through the thigh and had repeated secondary
hemorrhages from the femoral artery. Dr. Balfour stated, "It was
supposed that the profunda or one of its branches was injured, and
it was determined to tie the femoral artery above and below the
origin of the profunda; - this was done while the man was under
chloroform. In the course of the operation the man, who had had
his dinner previously, became sick and vomited. He subsequently
sank and shortly died from exhaustion, as was supposed. On examination of the body, the profunda was found cut across by the ball and
a false aneurysm formed at the seat of injury, and the trachea was
found filled with vomited matters."

Among those who commented on Balfour's paper to the Edinburgh
Obstetrical Society that evening was a Mr. Pridie who told of a
case many years before where his patient fainted under the use of
chloroform. After being resuscitated the patient argued that he
should have been dosed with brandy during his syncope but was soon
satisfied to the contrary when a day or two after, he read of a boy

at Newcastle who died while brandy was being poured down his throat
during syncope when anesthetized. Pridie's patient may have been
referring to a girl, Hannah Greener, the first reported anesthetic
death, who died on Friday, January 28, 1848. The inquest as reported
in the Lancet of Saturday, February 5, 1848,[51] brought forth a ver-
dict from the coroner's jury of death due to congestion of the
lungs from the effects of chloroform. James Young Simpson in a
communication the following Saturday[106] attributed the death to
aspiration of water and brandy which were poured into her mouth.
Simpson wrote, "The girl died, then, as I conceive, choked or asphyxiated by the very means intended to give her life."

An early American case of aspiration in a trauma victim was
presented by C. K. Briddon at a meeting of the New York Surgical
Society, November 24, 1897.[17] The patient, a six-year-old boy who
had been run over by a cable car, was anesthetized with ether for
a leg amputation three and one-half hours after the ingestion of
his dinner. Shortly after being returned to bed and while still
under the influence of ether he had a sudden attack of vomiting
and died. At autopsy a mass of spinach-green material filled
the trachea and extended down the bronchi. In this instance anesthesia had been purposely delayed for two hours so the patient could
digest his dinner. It was not yet recognized that gastric emptying
was delayed or halted by trauma.

Sporadic reports of aspiration continued to appear but it was
not until Mendelson's classic report,[83] which provided a name for
the syndrome, that the importance of the problem was realized. He
reviewed 66 cases of aspiration of stomach contents during obstetrical anesthesia and distinguished between aspiration of solids
and liquids. Three years after this the Council of the Association
of Anaesthetists of Great Britain and Ireland established a committee
to investigate deaths associated with anesthesia. In 1951[89] and
again in 1956[49] reports from the committee indicated the significant contribution to morbidity and mortality of aspiration. Of
the first 598 anesthetic deaths 110 were due to regurgitation or
vomiting and 5.5 percent of these occurred in traumatized patients.
Marshall and Gordon[77,78] who extensively reviewed the matter in
1958, described the clinical and pathological features, presented
case histories and discussed explicit therapy.

Impressed with the increasing number of aspirations among injured patients during anesthesia physicians studied its pathogenesis and devised methods of prevention and therapy.[138,37,7,14,23]
Most agreed that the safest anesthesia for the patient in control
of his airway was local infiltration or regional anesthesia. If
general anesthesia was required it should be induced only after the
trachea was intubated with a cuffed endotracheal tube under topical
anesthesia. Topical application involved spraying the agent from

a nebulizer on the tongue and soft palate and, by direct laryngoscopy, on the epiglottis and vocal cords. An additional method involved translaryngeal injection of the topical anesthetic via a
needle inserted through the cricothyroid membrane into the larynx
after confirming the position of the needle's tip by withdrawing
air. It was emphasized that a larynx so anesthetized often was
unable to prevent aspiration should regurgitation occur[132,27] and
that the severely injured patient even under local or regional
anesthesia might aspirate regurgitated material especially in the
supine position.

Special induction techniques were needed when "awake" intubation
was contraindicated as in the patient with an eyeball laceration or
intracranial hemorrhage where any coughing or straining must be
avoided. Large-bore nasogastric tubes were used[64] to remove gastric
contents and reduce intragastric pressure but they could not be
relied upon to completely empty the stomach and the insertion itself often provoked undesired coughing and straining. Cuffed stomach[76] or esophageal[60] tubes placed to prevent stomach contents from
entering the pharynx during anesthesia again were not always reliable.

Snow and Nunn[114] described a method for the rapid induction of
anesthesia and endotracheal intubation in patients with full stomachs. They placed the patient in a 40 degree head-up tilt so that,
in the presence of complete muscle paralysis from succinylcholine,
the larynx was high enough above the gastro-esophageal junction to
prevent intragastric pressure from forcing material from the stomach into the pharynx. Patients who were debilitated or hypovolemic
had their lower extremities raised to prevent further pooling of
blood as anesthesia depressed cardiovascular function. Sellick[105]
prevented aspiration by having an assistant press the cricoid cartilage backward against the bodies of the cervical vertebrae to
close the esophagus until endotracheal intubation was completed.
He conjectured that his maneuver might cause esophageal tearing
if forceful vomiting occurred but his fear was not borne out by
time. Miller and Way[84] measured intragastric pressure during succinylcholine-induced muscle fasciculations in men who had sustained
acute war wounds and demonstrated a several-fold increase which could
be prevented by the prior administration of gallamine or tubocurarine. In a subsequent study[85] they concluded that pretreatment with
these nondepolarizing relaxants had little effect on the magnitude
or type of the blockade produced by succinylcholine as indicated by
twitch-height response to nerve stimulation. However, Cullen[34]
reported difficulties during the intubation of some patients when
gallamine or tubocurarine preceded the customary succinylcholine
dose of 1 mg per kg body weight. When he increased the dose to
1.5 mg per kg and pretreated with 10 - 20 mg gallamine conditions
for intubation were satisfactory. Takki, Kauste and Kjellberg[119]

provided optimal conditions for intubation by administering 0.05 mg
per kg of succinylcholine. Pretreatment with gallamine or tubocurarine in these doses did not increase the time of onset of the
succinylcholine block. This was vitally important because assisted
or controlled ventilation often forced inspired gas into the esophagus and permitted regurgitation. The patient breathed 100 percent oxygen for several minutes before the administration of succinylcholine so that arterial oxygen tension remained adequate
when intubation was attempted later, after some 60 to 90 seconds
of apnea.

In a useful, critical review of the anesthetic management of
the patient with a full stomach Salem[102] listed the common errors
in the performance of the technique of rapid induction and intubation. The errors included the application of positive pressure to
the airway before intubation, accidental placement of the endotracheal tube in esophagus, failure to combine the cricoid pressure
maneuver or the head-up tilt with the thiopental-succinylcholine
sequence and the use of inadequate doses of relaxants resulting
in partial paralysis.

Means of avoiding such errors have been outlined by Stept and
Safar[115] in their discussion of the technique in preventing aspiration during induction. They emphasized the need for a well-trained
assistant to provide cricoid pressure, a well-secured intravenous
infusion, and adequately prepared equipment. They inserted a wide-
bore nasogastric tube to decompress the stomach but were undecided
as to its role in holding the esophagus open and possibly, permitting
regurgitation if it were not removed before induction. They placed
the patient in a semi-sitting, V-position with the trunk elevated
about 30 degrees to counteract regurgitation by gravity and with the
feet elevated to maintain blood pressure. Their patients breathed
100 percent oxygen to denitrogenate the lungs and oxygen administration was continued until removal of the mask just prior to intubation. Tubocurarine was injected intravenously and a predetermined
dose of thiopental was administered intravenously about two minutes
after the curare. Then the head was tilted back and the assistant
maintained firm cricothyroid pressure to close the esophagus until
the endotracheal tube's cuff was inflated. Succinylcholine was
injected after thiopental approximately 2 to 3 minutes after tubocurarine administration. Respirations were allowed to cease spontaneously and complete paralysis ensued before beginning laryngoscopy
for intubation. In the injured patient who was unconscious a nasogastric tube was not inserted until after intubation of the trachea
because of the likelihood of aspiration during or after insertion of
the gastric tube.

REFERENCES

1. Alexander, R. S.: Tonic and reflex functions of medullary sympathetic cardiovascular centers. J. Neurophysiol. 9: 205-217, 1946.

2. Allan, C. M., Cullen, W. G., and Gillies, D. M. M.: Ventricular fibrillation in a burned boy. Canad. Med. Ass. J. 85: 432-434, 1961.

3. Axelsson, J., and Thesleff, S.: A study on supersensitivity in denervated mammalian skeletal muscle. J. Physiol. 145: 48P-49P, 1959.

4. Axelsson, J., and Thesleff, S.: A study on supersensitivity in denervated mammalian skeletal muscle. J. Physiol. 147: 178-193, 1959.

5. Baird, W. L. M., and Reid, A. M.: The neuromuscular blocking properties of a new steroid compound, pancuronium bromide. Brit. J. Anaesth. 39: 775-780, 1967.

6. Balfour, G. W.: New cause of death under chloroform. Edinburgh Med. J. 8: 194-195, 1862.

7. Bannister, W. K., Sattilaro, A. J., and Otis, R. D.: Therapeutic aspects of aspiration pneumonitis in experimental animals. Anesthesiology 22: 440-443, 1961.

8. Baraka, A., Harrison, T., and Kachachi, T.: Catecholamine levels after ketamine anesthesia in man. Anesth. Analg. 52: 198-200, 1973.

9. Bart, A. J., Homi, J., and Linde, H. W.: Changes in power spectra of electroencephalograms during anesthesia with fluroxene, methoxyflurane and ethrane. Anesth. Analg. 50: 53p63, 1971.

10. Belin, R. P., and Karleen, C. I.: Cardiac arrest in the burned patient following succinylcholine administration. Anesthesiology 27: 516-518, 1966.

11. Bennett, E. J., Daughety, M. J., Bowyer, D. E., and Stephen, C. R.: Pancuronium bromide: experiences in 100 pediatric patients. Anesth. Analg. 50: 798-807, 1971.

12. Bennike, K. A., and Jarnum, S.: Myoglobinuria with acute renal failure possibly induced by suxamethonium, a case report. Brit. J. Anaesth. 36: 730-736, 1964.

13. Birch, A. A., Mitchell, G. D., and Cameron, J. L.: Aspiration pneumonia: pulmonary arteriography after experimental aspiration. J. Surg. Res. 12: 48-52, 1972.

14. Booth, D. J., Zuidema, G. D., Cameron, J. L.: Aspiration pneumonia: pulmonary arteriography after experimental aspiration. J. Surg. Res. 12: 48-52, 1972.

15. Botty, C., Brown, B., Stanley, V., and Stephen, C. R.: Clinical experiences with compound 347, a halogenated anesthetic agent. Anesth. Analg. 47: 499-505, 1968.

16. Brandus, V., Joffe, S., Rubin, J. M.: Histamine-like reaction to tubocurarine. Brit. J. Anaesth. 45: 108-110, 1973.

17. Briddon, C. K.: Death during anaesthesia resulting from the entrance of food into the larynx and trachea. Ann. Surg. 27: 372-373, 1898.

18. Buckett, W. R., and Bonta, I. L.: Pharmacological studies with NA97 (2β, 16β - Dippiperidino - 5α androstane - 3α, 17β - Diol diacetate dimethobromide). Fed. Proc. 25: 718, 1966.

19. Bullough, J.: Intermittent suxamethonium injections. Brit. Med. J. 5124: 786-787, 1959.

20. Bush, G. H., Graham, H. A. P., Littlewood, A. H. M., and Scott, L. B.: Danger of suxamethonium and endotracheal intubation in anaesthesia for burns. Brit. Med. J. 5312: 1081-1085, 1962.

21. Byles, P. H., Dobkin, A. B., Ferguson, J. H., and Levy, A. A.: Forane (compounc 469): cross-over comparison with enflurane (ethrane), halothane, and methoxyflurane in dogs. Canad. Anaesth. Soc. J. 18: 397-407, 1971.

22. Byles, P. H., Dobkin, A. B., and Jones, D. B.: Forance (compound 469): 3. comparative effects of prolonged anaesthesia on mature beagle dogs and young rhesus monkeys. Canad. Anaesth. Soc. J. 18: 397-407, 1971.

23. Cameron, J. L., Zuidema, G. D.: Aspiration pneumonia: magnitude and frequency of the problem. JAMA 219: 1194-1196, 1972.

24. Chasapakis, G., Kekis, N., Sakkalis, C., and Kolios, D.: Use of ketamine and pancuronium for anesthesia for patients in hemorrhagic shock. Anesth. Analg. 52: 282-287, 1973.

25. Chase, R. E., Holaday, D. A., Fiserova-Bergerova, V., Saidman,
         L. J., Mack, F. E.: The biotransformation of ethrane
         in man. Anesthesiology 35: 262-267, 1971.

26. Chen, G.: Sympathomimetic anesthetics. Canad. Anaesth. Soc.
         J. 20: 180-185, 1973.

27. Claeys, D. W., Lockhart, C. H., and Hinkle, J. E.: The effects
         of translaryngeal block and innovar on glottic competence.
         Anesthesiology 38: 485-486, 1973.

28. Coleman, A. J., Downing, J. W., Leary, W. P., Moyes, D. G.,
         Styles, M.: The immediate cardiovascular effects of pan-
         curonium, alcuronium and tubocurarine in man. Anaesthe-
         sia 27: 415-422, 1972.

29. Cooperman, L. H.: Succinylcholine-induced hyperkalemia in
         neuromuscular disease. JAMA 213: 1867-1871, 1970.

30. Corssen, G., Domino, E. F.: Dissociative anesthesia: further
         pharmacologic studies and first clinical experience with
         the phencyclidine derivative CI-581. Anesth. Analg. 45:
         29-40, 1966.

31. Corssen, G., Domino, E. F., and Sweet, R. G.: Neuroleptanal-
         gesia and anesthesia. Anesth. Analg. 43: 748-763, 1964.

32. Corssen, G., Miyakako, M., and Domino, E. F.: Changing con-
         cepts of pain control during surgery: dissociative
         anesthesia with CI-581. Anesth. Analg. 47: 746-759, 1968.

33. Cromwell, T. H., Stevens, W. D., Eger, E. I., Shakespeare, T.
         F., Halsey, M. J., Bahlman, S. H., and Fourcade, H. E.:
         The cardiovascular effects of compound 469 (forane) during
         spontaneous ventilation and carbon dioxide challenge in
         man. Anesthesiology 35: 17-25, 1971.

34. Cullen, D. J.: The effect of pretreatment with nondepolarizing
         muscle relaxants on the neuromuscular blocking action of
         succinylcholine. Anesthesiology 35: 572-578, 1971.

35. Dick, W., and Droh, R.: Pancuronium bromide: clinical exper-
         iences with a new steriod-like relaxant. Anaesthesist 19:
         173-176, 1970.

36. Dietzel, W., and Massion, W. H.: The prophylactic effect of
         innovar in experimental hemorrhagic shock. Anesth. Analg.
         48: 968-972, 1969.

37. Dines, D. E., Baker, W. G., and Scantland, W. A.: Aspiration
     pneumonitis-Mendelson's syndrome. JAMA 176: 229-231, 1961.

38. Dixon, S. H., Nolan, S. P., Stewart, S., and Morrow, A. G.:
     Neuroleptanalgesia: effects of innovar on myocardial
     contractility, total peripheral resistance, and capaci-
     tance. Anesth. Analg. 49: 331-335, 1970.

39. Dobkin, A. B., Byles, P. H., Arandia, H. Y., Ghanooni, S.,
     Nishioka, K., and Levy, A. A.: Comparative metabolic
     responses to halogenated anaesthetics. Acta Anaesth.
     Scand. 16: 69-75, 1972.

40. Dobkin, A. B., Byles, P. H., Ghanooni, S., and Valbuena, D. A.:
     Clinical and laboratory evaluation of a new inhalation
     anesthetic: forance (compound 469) CHF_2-O-CH Cl CF_3.
     Canad. Anaesth. Soc. J. 18: 264-271, 1971.

41. Dobkin, A. B., Byles, P. H., and Neville, J. F.: Neuroendo-
     crine and metabolic effects of general anaesthesia during
     spontaneous breathing, controlled breathing, mild hypoxia
     and mild hypercarbia. Canad. Anaesth. Soc. J. 13: 130-
     171, 1966.

42. Dobkin, A. B., Evers, W., Ghanooni, S., Levy, A. A., Thomas,
     E. T.: Pancuronium bromide: evaluation of its clinical
     pharmacology. Canad. Anaesth. Soc. J. 18: 512-535, 1971.

43. Dobkin, A. B., Kwang, P., Lee, Y., Byles, Ph. H., Israel, J. S.:
     Neuroleptanalgesics: a comparison of the cardiovascular,
     respiratory and metabolic effects of innovar and thiopen-
     tone plus methotrimeprazine. Brit. J. Anaesth. 35: 694-
     705, 1963.

44. Dobkin, A. B., Nishioka, K., Gengaze, D. B., Kim, D. S., Evers,
     W., Israel, J. S.: Ethrane (compound 347) anesthesia: a
     clinical and laboratory review of 700 cases. Anesth.
     Analg. 48: 477-494, 1969.

45. Domino, E. F., Chodoff, P., and Corssen, G.: Pharmocologic
     effects of CI-581, a new dissociative anesthetic, in man.
     Clin. Pharmacol. Ther. 6: 279-291, 1965.

46. Doughty, A. G., and Wylie, W. D.: An assessment of gallamine
     triethiodide. Proc. Roy. Soc. Med. 44: 375-388, 1951.

47. Dowdy, E. G., Kaya, K.: Studies of the mechanism of cardio-
     vascular responses to CI-581. Anesthesiology 29: 931-
     943, 1968.

48. Dundee, J. W., Bovill, J. G., Clarke, R. S. F., Pandit, S. K.: Problems with ketamine in adults. Anaesthesia 26: 86, 1971.

49. Edwards, G., Morton, H. J. V., Pask, E. A., Wylie, W. D.: Deaths associated with anaesthesia. Anaesthesia 11: 194-220, 1956.

50. Eger, E. I., Dolan, W. M., Stevens, W. C., Miller, R. D., Way, W. L.: Surgical stimulation antagonizes the respiratory depression produced by forane. Anesthesiology 36: 544-549, 1972.

51. Fatal application of chloroform. Lancet 1: 161-162, 1848.

52. Ferrer-Allado, T., Brechner, V. L., Dymond, A., Cozen, H., Crandall, P.: Ketamine-induced electroconvulsive phenomena in the human limbic and thalamic regions. Anesthesiology 38: 333-344, 1973.

53. Finer, B. L., Nylen, B. O.: Double cardiac arrest with survival. Brit. Med. J. 5122: 624-625, 1959.

54. Finer, B. L., Nylen, B. O.: Cardiac arrest in the treatment of burns, and report on hypnosis as a substitute for anesthesia. Plast. Reconstr. Surg. 27: 49-55, 1961.

55. Fleming, W. B., Hueston, J. T., Stubbe, J. L., Villiers, J. D.: Two episodes of cardiac arrest in one week. Brit. Med. J. 5167: 157-160, 1960.

56. Foldes, F. E., McNall, P. G., Borrego-Hinojosa, J. M.: Succinylcholine: a new approach to muscular relaxation in anesthesiology. New Eng. J. Med. 274: 596-600, 1952.

57. Forrest, T.: A report on two cases of cardiac arrest. Brit. J. Anaesth. 31: 277-279, 1959.

58. Fourcade, H. E., Stevens, W. C., Larson, C. P., Cromwell, T. H., Bahlman, S. H., Hickey, R. F., Halsey, M. J., Eger, E. I.: The ventilatory effects of forane, a new inhaled anesthetic. Anesthesiology 35: 26-31, 1971.

59. Gardner, A. E., Olson, B. E., Lichtiger, M.: Cerebrospinal-fluid pressure during dissociative anesthesia with ketamine. Anesthesiology 35: 226-228, 1971.

60. Guiffrida, J. G., Bizzarri, D.: Intubation of the esophagus. Amer. J. Surg. 93: 329-334, 1957.

61.   Greene, M. J.: Some aspects of the pharmacology of droperidol.
      Brit. J. Anaesth. 44: 1272-1276, 1972.

62.   Halsey, M. J., Sawyer, D. C., Eger, E. I., Bahlman, S. H., Im-
      pelman, D. M. K.: Hepatic metabolism of halothane, meth-
      oxyflurane, cyclopropane, ethrane and forane in miniature
      swine. Anesthesiology 35: 43 - 47, 1971.

63.   Holderness, M. C., Chase, P. E., Dripps, R. D.: A narcotic
      analgesic and a butyrophenone with nitrous oxide for
      general anesthesia. Anesthesiology 24: 336-340, 1963.

64.   Inkster, J. S.: The induction of anaesthesia in patients likely
      to vomit with special reference to intestinal obstruction.
      Brit. J. Anaesth. 35: 160-167, 1963.

65.   Innes, R. K. R., Stromme, J. H.: Rise in serum creatine phos-
      phokinase associated with agents used in anaesthesia.
      Brit. J. Anaesth. 45: 185-190, 1973.

66.   Janssen, P. A. J., Neimegeers, C. J. E., Shellekens, K. H. L.,
      Verbruggen, F. J. and Van Nueten, J. M.: The pharmacology
      of dehydrobenzperidol, a new potent and short acting neu-
      roleptic agent chemically related to haloperidol. Arznei-
      mittelforschung 13: 205-211, 1963.

67.   Joas, T. A., Stevens, W. C.: Comparison of the arrhythmic doses
      of epinephrine during forane, halothane and fluroxene anes-
      thesia in dogs. Anesthesiology 35: 48-53, 1971.

68.   Katz, R. L. : Pancuronium. 22nd Annual Refresher Course Lec-
      tures. A. A. A., Atlanta, 1971, p. 220A, 1-7.

69.   Kelman, G. R., Kennedy, B. R.: Cardiovascular effects of pan-
      curonium in man. Brit. J. Anaesth. 43: 335-338, 1971.

70.   Klupp, H., Kraupp, O.: The liberation of potassium from muscles
      under the influence of muscle relaxants. Arch. Int.
      Pharmacodyn. 98: 340-354, 1954.

71.   Laborit, H., Huguenard, P.: Pratique de l'hibernotherapie en
      chirurgie et en medicine. Paris, Masson, 1954.

72.   Larson, C. P., Eger, E. I., Muallem, M., Buechel, D. R., Munson,
      E. S., Eisele, J. H.: The effects of diethyl ether and
      methoxyflurane on ventilation: 11. a comparative study in
      man. Anesthesiology 30: 174-184, 1969.

73.   Leigh, M. D., McCoy, D. D., Belton, M. K., Lewis, G. B.: Brady-

cardia following intravenous administration of succinylcholine chloride to infants and children. Anesthesiology 18: 698-702, 1957.

74. List, W. F.: Serum potassium changes during induction of anaesthesia. Brit. J. Anaesth. 39: 480-484, 1967.

75. Lowenstein, E.: Succinylcholine adminis-ration in the burned
patient. Anesthesiology 27: 494-496, 1966.

76. MacIntosh, R. R.: A cuffed stomach tube. Brit. Med. J. 2: 545,
1951.

77. Marshall, V. M., Gordon, R. A.: Vomiting, regurgitation and
aspiration in anaesthesia I. Canad. Anaesth. Soc. J. 5:
274-281, 1958.

78. Marshall, V. M., Gordon, R. A.: Vomiting, regurgitation and
aspiration in anaesthesia II. Canad. Anaesth. Soc. J. 5:
438-447, 1958.

79. Masters, F. W., Hansen, J. M., Robinson, D. W.: Anesthetic
complications in plastic surgery. Plast. Reconstr. Surg.
24: 472-480, 1959.

80. McCaughey, T. J.: Hazards of anaesthesia for the burned child.
Canad. Anaesth. Soc. J. 9: 220-233, 1962.

81. McCaughey, T. J., Lowenstein, E.: Use of succinylcholine in
burned patients (correspondence). Anesthesiology 28:
234-235, 1967.

82. McDowall, S. A., Clarke, R. S. J.: A clinical comparison of
pancuronium with d-tubocurarine. Anaesthesia 24: 581-
590, 1969.

83. Mendelson, C. L.: Aspiration of stomach contents into lungs
during obstetric anesthesia. Amer. J. Obstet. Gynec. 52:
191-205, 1946.

84. Miller, R. D., Way, W. L.: Inhibition of succinylcholine-induced increased intragastric pressure by nondepolarizing
muscle relaxants and lidocaine. Anesthesiology 34: 567-
571, 1971.

85. Miller, R. D., Way, W. L.: The interaction between succinylcholine and sub-paralyzing doses of d-tubocurarine and
gallamine in man. Anesthesiology 35: 567-571, 1971.

86. Moncrief, J. A.: Complications of burns. Ann. Surg. 147:
        443-475, 1958.

87. Moran, J. E., Rusy, B. F., Vongvisces, P., Lattanand, S.:
        Effects of halothane-oxygen and innovar-nitrous oxide-
        osygen on the maximum acceleration of left ventricular
        ejection and the tension-time index in dogs. Anesth.
        Analg. 51: 350-354, 1972.

88. Mori, K., Kawamata, M., Mitani, H., Yamazaki, Y., Fuzita, M.:
        Aneurophysiologic study of ketamime anesthesia in the cat.
        Anesthesiology 35: 373-383, 1971.

89. Morton, H. J. V., Wylie, W. D.: Anaesthetic deaths due to
        regurgitation or vomiting. Anaesthesia 6: 190-201, 1951.

90. Neigh, J. L., Garman, J. K., Harp, J. R.: The electroencepha-
        lographic pattern during anesthesia with ethrane: effects
        of depty of anesthesia, $PaCO_2$, and nitrous ox ide, Anes-
        thesiology 35: 482-487, 1971.

91. Nettles, D. C., Herrin, T. J., Mullen, J. G.: Ketamine induc-
        tion in poor-risk patients. Anesth. Analg. 52: 59-64,
        1973.

92. Nightingale, D. A., Bush, G. H.: A clinical comparison between
        tubocurarine and pancuronium in children. Brit. J. Anaes-
        th. 45: 63-70, 1973.

93. Nilsson, E.: Origin and rationale of neurolept-analgesia.
        Anesthesiology 24: 267-268, 1963.

94. Nilsson, E., Janssen, P.: Neurolept-analgesia, an alternative
        to general anesthesia. Acta Anaesth. Scand. 5: 73-84,
        1961.

95. Oyama, T., Matsumoto, F., Kudo, T.: Effects of ketamine on
        adrenocortical function in man. Anesth. Analg. 49:
        697-700, 1970.

96. Paton, W. D. M.: Mode of action of neuromuscular blocking
        agents. Brit. J. Anaesth. 28: 470-480, 1956.

97. Paton, W. D. M.: The effects of muscle relaxants other than
        muscular relaxation. Anesthesiology 20: 453-463, 1959.

98. Penrose, B. H.: Aspiration penumonities following ketamine
        induction for general anesthesia. Anesth. Analg. 51:
        41-43, 1972.

99. Puddy, B. R.: Effects of droperidol on the vasoconstriction produced by noradrenaline, histamine, sympathetic nerve stimulation and potassium ions in the isolated rabbit auricular artery. Brit. J. Anaesth. 43: 441-444, 1971.

100. Roth, F., Wuthrich, H.: The clinical importance of hyperkalemia following suxamethonium administration. Brit. J. Anaesth. 41: 311-316, 1969.

101. Ryan, J. F., Kagen, L. J., Hyman, A. I.: Myoglobinemia after single dose of succinylcholine. New Eng. J. Med. 285: 824-287, 1971.

102. Salem, M. R.: Anesthetic management of patients with "a full stomach." Anesth. Analg. 49: 47-55, 1970.

103. Sawyer, D. C., Eger, E. I., Bahlman, S. H., Cullen, B. F., Impelman, D.: Concentration dependence of hepatic halothane metabolism. Anesthesiology 34: 230-235, 1971.

104. Schaner, P. J., Brown, R. L., Kirksey, T. D., Gunther, R. C., Ritchey, C. R., Gronert, G. A.: Succinylcholine-induced hyperkalemia in burned patients. Anesth. Analg. 48: 764-770, 1969.

105. Sellick, B. A.: Cricoid pressure to control regurgitation of stomach contents during induction of anaesthesia. Lancet 2: 404-405, 1961.

106. Simpson, J. Y.: Remarks on the alleged case of death from the action of chloroform. Lancet 1: 175-176, 1848.

107. Skovsted, P., Price, H. L.: The effect of methoxyflurane on arterial pressure, preganglionic sympathetic activity and barostatic reflexes. Anesthesiology 31: 515-521, 1969.

108. Skovsted, P., Price, H. L.: Central sympathetic excitation caused by diethyl ether. Anesthesiology 32: 202-209, 1970.

109. Skovsted, P., Price, H. L.: Central sympathetic excitation caused by fluroxene. Anesthesiology 32: 210-217, 1970.

110. Skovsted, P., Price, H. L.: The effects of ethrane on arterial pressure, preganglionic sympathetic activity and barostatic reflexes. Anesthesiology 36: 257-262, 1972.

111. Skovsted, P., Price, M. L., Price, H. L.: The effect of halothane on arterial pressure, preganglionic sympathetic activity and barostatic reflexes. Anesthesiology 31:

507–514, 1969.

112. Slogoff, M. D., Allen, G. W., Mendenhall, M. K.: Use of halo-thane in the management of burns. Clin. Aspects of Anesthesiology (Ayerst) Dec., 1972.

113. Smith, T. N., Corbascio, A. N.: The hemodynamic effects of potassium infusion in dogs. Anesthesiology 26: 633;641, 1965.

114. Snow, R. G., Nunn, J. F.: Induction of anaesthesia in the foot-down position with a full stomach. Brit. J. Anaesth. 31: 493–497, 1959.

115. Stept, W. J., Safar, P.: Rapid induction/intubation for prevention of gastric-content aspiration. Anesth. Analg. 49: 633–635, 1970.

116. Stevens, W. C., Cromwell, T. H., Halsey, M. J., Eger, E. I., Shakespeare, T. F., Bahlman, S. H.: The cardiovascular effects of a new inhalation anesthetic, forane, in human volunteers at constant arterial carbon dioxide tension. Anesthesiology 35: 8–16, 1971.

117. Stoelting, R. K.: The hemodynamic effects of pancuronium and d-tubocurarine in anesthetized patients. Anesthesiology 36: 612–615, 1972.

118. Stovner, J., Endresen, R., Bjelke, E.: Suzamethonium hyper-kalemia with different induction agents. Acta Anaesth. Scand. 16: 46–50, 1972.

119. Takki, S., Kauste, A., Kjellberg, M.: Prevention of suxameth-onium-induced fasciculations by prior dose of d-tubocura-rine. Acta Anaesth. Scand. 16: 230–234, 1972.

120. Takki, S., Nikki, P., Jaattela, A., Tammisto, T.: Ketamine and plasma catecholamines. Brit. J. Anaesth. 44: 1318–1322, 1972.

121. Tammisto, T., Airaksinen, M.: Increase of creatine kinase activity in serum as a sign of muscular injury caused by intermittently administered suxamethonium during halothane anaesthesia. Brit. J. Anaesth. 38: 510–515, 1966.

122. Tarhan, S., Moffitt, E. A., Lundborg, R. O., Frye, R. L.: Hemodynamic and blood-gas effects of innovar in patients with acquired heart disease. Anesthesiology 34: 250–255, 1971.

123. Thomas, E. T.: Circulatory collapse following succinylcholine.
      Anesth. Analg. 48: 333-337, 1969.

124. Tobey, R. E.: Paraplegia, succinylcholine and cardiac arrest.
      Anesthesiology 32: 359-364, 1970.

125. Tobey, R. E., Jacobsen, P. M., Kahle, C. T., Clubb, R. J.,
      Dean, M. A.: The serum potassium response to muscle re-
      laxants in neural injury. Anesthesiology 37: 332-337,
      1972.

126. Tolmie, J. D., Joyce, T. H., Mitchell, G. D.: Succinylcholine
      danger in the burned patient. Anesthesiology 28: 467-470,
      1967.

127. Traber, D. L., Wilson, R. D.: Involvement of the sympathetic
      nervous system in the pressor response to ketamine.
      Anesth. Analg. 48: 248-252, 1969.

128. Traber, D. L., Wilson, R. D., Priano, L. L.: Blockade of the
      hypertensive response to ketamine. Anesth. Analg. 49:
      420-426, 1970.

129. Traber, D. L., Wilson, R. D., Priano, L. L.: The effect of
      alpha-adrenergic blockade on the cardiopulmonary response
      to ketamine. Anesth. Analg. 50: 737-742, 1971.

130. Virtue, R. W., Lund, L. O., Phelps, M., Vogel, J. H. K.,
      Beckwitt, H., Heron, M.: Difluoromethyl 1, 1, 2 - tri-
      fluoro - 2 - chloroethyl ether as an anaesthetic agent:
      results with dogs, and a preliminary note on observations
      with man. Canad. Anaesth. Soc. J. 13: 233-241, 1966.

131. Vitcha, J. F.: A history of forane. Anesthesiology 35: 4-7,
      1971.

132. Walts, L. F.: Anesthesia of the larynx in the patient with a
      full stomach. JAMA 192: 705-706, 1965.

133. Waters, D. J., Mapleson, W. W.: Suxamethonium pains: hypo-
      thesis and observation. Anaesthesia 26: 127-141, 1971.

134. Weintraub, H. D., Heisterkamp, D. V., Cooperman, L. H.:
      Changes in plasma potassium concentration after depolar-
      izing blockers in anaesthetized man. Brit. J. Anaesth. 41:
      1048-1052, 1969.

135. Westgate, H. D., Van Bergen, F. H.: Changes in histamine blood
      levels following d-tubocurarine administration. Canad.

Anaesth. Soc. J. 9: 497–503, 1962.

136. Whitwam, J. G., Russell, W. J.: The acute cardiovascular
     changes and adrenergic blockade by droperidol in man.
     Brit. J. Anaesth. 43: 581–591, 1971.

137. Wyant, G. M.: Editorial: new concepts in anaesthesia.
     Canad. Anaesth. Soc. J. 18: 1–2, 1971.

138. Wykoff, C. C.: Aspiration during induction of anesthesia:
     its prevention. Anesth. Analg. 38: 5–13, 1959.

139. Zauder, H. L., Del Guercio, L. R. M., Feins, N., Barton, N.
     Wollman, S.: Hemodynamics during neuroleptanalgesia.
     Anesthesiology 26: 266, 1965.

SPINAL CORD INJURY AND TRAUMA

Theodore M. Cole

University of Minnesota

HISTORY

Trauma of the spinal cord has almost always been associated with feelings of hopelessness and foreboding. The first recorded reference to injury of the spinal cord was found in the Edwin Smith Surgical Papyrus, written about 2500-3000 B. C. "Thou shouldst say concerning him, one having a dislocation of the vertebrae of his neck while he is unconscious of his two legs and two arms and his urine dribbles. An ailment not to be treated."[1] Almost 5,000 years later, Robert Penn Warren described a quadriplegic in his novel All The Kings Men (1946). He describes a lifetime of paralysis, hopelessness, and impending death.

World War I, produced a plethora of patients with spinal cord injuries but did not produce a significant change in the attitude of the medical profession towards spinal cord injury. In his review of the literature, Dick points out that until the second world war paraplegic patients were in no way able to look after themselves nor were they encouraged to do so.[2] The prevailing attitude of hopelessness allowed many to become drug addicts. The fact that a few survivors were left as pathetic wrecks, dependent and helpless, to be cared for in homes and hospitals for the chronically sick, largely escaped notice.

World War II brought about improvement. Britain decided to organize spinal cord treatment centers and to staff them with neurosurgical teams. In 1940 the first center was established in Winwick, in Northern England. Generally however, patients were admitted to the hospital long after their injury and in poor condi-

tion. Although treatment was more aggressive, there did not seem
to be a definite plan of rehabilitation even for the more fit
patients.

At about this time a new era in the rehabilitation of the
paraplegic patient began at Stoke Mandeville Hospital in England.

Ludwig Guttmann operated a well equipped and organized unit on
the concept that the paraplegic, although disabled, is a healthy
and independent person with a productive future in society. This
concept permeated the entire institution and was the first real
breakthrough in new thinking for spinal cord injured adults.

DEMOGRAPHY

A variety of conditions may injure the spinal cord. However,
infections, metabolic diseases, neoplasms, and congenital anomalies
are sufficiently different from the traumatically induced spinal
cord injury that they will not be considered here. Neither will
injuries of the first three cervical segments be included here.
Patients with such injuries are unable to sustain themselves independently from mechanical respiratory assistance and specially
equipped medical units staffed by highly trained personnel.

Most spinal cord injured adults are males and the preponderant
majority of them are young. According to Carter's survey of over
700 patients, the average age of onset was 25 years.[3] Excluding
industrial accidents in which the average age was 40 years, the
average age is probably closer to 21 or 22 years. This is confirmed
by a large study reported by Wilcox et al.[4] The peak distribution
by age for quadriplegia was found in the age group 15 to 19 years
and in the age group 20 to 24 years for paraplegia.

Most spinal cord injuries are caused by a relatively small number
of circumstances. In Switzerland 36 percent are caused by road traffic, 35 percent occur at work and 29 percent result from sporting
and home accidents.[5] In Australia 17 years of experience with 625
patients revealed that 38 percent were caused by road traffic accidents, 9 percent by sports and games, 4 percent by assaults and 5
percent occurred in the armed forces.[6] In Japan a small study of
60 cases revealed that automobile accidents accounted for 49 percent
and industrial accidents 25 percent.[7] Two studies done in the United
States,[3,8] showed that automobile accidents accounted for 37 percent
to 43 percent, diving accidents from 13 to 17 percent, and gunshot
wounds 8 to 15 percent. Industrial accidents in one study accounted
for 13 percent of spinal cord injuries.

The total number of adults with spinal cord injury in the United

States is unknown. However, estimates range between 60,000 and
125,000, and it has been estimated that 6,000 new cases occur each
year.[9]

The prevalence of spinal cord injury is variable depending upon
the location of the study. In Canada the prevalence has been found
to be 16 cases per million population. In Switzerland, which is a
highly organized small country, the prevalence of spinal cord injury has been found to be 15 cases per million population and the
annual increase has been 1.7%.[5] Two separate studies in the United
States show the prevalence to range between 26 and 50 cases per
million population.

Freed[10] reviewed mortality rates reported from a number of
studies done between World War I and 1966. Among American troups
during World War I, 80% of the 2,324 men who had received spinal
cord injury died before they could be returned from overseas. Of
the remainder who were successfully evacuated to the United States,
only 10 percent survived the first year. In 1946 it was estimated
that less than 1 percent of those who had survived the first year
were still living. Hoffman and Bunts did a 15 year follow-up of
170 World War II paraplegic patients.[11] They recorded an overall
mortality rate of 25.2%. In Freed's study of 243 civilian patients
with traumatic spinal cord injury and permanent neurological impairment, over 23 years of follow-up time is included. Ten patients
died within the first two months after injury and were classified
as early deaths. Forty-four patients died at an interval greater
than two months and were classified as late deaths. Of the ten
early deaths, three resulted from pulmonary embolism, three from
cardiac arrest during surgical procedures, one from bed sores and
one from anterior spinal artery thrombosis. One death was unrelated
to the spinal cord injury. Of the 44 who died after two months,
24 could be said to be related to the spinal cord injury and 16
unrelated. Renal disorder accounted for 14 of the 24 deaths attributed to the spinal cord injury. Carter estimates 1,800 to 3,000
deaths per year based upon 50,000 to 100,000 total estimated cases
in the United States.[3] In Australia, Cheshire[6] found the crude
death rate in 325 cases of acute spinal cord trauma to be 9.2% of
all spinal cord injured patients. Of those who died, only 66.1% of
the deaths were directly related to the injury. Forty-six percent of
the deaths occured more than 60 days after the accident. Of these,
84 percent were related to the spinal injury.

In the recent Vietnamese war, Jacobson and Bors found a mortality
rate of 1.7% over the short follow-up period which they studied.[12]
They stressed that combat injuries differ significantly from civilian
injuries in their severity, problems encountered in first aid and
evacuation, and in regard to complications and course. However,
Jacobs points out[13] that neurosurgical casualties were sometimes

admitted within 15 to 20 minutes of injury and seldom longer than
several hours post trauma. The availability of a variety of antibiotics as well as speed and efficiency of the system of evacuation
are unique to modern war injuries.

In comparison to other neurological conditions provocative of
disability, a spinal cord injury accounts for relatively few cases.
The number of deaths is even smaller.

Comparative Mortality Prevalence
and Cost of Neurological Conditions
Provocative of Disability[3]

Disability	Mortality	Estimated Total Cases	Estimated Cost of Care
Cerebral Palsy	1,161	600,000	$500,000,000
Epilepsy	2,043	1,000,000–2,000,000	$750,000,000
Multiple Sclerosis	1,533	250,000	$250,000,000
Muscular Dystrophy	921	200,000	$125,000,000
Parkinson's Disease	3,108	500,000	$400,000,000
Spinal Cord Injury	1,800–3,000 (estimated)	60,000–100,000	$2,400,000,000
Stroke	204,841	2,000,000	$440,000,000
Tumors of the Brain and other parts of the Nervous System	8,328	140,000	$272,000,000

Because the person most likely to sustain spinal cord injury
is a young male in his second or third decade, the cumulative cost
to society is enormous. According to Carter[3] the average expense
of the initial hospitalization for the traumatic paraplegic is
$18,000. Employing a 50% life expectancy of approximately 20 years,
annual expenses would total almost $47,000. Added to the cost of
the initial cost of hospitalization, such a patient would have
required an expenditure of $65,000 for medically related costs alone.

If one adds the expense of lost wages prior to age 65 and tax loss
calculated at 10 percent of gross annual income, the total economic
impact of the average paraplegic patient is $315,800.

The spinal cord injury causes the patient chronic medical,
psychosocial and vocational problems which often lengthen the
hospital stay. In her study, Wilcox[4] found that the mean duration
of hospitalization for a mixed group of paraplegics and quadriplegics
varied between 130 and 150 days over three years time. Whereas some
patients may be admitted and discharged within a two or three week
period, some others stayed well over a year and a half to control
medical complications.

A study done for the decade ending in 1971 at the Texas Institute
for Rehabilitation and Research showed that annual mean duration of
hospital stay varied between 77 and 159 days.[3] A relationship was
seen between the time which elapsed from the injury to the onset of
admission and the patient's arrival at rehabiliation center. The
longer the wait before rehabilitation treatment, the longer and
more costly the rehabilitation tended to be. For quadriplegics
treated at the same center during the same decade, the mean duration
of hospitalization varied between 97 and 188 days. A similar correlation between elapsed time before rehabilitation and duration of
rehabilitation treatment was noted.

Certain complications are commonly seen in spinal cord injury
and prolong hospitalization and increase cost. Carter[3] demonstrated
in a group of C5-C6 complete spinal cord injuried adults that program
cost varied whether or not they had ischemic decubitus ulcers. Those
who had no ulcers generated a program cost of $9,000, while those
with ulcers tended to cost in excess of $15,000. The duration of
hospitalization for the two groups was 96 and 138 days respectively.

THE PHYSIOLOGIC BASIS OF ACUTE CARE

The correct selection of the surgical approach to the acutely
traumatized spinal cord must be individualized to each case. There
continues to be controversy among experienced clinicians working
with spinal cord injury as to whether or not the risks of laminectomy, decompression or debridement out-weigh the advantages. Clinicians agree that an exploratory laminectomy is required in patients
with penetrating wounds of the spine. In cases of recent closed
trauma a wide difference of opinion exists concerning the indications
for laminectomy. Some surgeons wish to make the decision on the
basis of myelography, while others decide on the basis of spinal
manometrics. In the absence of a block of spinal fluid flow, some
feel that the neural structures are not being continuously or
seriously compressed. In these cases maximum destruction may

already have occurred and there may be nothing to gain by laminectomy. In one study,[14] 16 percent of laminectomized patients had
some degree of neurological improvement as against 29 percent of
non-laminectomized patients who also showed improvement. In another
study, 24 percent of patients who received laminectomy improved in
comparison to 60 percent who improved without surgical intervention.

In patients with complete blocks further improvement without
laminectomy is unlikely. Yet only few will improve noticeably with
laminectomy even in the presence of a complete block. Neither is
the X-ray appearance of the vertebrae a sufficient criterion by
itself to indicate need for surgical intervention. Some patients
with excellent bony alignment show no neurological improvement with
time. Others with significant bony abnormalities may improve neur-
ologically despite continued malalignment of the spine. Most
workers agree, however, that progressive neurological deficit is
an indication for laminectomy and progressive neurological recovery
contraindicates it. Commar[14] states that even when laminectomy
is indicated it need not necessarily be done immediately. Some
patients benefited from laminectomy even up to one year after injury
but patients receiving laminectomy longer than one year after injury
received no practical benefit other than psychological. He argues
that a conservative attitude toward immediate laminectomy serves
three purposes: 1) To provide time in which edema may recede,
thereby releasing the spinal block detected on manometric study and
making laminectomy unnecessary. 2) Allowing time for more thorough
examination of the patient and delineation of associated injuries.
3) Better preparation of the patient for surgery by stabilizing
associated metabolic abnormalities.

Tarlov[15] argues that mechanical deformation rather than spinal
cord anoxia is the main factor in acute paralysis caused by spinal
cord compression. In man, the rich vascular network of the cervical
spinal cord may permit adequate oxygenation of ischemic segments
through adjacent non-ischemic parts. Mechanical compression or
deformation of the cord may interfer with this blood flow, however,
and may need to be corrected. Some clinicians have reported patients
in whom partial or complete recovery from what appeared to be an
acute and total sensorimotor paralysis has occurred when compressive
forces were removed within 24 hours after the injury. Intramedullary
hemorrhages or pressure from edema or dislocated bone may not always
be differentiated from external compression of the spinal cord. In
such cases the surgeon must consider laminectomy or myelotomy. The
significant facts must be kept in mind.

Exposure, debridement, and decompression of the spinal cord may
offer the possibility of removing extrinsic or intrinsic pressure
and thereby promoting improved circulatory dynamics and oxygenation
of the nervous tissue. On the other hand, surgery is not conducted

without risk and the clinician must evaluate each patient individually.

Recently increased attention has been paid to the microvasculature of the spinal cord and changes observed within it and the
surrounding nervous tissue following injury.[16,17,18,19,20] Most
of the spinal cord gray matter receives its blood supply from penetrating central arteries which arise from the anterior spinal artery.
Five to eight central arteries arise from each centimeter length of
the anterior spinal artery in the cervical region. The thoracic
anterior spinal artery gives rise to two to five central arteries
per centimeter; and the lumbosacral anterior spinal artery, five
to twelve. Terminal branches of the central arteries extend up
and down to overlap one another.

The periphery of the cord receives its blood supply from pene=
trating braches from the pial artery plexus. Thus the posterior
horns, the lateral white matter, and the periphery of the lateral
gray matter receive their blood supply from this plexus. Capillary
networks are much more numerous in the gray matter than in the
white matter which probably reflects the greater metabolic requirement of the cell bodies as opposed to the axons. At all levels of
the spinal cord there is an intermediate zone which receives its
blood supply both from the central arteries and the penetrating
branches of the pial arteries. This doughnut shaped area includes
the outer margins of the gray matter and the inner margins of the
white matter.

When the spinal cord is exposed and subjected to experimental
acute compression, an ischemic syndrome is seen restricted to the
traumatized segment of the cord. Diffusion remains adequate above
and below the experimentally injured area. Narrowing of the central
arteries can be seen as early as 20 minutes after injury and may not
be restricted to the traumatized segment. Narrowing can be found
two centimeters proximal and distal to the experimental lesion. As
early as five minutes post contusion erythrocytes can be seen
distending the central venules of the gray matter, and within 15 to
30 minutes they can be seen in the perivascular spaces. Hemorrhage
into the gray matter can be seen one hour post injury.

In the white matter a marked decrease in the number of vessels
perfused can be seen within 15 minutes. However, half an hour after
injury evidence of renewed blood flow in the white matter begins to
be seen. Between one and eight hours post contusion circulation
remains considerably decreased in the white matter. At eight hours
only the peripheral half of the white matter shows evidence of significant perfusion. However, by 24 hours most of the vessels in the
white matter are receiving flow but the gray matter remains without
perfusion.

Microangiographs can define two zones in the injured spinal cord. The first zone is located in the posterior central part of the cord. Within it, degenerative changes in the neurons are visible about one hour after injury and capillaries progressively lose their ability to conduct blood over the first four hours. Necrosis of all elements within this zone follows soon thereafter. Zone two surrounds zone one. Although axonal and neuronal degeneration is severe in zone two, the microvascular patterns appear normal. The evidence obtained by Fairholm[18] indicates that at all times in the pathogenesis of the spinal cord injury the microvasculature in zone two is capable of perfusion. Degeneration of neural structures either precedes microvascular breakdown or occurs in the absence of microvascular disruption.

Ducker,[21] using an Allen's methodology of experimental impaction of the exposed spinal cord, studied the pathology of the cord and correlated it with the severity of the impact as well as time after injury. The development of demonstrable spinal cord pathology did not parallel the clinical neurological condition in the monkeys studied. Evidences of spinal cord injury were more prominent in the center of the cord even when the traumatic blow was delivered to the surface. Generally, the pathological changes began in the gray matter as small hemorrhages and edema and progressed to central necrosis and demyelinization of the adjacent white matter.

Dohrmann[22] examined the white matter of the monkey spinal cord by electron microscopy during the first four hours following a contusion sufficient to produce a transitory paraplegia. Electron microscopic changes could be noticed as early as 15 to 30 minutes after injury and consisted of moderately enlarged periaxonal spaces. Within an hour following trauma the myelin sheaths could be seen to be attenuated and the lamellae were splayed. By four hours after the contusion approximately 25 percent of the fibers showed breaks in the myelin sheaths, denuding of axons, attenuation of the myelin sheaths and greatly enlarged periaxonal spaces.

The role of norepinephrine in the pathogenesis of spinal cord injury has been studied.[23] Chemical analysis of spinal cord following experimental contusion shows a doubling of norepinephrine within 30 minutes and a quadrupling within 60 minutes after the injury. A slow decline within the tissue occurs over the next four hours. A direct correlation is seen between tissue injury and the amount of increased norepinephrine present. Massive-central hemorrhages are inversely associated with increases in tissue norepinephrine. It is hypothesized that norepinephrine in toxic amounts induces intense vasospasms and thereby perfusion in the microcirculation.

Alpha methyl tyrosine which blocks norepinephrine synthesis has been administered experimentally to cats who are subjected to spinal cord injury.[24] The expected accumulations of norepinephrine in the injury sites is reduced, suggesting that the drug has a protective effect against hemorrhagic necrosis.

Experimental extravascular surface cooling of the acutely injured spinal cord[25] has also been studied. Experimental dogs have been found to tolerate extravascular surface cooling of the cord to 17 to 19 degrees centigrade. Cooling may reduce the oxygen demand of the nervous tissue which is in a state of vascular compromise due to trauma. The technique may be advantageous in the treatment of trauma to these regions by reducing edema and ischemic changes.

Tissue pO_2 in the healthy canine spinal cord can be modified by ventilating the animals with hyperbaric oxygen. Using hyperbaric oxygen at two atmospheres, Kelly[26] treated traumatic paraplegic dogs and noted recovery to a greater degree than in the untreated control group. In earlier work, Kelly has shown that spinal cord pO_2 decreased rapidly to near zero following experimental non-disruptive trauma in dogs. The hypoxic area was confined to a short segment of cord at the site of injury. In animals experimentally treated with two atmospheres of 100 percent oxygen for four hours shortly after trauma, pO_2 at the traumatized site was greatly increased compared to animals who were allowed to breath 100 percent oxygen at one atmosphere only. Furthermore, the treated animals' functional outcome was noticeably better than the untreated animals.

In all cases of spinal cord injury with a neurological deficit, the patient will wonder if there is any reasonable hope for useful regeneration of the spinal cord. In February of 1970, the National Paraplegic Foundation, with partial support from the Paralyzed Veterans of America, promoted and financed a three day seminar on "The Enigma of Spinal Cord Regeneration." A summary of the proceeding of that meeting appears in the fifth supplement of Experimental Neurology, September, 1970.[27] The conferees agreed that continued research was necessary in several directions: 1) the process of colateral sprouting must be investigated; 2) the mechanisms which regulate protein synthesis, transport, and degradation in neurons must be studied; 3) neuronal specificities must be further investigated by biochemical studies in tissue culture; 4) the embryonic development and the adult state of the nervous tissue must be studied for changes in nerve specificities; 5) all growth and metabolism between elements of the central nervous system and its circulation must be more thoroughly understood. Although there is no assurance that further study will resolve the enigma of regeneration in the human central nervous system, certainly the problem cannot be avoided.

EARLY AND LATE EVALUATION OF INJURIES

As one might expect, the trauma that leads to injury of the spinal cord, enclosed as it is in a protective shell of bone and soft tissue, also frequently leads to multiple associated injuries. Associated injuries are seen most often in those who have been injured under combat conditions. Jacobson[12] and Jacobs[13] have reported on the associated injuries that they have observed in soldiers injured during the Vietnam war. Bullet wounds and shell fragments accounted for the majority of injuries. Chest injuries were the most common of the associated injuries, being seen often in patients with cervical and thoracic injuries. Gastrointestinal injuries were second most common, occurring most frequently in patients with lumbosacral spinal cord injuries. The emergency techniques most frequently employed were thoracotomy and/or tube insertion followed by bowel resection and fracture alignment of extremity bones.

Because many of the injuries were received from missiles and penetrating wounds, most of these patients underwent laminectomy and exploration of the spinal canal. Decision for non-operative treatment continued to depend upon either spontaneous improvement or absence of radiologic evidence of bone injury in the presence of normal cerebrospinal fluid manometrics. A variety of spinal cord injuries were found at surgery. These included transsection, laceration or contusion of the cord, and/or the roots. Some cords appeared grossly normal.

Perhaps more than any other medical or surgical condition, spinal cord injury is capable of abrupt and long term alteration of the body's organ systems. Virtually no organ system is spared if the spinal cord is injured in the upper thoracic or cervical segments. Few are spared if the injury is below that level.

Spinal shock is that state of hypoexcitability or absence of excitability of the spinal cord after transsection. It exists immediately after injury and continues for variable periods of time. The modern concept of spinal shock is based upon the work of Sherington. Transient reflex depression in the spinal segments below the transsection results from a sudden withdrawal of facilitating impulses from supraspinal areas. The end result is a disruption of transmission at the synapse which makes nerve conduction difficult if not impossible.

Frequent neurological examinations by the same examiner during and immediately after spinal transsection is invaluable in determining the intensity and duration of spinal shock in patients with incomplete as well as complete traumatic lesions. Some clinicians have found that the ankle jerk, plantar response, and bulbocavernosus reflex are still present immediately after cord transsection in man.[28]

The intensity and duration of spinal shock varies with the species under consideration. As one progresses up the phylogenetic scale, spinal shock is more intense and lasts longer. In man reflex activity of the skeletal muscles may appear from three days to two months after injury. Repression of reflexes is usually more pronounced in those spinal segments closest to the transsection but it is not unusual for segments of the spinal cord above the level of the lesion to show initial transient depression of cord function. Segments more removed usually recover from spinal shock somewhat earlier.

The most life threatening effect of transsection of the spinal cord is upon the cardiovascular and respiratory systems. Denervation of vasomotor control leads to sudden hypotension. Peripheral vasoconstriction does not occur reflexly and pooling of the blood in the splanchnic bed and dependent portions of the body causes a marked reduction of cardiac filling and cardiac output. As the heart rate increases to compensate for the decreased diastolic filling, stroke volume decreases even though cardiac output may remain unchanged. The quadraplegic is therefore relatively defenseless to compensate.

Meyer[29] studied a small group of military personnel during the first few hours after the onset of traumatic quadriplegia. He points to the danger of excessive fluid replacement in such patients. Because of the acute somatic and autonomic denervation, the vascular bed and the heart cannot respond normally to over hydration. Nor can the quadriplegic produce sufficient adjustments in respiratory rate, volume and flow. Reduction of inspiratory capacity, expiratory reserve volume, and vital capacity may all continue to be reduced by about 2/3 of the normal values. Expiratory reserve volume is particularly limited by weak abdominal muscles and accessory muscles of respiration. Fulminating and unexplained pulmonary edema has been the only explanation for sudden unexpected death in quadriplegic patients as long as six years after their injuries.

When autonomic tone returns, the patient with a spinal cord injury about the T4 segment may experience another cardiovascular challenge, autonomic hyperreflexia with hypertension. In the fully developed syndrome, autonomic hyperreflexia may consist of hypertension, bradycardia, sweating, pilomotor activity, flushing, dilated pupils, nasal stuffiness, blurred vision and headache.[30] Hypertension is the most serious of all of these symptoms since generalized seizures and subarachnoid hemorrhage have been reported.

The initiating stimulus for the massive sympathetic discharge is usually distention of a hollow viscus below the level of the

spinal cord injury. However, a variety of stimuli have been reported to cause the response. Stimulation of the urinary bladder,
rectum, pregnant uterus, gallbladder, bile ducts and duodenum
have all been reported to initiate the syndrome. Afferent impulses
from these organs enter the posterior horn of the spinal cord and
may initiate segmental reflexes. They may also ascend the spinal
cord to synapse with neurons in the intermediolateral columns of
the thoracic cord, thus initiating autonomic vasoconstrictor reflexes. In the normal person inhibitory impulses from higher centers in the brain are able to regulate these vasocontrictor reflexes. In the quadriplegic patient, however, the descending inhibitory impulses are blocked at the level of the spinal cord
injury leaving the spinal autonomic reflexes uncontrolled.

The remaining defense against uncontrolled hypertension is
bradycardia which is initiated by baroceptors in the aorta and
carotid sinus and mediated through the medullary vasomotor centers.
Bradycardia is a common feature of the fully developed attack of
autonomic hyperreflexia. Severe and disabling headache is however,
the most distressing symptom to the patient.

Claus-Walker et al.[31] demonstrated that the quadriplegic's
sympathetic nerve endings are partially depleted of norepinephrine.
Quadriplegics have an increased reflex section of adrenal catecholamines and the circulating norepinephrine is taken up by the
depleted neurons. Some may leak passively from the neurons to maintain normal blood pressure. Hypertension is caused when a reflex
stimulation spreads to the sympathetic nerves causing massive release
of neuronal norepinephrine. The stress affects the adrenal medulla
which then releases more epinephrine leading to a sustained episode
of hypertension which ceases when the adrenal medulla is exhausted.
Sell et al.[32] were able to demonstrate that the excretion of all
metabolites of catecholemines increased significantly after the
hypertensive crisis. The values obtained were comparable to values
seen only in patients with pheochromocytoma. Cole[30] was able to show
that the cardiovascular system of the quadriplegic patient is well
able to tolerate the usual activities of daily living.

Pulmonary compliance is lower in the quadriplegic than in the
normal person. Respiratory reserve volume is particularly limited
by weak abdominal muscles and reduction in pulmonary function is
more prominent in the sitting than in the supine position. Consequently the quadriplegic must expend more energy to breathe than
a normal person. Since the diaphragm normally delivers only 40 percent to 60 percent of inspiratory volume, a significant portion of
the lung remains unserved and unused due to weak chest and abdominal
muscles. Some clinicians have recommended that such patients learn
the technique of glossopharyngeal breathing. This technique may
add 700 to 1,000 milliliters to the patient's vital capacity. The

increased inspired air can be utilized to augment gas exchange as well as to enhance the effectiveness of coughing. One can expect such a patient to learn to increase his breath holding time to about 90 percent of predicted, maximum breathing capacity from about 30 to 50 percent of predicted, vital capacity from about 55 to 65 percent of predicted, and maximum expiratory flow from 40 to 90 percent of the predicted value.

Loss of motor function of the trunk and extremities produces rapid calcium mobilization from bones. With time bones which are no longer subjected to mechanical stresses become atrophied, a common complication of spinal cord injury.

Claus-Walker et al.[33] demonstrated that this effect was principly related to the duration of paralysis. Calciuria increases within ten days after the onset of paralysis, becomes maximum between one and six months, and then diminishes slowly. After one year calcium excretion is normal. At its peak, maximum calcium excretion can be two to four times greater than that in healthy subjects during long term bed rest. Unfortunately, rehabilitation exercises such as weight bearing on the long bones or prolonged wheelchair activity, do not alter this pattern. Furthermore, in the stabilized quadriplegic a period of three days of complete bed rest does not increase the calcium excretion as it would be expected to in normal subjects. It appears that this effect can be significantly lowered by reducing extracellular volume by sodium depletion. With time compensatory hormonal changes begin to conserve calcium but the spinal cord transsection permanently renders a large portion of the skeleton osteoporotic.

In the past, the most common cause of death in the chronic spinal cord injured adult has been renal failure and sepsis. Since normal kidney function is essential to the preservation of health, much attention has been paid to preservation of renal function and the lower urinary tract. Among other things the importance of fastidious urinary hygiene has been recognized. Medical and surgical management of these chronic urinary problems has been greatly improved over the past two decades.

A more aggressive approach to management of the neurogenic bladder probably began in the early part of World War II when the British government made a conscious decision to recommend suprapubic cystotomies as a method of choice for bladder drainage in those patients whose bladder paralysis lasted beyond 48 hours.[2] Recently, Price[34] reported on a longitudenal study done on 87 patients. Only 4 percent of her patients showed serious decline in kidney function, 19 percent showed mild deterioration, and good function was maintained by 76 percent. She reported that quadriplegic patients are more likely to suffer renal deterioration than paraplegics. Vesico-ureteral reflux

was more prominent in patients with depressed function than it was
in those with no deterioration. However, the patient with spinal
injury who is most likely to maintain renal integrity appears to
be appliance free paraplegic with sterile urine and no history of
vesicoureteral reflux, kidney stones, or bladder calculi.

Recently there has been a flurry of interest in the technique
of intermittent catheterization for people with neurogenic vesicle
dysfunction.[35,36] Prolonged intermittent catheterization following
removal of an indwelling catheter appears to be a safe and effective
method of treating the spastic neurogenic bladder. Some patients
so managed can become catheter free who would not be allowed to be
catheter free on the basis of occasional residual urine determinations. A high residual volume in the bladder is the most frequent
cause of failure. Occasional urinary tract infections and instances
of autonomic hyperreflexia complicate some of the attempts. The
duration of prior indwelling catheters does not seem to influence
the success of the intermittent catheter program. Low mean residual
volumes and duration of intermittent catheterization trial of longer
than 16 days seems to be statistically associated with success.

Changes also occur in other organ systems. Denervation of the
gastrointestinal tract during the stage of spinal shock leads to
gastric distention and small and large bowel hypotonia. Generalized
abdominal distension in turn may limit diaphragmatic excursion and
embarrass respiratory function. Muscular atrophy, spasticity and
occasional heterotopic bone formation within muscles leads to problems of positioning and nursing care. Skin atrophy commences as
soon as the patient is put to bed. Ischemic necrosis of the skin
and subcutaneous tissue becomes a major medical complication which
threatens the patient's physical health and his economic resources.
Alterations in autonomic control of the sweating mechanism may lead
to excessive moisture on the skin, setting the stage for maceration
and skin breakdown. In other areas the sweating response may be
diminished, limiting the ability of the body to regulate its temperature through head dissipation. Peptic ulcerations of the stomach
and duodenum are seen especially in those who are further stressed
by subsequent surgery for associated injuries or spinal stabilization. Above all is the constant threat of connective tissue proliferation and contracture contribued to by immobility and soft tissue
injuries.

The level at which the spinal cord is injured will determine
the potentiality for function in the paraplegic and quadriplegic
patient. In the absence of medical and psychological limitations
the most important factor controlling functional ability is the
amount of muscle power remaining. This in turn is determined by
the spinal segments at which important muscle groups are innervated.

Most clinicians recognize about seven critical levels of spinal cord severence. Beginning with the fourth cervical segment the addition of each successive critical segment adds an important increment to the muscle power of the patient.[37,38] The quadriplegic who retains the fourth cervical segment will continue to have function in the sternomastoids, upper cervical paraspinal muscles and the trapezius muscle. Voluntary use of the arms, trunk, and lower extremities are absent but he may benefit from training in the use of externally powered devices to achieve some useful function of the arms. He can control his head position which in turn assists in balancing the trunk while sitting in the wheelchair. Many such patients find an adaptive tool, held in the mouth for typing, writing, page turning, etc., to be more useful than adaptive devices attached to the denervated arms.

When the fifth cervical segment is preserved the deltoid and biceps muscles continue to function together with the external rotators of the shoulder. These muscles usually remain weak however, and frequently require external assistance in order to accomplish useful work. Elbow, forearm, and hand control remain absent and continued use of assistive devices is necessary if the patient is to engage in useful upper extremity activities. Such a patient may be able to do some self feeding and grooming and to assist in dressing activities. He will be unable to push his wheelchair enough to avoid use of a motorized wheelchair. He will also be unable to voluntarily assist in transferring himself to or from the chairs, beds, and toilets.

When the sixth cervical segment is retained the patient may gain some voluntary strength in shoulder depression, scapular stabilization, shoulder adduction and internal rotation, and radial wrist extension. These added muscles greatly enhance the ability to perform activities of daily living. Especially important is the wrist extensor mechanism which can be harnessed through a special "tenodesis" splint which drives the fingers into flexion. He is more capable at pushing his wheelchair if the wheels have adaptive devices to allow his hand to grasp the rim. With training he can often learn to lean on his arms to periodically shift weight off his buttocks. His sitting tolerance may thus increase to six to eight hours. He can be more helpful in dressing himself and asisting others in transferring himself from bed to chair, etc. Some patients who retained sixth cervical segment function are able to achieve safety and confidence in operating an automobile with special adaptive equipment and hand controls.

When the seventh cervical segment is retained the patient now gains some voluntary control over the triceps muscle for elbow extension. Wrist and finger extension becomes somewhat stronger and he may gain some use of his finger flexes. He is now able to do

pushups on the arms of his wheelchair and therefore his sitting
tolerance can be expected to be increased to the limit of his
overall physical endurance. Most importantly, hand function is
sufficient to allow him to grasp and release voluntarily. This
adds important control over activities of daily living as well as
greatly increased potential for vocational activity. He is independent in his wheelchair and should need no motorized assistance.
whatsoever. Often he can be trained to use his hand control to
irrigate his catheter or stimulate the rectum for bowl movements.

The individual whose lesion is at the upper thoracic level
has full control of his upper extremities and is therefore more
properly called paraplegic than quadriplegic. Although trunk and
abdominal innervation are absent, he can be trained to be completely
independent in his wheelchair. With sufficient skills and motivation he is able to work full time in competitive employment and
can operate independently in a community without major architectural
barriers to the wheelchair. He can drive, propel himself for long
distances in his wheelchair and manage to maneuver the wheelchair
up and down or over a single small step or curb. He is also independent in managing his urinary drainage apparatus and inserting
rectal suppositories or a gloved finger for stool extraction.

The next critical level is at the lower thoracic or upper lumbar segments. This paraplegic has full abdominal, upper back, and
respiratory control. He can do all of the activities listed above
but because of his increased trunk innervation, his sitting balance
is enhanced. This allows him greater capability for wheelchair
operating skills and athletics.

The patient with a low lumbar spinal cord transsection (L4) has
the use of his hip flexors and knee extensors although he lacks
voluntary control over hip extensors and abductors. He is able to
walk with proper bracing and training but the energy required for
functional ambulation is so high as to make it necessary for many
such patients to retain the use of the wheelchair on certain occasions.

It is not within the scope of this paper to discuss the poten-
tialities for surgical intervention to provide greater upper extremity function in quadriplegic patients. The reader is referred
to Zancolli's text for a complete discussion of this.[39] Orthopedic
surgical procedures such as tendon transfers and joint stabilization
procedures can sometimes produce significant increase in hand function. For carefully selected patients the increased function may
make a difference between independence and semi-dependence, vocational success or unemployment.

OUTCOME

The long term outcome of spinal cord injured patients is as
varied as the causes which produce the injury. Until recently
however, the literature has been imprecise in specifying outcomes
so that the processes which lead to them can be analyzed.

Cheshire[40] suggests a detailed description of outcome for purposes of comparing results between various centers. He postulates
three overall groupings: complete recovery, incomplete recovery,
and no recovery. As the term implies, the completely recovered
case shows no neurological deficits whatever. The case with no
recovery shows complete and motor sensory paralysis below the
level of the lesion and the patient is confined to a bed and wheelchair except for therapeutic standing. He has a neurogenic bladder
and is rarely employable outside of the home. Cheshire divides
the incomplete recovery group into five subgroups which differ from
one another according to motor and sensory function, spasticity,
upper and lower extremity function, ambulation ability, bladder
function, and employability.

Deyoe[41] interviewed 219 recently injured spinal cord veterans.
Their injuries were of long duration, had all received comprehensive
rehabilitation at the Veterans Administration hospitals, and all
had been living at home for periods up to 25 years. He found that
the majority were married and living in the suburbs in accomodations
ranging from specially built homes to apartments and mobile homes.
Travel was generally not considered a problem for these successfully
rehabilitated paraplegics and quadriplegics. Many were traveling
more miles by car and plane than the general population. Relatively
few, however, were employed full time. Employment often was not
required for income, since many of the veterans received pensions.
However, he reported that many were participating in community and
service activities and argued that this group must be considered
self-sufficient and productive members of society. It should be
pointed out however, that his study was limited to those who had
succeeded in living outside of the hospital environment and who
were returning to the spinal cord injury service of the Veteran's
Hospital for medical check-up or treatment during the period of
the study.

In contrast, Ooi[7] studied quadriplegic patients in Japan. Of
the 66 cases he evaluated, he found only three who found employment
while living in the wheelchair. He stated that the large industrial corporations were unwilling to hire patients with such
severe disabilities at that time.

Hallin[8] reported on follow-up studies done in approximately
half of the paraplegic and quadriplegic patients discharged from

the inpatient service between January, 1955 and November, 1967
who were willing to return for follow-up. The majority of
the patients did very well. Motivation was high, marriages were
remaining intact and they were living in a fairly normal manner
maintaining their health in self-directed activities of some sort.
Although the level and completeness of the spinal cord injury certainly affected the outcome, good psychosocial adjustment was found
in even the most severely injured persons. He classified patients
as having less than 10 percent independence, between 10 and 90 percent independence, and more than 90 percent independence. He found
57 percent of the patients could be classified as more than 90 percent independent while only 25 percent less than 10 percent independent. Almost half of his patients were directing more than
50 percent of their own activities whereas approximately a third
were able to direct less than 10 percent of their own activities.

   Neurological impairment, medical management, and residual functional capacity not withstanding, the long range outcome of the
patient will depend largely upon his ability to accept the disability
of the lightening-bolt-like trauma of spinal cord injury. Kerr[42]
concluded that the adjustment process takes at least two years
to complete and 100 percent acceptance of a spinal cord injury is
probably impossible. Although acceptance or rejection of disability
are terms commonly used, they may give rise to the false impression
that the patient is voluntarily able to exercise his free will of
acceptance or rejection. However, many patients make successful
mental adjustments to their disability and continue productively
and meaningfully for themselves and the significant other people
in their lives. The most important factors in achieving a good
adjustment are the previous background of the patient, both personal
and environmental, and the patient's age. The young usually adjusted
better than the old.

   Until recently sexual rehabilitation has been viewed guardedly
by physicians and hospital staff. However, within the last few years
taboos and biases against dealing directly with sexual adjustment
have gradually diminished. Men and women with traumatic spinal cord
injuries are making increased demands upon the health care system
to do more than simply manage them through their immediate post
injury period. The reluctance to aggressively study self help
methods to increase the patient's sexual capabilities stems from
a general uneasiness with the whole topic of human sexuality. Irrespective of professional credentials it is difficult to speak openly
of the personal feelings associated with sexual expression. Talbot[43]
states that of all the problems confronting the spinal cord injured
patient there is none from which it has taken longer to dispell the
mists of ignorance than the matter of sexual function. The unwill-
ingness of the medical profession to inform itself on the subject
has been incomprehensible and inexcusable.

Loss of somatic sensation if frequent in paraplegia but in the vast majority of patients, psychosexual content remains substantially normal in spite of loss of sensation over primary erogenous zones of the body. Reflex penile erection may still occur as a result of visceral or external stimulation. Most patients with complete upper motor neuron lesions at any level can achieve reflex erection for puposes of copulation. However, the higher the spinal cord transsection, the more likely it is that the patient will be able to regularly elicit the reflex erection. The sensation of orgasm may not occur in the same sense that it did prior to injury. Since the genital and pelvic areas may be denervated, ejaculation may also be absent or retrograde into the urinary bladder.

Fantasy plays an important role in sexual expression for both men and women, whether able bodied or disabled. Spinal cord injured adults report the importance of fantasy for enhancement of sexual sensation and satisfaction. Some report a body awareness or orgasm which, through fantasy, may be as intense as orgasm experienced prior to the spinal cord injury. Others report no sense of orgasm but obtain great personal satisfaction from sexual activity with their partners.

The Minnesota program reported by Cole[44,45] is an intensive two day sight and sound program intended to desensitize the participant to explicit sexual stimuli, thereby allowing resensitization to an open, warm and experimental attitude toward sexual activity. The format includes explicit slides, films and talks coupled with periodic small group discussions lead by trained group leaders. His results suggest that continued sexual interest, activity and satisfaction can be expected for many paraplegic and quadriplegic adults.

NEW SYSTEMS OF SPINAL CORD CARE

At the end of World War II many energies were brought to bear on the problems of the spinal cord injured veteran. As a result organizations were formed which had as their prime objective improvement of health care for spinal cord injured people. A group of veterans with service connected disabilities, largely paraplegics and quadriplegics, first organized the Paralyzed Veterans of America. Shortly afterward the National Paraplegia Foundation was also organized. These organizations have continued to grow and now the National Paraplegia Foundation sponsors an annual conference on medical and consumer problems of paraplegics and quadriplegics.

In England, the Stoke Mandeville Hospital was taken over by the National Health Service and became a 195 bed national center admitting patients from all over the United Kingdom. More than 4,000 patients have been treated there under the eminent leadership of

Sir Ludwig Guttmann,[46] who was instrumental in founding the Stoke
Mandeville games.   Since 1952 these games have developed into the
first organized annual sports festival for severely disabled people
in the world.

In several countries outside of the United States treatment of
spinal cord injury has been centralized.   Several centers have
sprung up in Australia.   Dr. G. M. Bedbrook of the Royal Perth
Hospital in Perth, Western Australia, described his center at the
1967 annual scientific meetings of the International Medical Society
of Paraplegia.

At the Rehabilitation Institute of Warsaw University, Poland,
Professor Marian Weiss described the Rehabiliation Institute of
Warsaw and stated that the organization of similar centers was
being suggested for other regions of his country.[48]

The notion of creating pilot centers to pioneer regional care
of spinal cord injury was stated by Dr. Paul Bucy in his conversa-
tions with Representative Melvin Laird, at that time Minority Leader
in the House of Representatives, United States Congress.   Encourag-
ingly, funds have finally been provided by the National Institute
of Neurological Disease and Stroke of the National Institutes of
Health for planning grants for pilot centers.

In September, 1969, the report, "Interdisciplinary Clinical
Educational and Research Aspects for a Regional Center for the
Rehabilitation of Spinal Cord Injured Persons" was published on
the basis of the work done at Rancho Los Amigos Hospital in Downey,
California.   The publication reports on a pilot study made in California and adjacent states to "determine the need and feasibility
of a regional center to provide rehabilitative care for spinal cord
injured civilians."   The Report deals with the clinical, adminis-
trative, educational and research aspects of the development and
operation of such a center and provides guidelines to help pro-
fessionals in providing treatment for patients.

Other centers in the United States have pioneered in the region-
alization of spinal cord care.   Based on cooperation between the
Barrow Neurological Institute and the Institute of Rehabilitation
Medicine in Phoenix, Arizona, Dr. John Young gives his plan for
the 70's, a national network of regional spinal injury care systems.
After two years of preliminary organization, the Southwest Regional
System for Treatment of Spinal Cord Injury, funded by Social and
Rehabilitation Services, demonstrated a system of spinal injury
care.   As a result of this work, a national network of regional
systems for the treatment of spinal cord injury was proposed.   It
was suggested that the "federal government define broad specifica-
tions for these systems and let contracts to private or public

organizations for facilities to develop and implement the systems."
As Dr. Young saw it, the plan would create a network of regional
centers within a decade. It would represent a cooperative effort
between federal and local medical service programs and would be
financed by direct revenue sharing. It could provide the best in
spinal cord injury care at a cost which the country could afford
and is already spending on its fragmented, inefficient system of
spinal cord injury care.

Subsequently, the Social and Rehabilitation Services published
grant guidelines for model regional systems of spinal cord injury
rehabilitation based upon many of the principles learned from
Rancho Los Amigos and the Institute of Rehabilitation Medicine.
They recommended that model regional systems of spinal cord injury
rehabilitation be created to demonstrate, within a defined region
of the country, a multi-disciplinary system of comprehensive rehabilitation services extending from the point of injury through
acute care and rehabilitation management to community and job
placement and long term follow-up.

A national network of new centers is not being established which
will demonstrate cost effectiveness of regional systems. They will
establish a rehabilitation research environment, provide evaluation
and application of new methods and equipment, demonstrate the feasibility of expanding and improving community resources, and demonstrate methods of community outreach for the spinal cord injured
population.

REFERENCES

1. Elsberg, C. A.: The Edwin Smith Surgical Papyrus, and the diag-
    nosis and treatment of injuries to the skull and spine
    5,000 years ago. Ann. Med. Hist. 3:271-279 (May) 1931.

2. Dick, T. B. S.: Traumatic paraplegic pre-Guttmann. Paraplegia
    7:173-177 (Nov) 1969.

3. Carter, R. E.: Research and training center accomplishments
    in spinal cord injury. Presented at the Conference of
    Rehabilitation, Research and Training Centers, February
    1972, Temple University, Philadelphia, Pa.

4. Wilcox, N. E., Stauffer, E. S., Nickel, V. L.: A statistical
    analysis of 423 consecutive patients admitted to the
    spinal cord injury center, Rancho Los Amigos Hospital,
    1 January, 1964, through 31 December 1967. Paraplegia
    8:27-35 (May) 1970.

5. Gehrig, R., Michaelis, L. S.: Statistics of acute paraplegia

and tetraplegia on a national scale. Paraplegia 6:93-95
(Aug) 1968.

6. Cheshire, D. J. E.: The complete and centralised treatment of
paraplegia. Paraplegia 6:59-73 (Aug) 1968.

7. Ooi, Y., Tateiwa, K., Uehara, M., Murai, S.: Quadriplegia in
Japan: a study of 60 cases. Arch. Phys. Med. Rehabil.
54:136-139 (March) 1973.

8. Hallin, R. P.: Follow-up of paraplegics and tetraplegics after
comprehensive rehabilitation. Paraplegia 6:128-134 (Nov)
1968.

9. Heyl, H. L.: Editorial: Spinal cord injuries. J. Neurosurg.
35:251-252 (Sept) 1971.

10. Freed, M. M., Bakst, H. J., Barrie, D. L.: Life expectancy,
survival rate, and causes of death in civilian patients
with spinal cord trauma. Arch. Phys. Med. Rehabil. 47:
457-463 (Jul) 1966.

11. Hoffman, C. A., Bunts, R. C.: Present urologic status of the
World War II paraplegic, 15 year follow-up; comparison
with status of 5 years Korean War paraplegic. J. Urol.
86:60-68 (July) 1961.

12. Jacobson, S. A., Bors, E.: Spinal cord injury in Vietnamese
combat. Paraplegia 7:263-281 (Feb) 1970.

13. Jacobs, G. B., Berg, R. A.: The treatment of acute spinal cord
injuries in a war zone. J. Neurosurg. 34: 164-167 (Feb)
1971.

14. Comarr, A. E.: The practical urological management of the patient with spinal cord injury. Br. J. Urol. 31:1-45
(March) 1959.

15. Tarlov, I. M.: Acute spinal cord compression paralysis. J.
Neurosurg. 36:10=20 (Jan) 1972.

16. Turnbull, I. M.: Microvasculature of the human spinal cord.
J. Neurosurg. 35:141-146 (Aug) 1971.

17. Dohrmann, G. J., Wagner, F. C., Bucy, P. C.: The microvasculature in transitory traumatic paraplegia. An electron
microscopic study in the monkey. J. Neurosurg. 35:263-271
(Sept) 1971.

18. Fairholm, D. J., Turnbull, I. M.: Microangiographic study of
         experimental spinal cord injuries. J. Neurosurg. 35:
         277-285 (Sept) 1971.

19. Fried, L. C., Goodkin, R.: Microangiographic observations of
         the experimentally traumatized spinal cord. J. Neurosurg.
         35:709-714 (Dec) 1971.

20. Dohrman, G. J., Wick, K. M., Bucy, P. C.: Spinal cord blood
         flow patterns in experimental traumatic paraplegia. J.
         Neurosurg. 38:52-57 (Jan) 1973.

21. Ducker, T. B., Kindt, G. W., Kempe, L. G.: Pathological find-
         ings in acute experimental spinal cord trauma. J.
         Neurosurg. 35:700-707 (Dec) 1971.

22. Dohrmann, G. J., Wagner, F. C., Bucy, P. C.: Transitory
         traumatic paraplegia: electron microscopy of early
         alterations in myelinated nerve fibers. J. Neurosurg.
         36:407-414 (Apr) 1972.

23. Osterholm, J. L, Mathews, G. J.: Altered norepinephrine meta-
         bolism following experimental spinal cord injury. Part I:
         Relationship to hemorrhagic necrosis and post-wounding
         neurological deficits. J. Neurosurg. 36:386-393 (Apr)
         1972.

24. Osterholm, J. L, Mathews, G. J.: Altered norepinephrine meta-
         bolism following experimental spinal cord injury. Part 2:
         Protection against traumatic spinal cord hemorrhagic
         necrosis by norepinephrine synthesis blockade with alpha
         methyl tyrosine. J. Neurosurg. 36:395-400 (Apr) 1972.

25. Selker, R. G.: Experimental extravascular surface-cooling of
         the brain stem, fourth ventricle, and high cervical
         spinal cord. J. Neurosurg. 35:432-436 (Oct) 1971.

26. Kelly, D. L., Lassiter, K. R. L., Vongsvivut, A., Smith, J. M.:
         Effects of hyperbaric oxygenation and tissue oxygen
         studies in experimental paraplegia. J. Neurosurg. 36:
         425-429 (Apr) 1972.

27. Guth, L, Windle, W. F.: The enigma of central nervous regen-
         eration. Exp. Neurol., Supp. 5 28:38-39 (Sept) 1970.

28. Guttmann, Sir L.: Spinal shock and reflex behavior in man.
         Paraplegia 8:100-109 (Aug) 1970.

29. Meyer, G. A., Berman, I. R., Doty, D. B., Moseley, R. V.,

Gutierrez, V. S.: Hemodynamic responses to acute quadriplegia with or without chest trauma. J. Neurosurg.
34:168-175 (Feb) 1971.

30. Cole, T. M., Kottke, F. J., Olson, M., Stradal, L., Niederloh:
     Alterations of cardiovascular control in high spinal
     myelomalacia. Arch. Phys. Med. Rehabil. 48:359-368
     (July) 1967.

31. Claus-Walker, J., Campos, R. J., Carter, R. E.: Hypertensive
     episodes in quadriplegic patients: neuroendocrine mech-
     anisms. Arch. Phys. Med. Rehabil. 53:47-50 (Feb) 1972.

32. Sell, G. H., Naftchi, N. E., Lowman, E. W., Rusk, H. A.:
     Autonomic hyperrefelxia and catecholamine metabolites in
     spinal cord injury. Arch. Phys. Med. Rehabil. 53:415-
     424 (Sept) 1972.

33. Claus-Walker, J., Campos, R. J., Carter, R. E., Vallbona, C.,
     Lipscomb, H. S.: Calcium excretion in quadriplegia.
     Arch. Phys. Med. Rehabil. 53:14-20 (Jan) 1972.

34. Price, M.: Renal function in patients with traumatic myelo-
     pathy. Arch. Phys. Med. Rehabil. 53:261-265 (June) 1972.

35. Stover, S. L., Miller, J. M., Nepomuceno, C. S.: Intermittent
     catheterization in patients previously on indwelling
     catheter drainage. Arch. Phys. Med. Rehabil. 53:25-30
     (Jan) 1973.

36. McMaster, W. C., Nicholas, J. J., Rosen, J. S.: Intermittent
     catheterization for spinal cord injury patients with
     chronic indwelling urethral catheters. Arch. Phys. Med.
     Rehabil. 53:563-567 (Dec) 1972.

37. Long, C., Lawton, E. B.: Functional significance of spinal
     cord lesion level. Arch. Phys. Med. Rehabil. 36:249-255
     (Apr) 1955.

38. Long, C.: Congenital and traumatic lesions of the spinal cord.
     In - Handbook of Physical Medicine and Rehabilitation.
     Krusen, F. H., Kottke, F. J., Ellwood, P. M., Jr. (Eds.):
     W. B. Saunders Company, Philadelphia, 1971, pp. 566-578.

39. Zancolli, E.: Functional restoration of the upper limbs in
     complete traumatic quadriplegia. In - Structural and
     Dynamic Bases of Hand Surgery: J. Lippincott Company,
     Philadelphia and Toronto, Chap. 11, 1968, pp. 155-174.

40. Cheshire, D. J. E.: A classification of the functional end-
         results of injury to the cervical spinal cord. Para-
         plegia 8:70-73 (Aug) 1970.

41. Deyoe, F. S.: Spinal cord injury: Long term follow-up of
         veterans. Arch. Phys. Med. Rehabil. 53:523-529 (Nov)
         1972.

42. Kerr, W. G., Thompson, M. A.: Acceptance of disability of
         sudden onset in paraplegia. Paraplegia 10:94-102 (May)
         1972.

43. Talbot, H. S. (1969) Proceeding of the seventh Veterans
         Administration Spinal Cord Injury Conference, Veterans
         Administration Hospital, Bronx, New York. pp. 222-223.

44. Sex and the paraplegia. Medical World News. 13:35-38 (Jan)
         1972.

45. Cole, T. M., Chilgren, R., Rosenberg, P.: A new program of
         sex education and counseling for spinal cord injured
         adults and health care professionals. Accepted for pub-
         lication in International Journal of Paraplegia.

46. Guttmann, Sir L.: History of the national spinal injury ser-
         vices, Stoke Mandeville Hospital, Aylesbury. Int. J.
         Paraplegia 5:115-126 (Nov) 1967.

47. Bedbrook, G. M.: The organization of a spinal injuries unit
         at Royal Perth Hospital. Paraplegia 5:150-158 (Nov)
         1967.

48. Weiss, M.: 15 years experience on rehabilitation of para-
         plegics at the Rehabilitation Institute of Warsaw Uni-
         versity, Poland. Paraplegia 5:1158-1166 (Nov) 1967.

PROBLEMS IN REHABILITATION FOLLOWING TRAUMA

A. Turkyilmaz Ozel

Frederic J. Kottke

University of Minnesota

Rehabilitation, which may be defined as "the treatment and training of the patient so that he may obtain his optimal potential for normal living physically, psychologically, socially, and vocationally," is a component of the management of most serious posttraumatic conditions. These conditions may include congenital and acquired brain injuries, spinal cord injuries, fractures, traumatic arthritis, peripheral nerve injuries, amputations, burns, frostbite injuries, radiation injuries, decompression arthropathy, and trauma to extra-articular structures, such as muscles, tendons, ligaments, and other connective tissue. While great importance is being given to the life-threatening conditions following severe trauma, simple procedures which will prevent future problems for the patient may be neglected during his early care. Consequently, disabilities secondary to dysmobility may develop which will interfere with the course of the post-traumatic condition. Therefore, initiation of rehabilitation at the beginning of the treatment of post-traumatic patients should be considered.

This chapter is concerned with the common problems of rehabilitation following trauma and the general principles for their treatment rather than an explantion of each disorder. However, the common problems and general principles for management are applicable in each individual case.

GENERAL CONSIDERATIONS

Repair Process in Collagen Systems: Although a detailed chapter about wound healing is found elsewhere in this book, we will touch

329

here on some important points related to the problems which are
encountered in the rehabilitation process. There is a continual
turnover of the components of connective tissue by breakdown and
replacement and by reorganization of the attachments of the various
components. Skin wounds become epithelialized by proliferation
and migration of non-keratinized epidermal cells from the wound
margins. Hyperplasia of epidermal cells occurs. DNA synthesis
may be detected in four hours by using H^3-thymidine (Hell and
Cruickshank, 1963). Mytotic activity begins twelve hours after
the injury (Bullough and Laurence, 1960). Full thickness skin
defects in mammals heal in part by gradual approximation of the
wound edges by a process of connective tissue reaction described
as contraction (Grillo et al, 1958). By about the tenth day
after injury the area has contracted to a minimal size in small
excised wounds (Abercrombie et al, 1954). The collagen content in
the area of healing increases during the phase of contraction
(Dunphy and Udupa, 1955) and continues to do so after the skin is
intact again (Abercrombie et al, 1954). The collagen content in
the area of healing increases during the phase of contraction
(Dunphy and Udupa, 1955) and continues to do so after the skin is
intact again (Abercrombie and James, 1957). Consequently, by fifty
days the concentration of collagen per unit of wet weight of scar
has caught up with that of control skin and by one hundred days the
scar clearly surpasses the normal skin in collagen per unit of wet
weight and by two hundred days in collagen per unit of area (Abercrombie and James, 1957).

   Buck (1953) sectioned the Achilles tendon of the rat and allowed
it to retract without suturing. The gap between the cut ends of the
tendon was filled within twenty-four hours by a fibrin clot oriented
longitudinally with the tendon. Fibroblasts began to migrate into
the coagulum from the periphery within three days. Reticulum and
collagen fibers were laid down within four days. The collagen fibers
were oriented parallel to the fibrin threads in the long axis of the
tendon. The entire length of the tendon was invaded by fibroblasts
within two weeks. When tenotomy was performed one month after denervation of the muscle, the result was the production of a much
thinner tendon. Spontaneous healing of sectioned Achilles tendon
without suturing was confirmed by Lipscomb and Wakim (1961). Since
effective repair depends on the invasion of the sutured area by fibroblasts from the surrounding tissue some degree of adherence to
promote the required vascularity of the region and to provide a
a pathway for the entry of cells is essential (McMinn, 1969). Within
four to five days, collagenous adhesions begin to form between the
sutured tendon and the surrounding structure. Potenza (1962) claimed
that associated trauma during suturing of a divided tendon was mainly
responsible for the adhesions and these adhesions could be prevented
or minimized by meticulous surgical techniques. It appears from
clinical evaluation of our patients that support of the ankle in the

desired position by a brace after Achilles tenotomy allows early
ambulation, and the limited motion associated with triceps surae
contraction stretches the newly-forming adhesions so that they do
not restrict the desired ankle motion.

MAINTENANCE OF NORMAL MOBILITY

Factors Which Decrease Normal Mobility: Following serious
trauma immobilization of part or all of the body may be necessary.
Changes in the function of the immobilized part may lead to additional disabilities not directly related to the initial injury.
Therefore, immobilization is one of the major concerns of rehabilitation medicine, since it is the source of many problems which are
extremely difficult or sometimes impossible to resolve.

When a part is immobilized, the collagenous and reticular networks become contracted and the distances between attachments in
the networks are shortened so that the tissue becomes dense and hard
and loses the suppleness of normal areolar tissue. Immobilization
in post-traumatic conditions is usually associated with an injury;
therefore, additional problems secondary to the wound's repair
process may be expected. Histological evidence of fibrosis may
occur within a few days in the injured site as is mentioned above.
Gross restriction of motion begins to occur in approximately four
days and develops progressively.

After prolonged immobilization contractures of both the muscle
and joint capsule are responsible for the restriction of motion with
the shortening of the muscle being primarily at fault (Evans et al,
1960). The intra-articular effects of prolonged immobilization
have been studied in animals (Evans et al, 1960; Hall, 1963; and
Thaxter et al, 1965) and in humans (Enneking and Horowitz, 1972).
Proliferation of intracapsular connective tissue and the formation
of adhesions occur. Later on, major cartilage alterations, such as
matrix fibrillation, cleft formation and erosions, as well as their
adjacent subchondral lesions, result from abnormal friction and
pressure in a joint compromised by limitation of motion. Enneking
and Horowitz (1972), studying ten human knees, confirmed the existence of progressive contractures of the capsular and paracapsular
structures and concomitant encroachment on the joint by intra-articular fibrofatty connective tissue with eventual obliteration of
the joint cavity by this tissue. When the articular surfaces are
in direct opposition to one another the cartilage becomes fibrillated
and small cystic defects appear in the superficial and deep layers.
Subsequently, in what appears to be a reparative effort, there is
general replacement of this defect by primitive mesenchymal tissue
which matures and may eventually ossify. Forceful remobilization
of the joints following prolonged immobilization results in tearing

of the connective tissue that is proliferated within the joint and in avulsion of pieces of articular cartilage (Evans et al, 1960; Enneking and Horowitz, 1972).

Factors promoting the formation of dense fibrosis are immobilization, edema, trauma, and impaired circulation (Kottke, 1971a).

In the normal relaxed standing posture, extension of the back, hip, and knee is maintained by positioning the center of gravity of the body above these joints, so that the weight of the body holds the joints extended against restricting ligaments and the extensor muscles are relaxed. Maintenance of balance is provided mainly by activity of the soleus muscle (Kottke and Kubicek, 1956).

During relaxed standing there is not free extension of the hip or knee beyond this position. The "overextension" of the hip joint is actually extension of the lumbar spine and flexion of the opposite hip. Maximum extension of the hip joint is 170 degrees in men and 165 degrees in women during relaxed standing. Further extension is possible only to the extent that connective tissue can be stretched (Mundale et al, 1956). During normal walking the iliofemoral, pubofemoral, and ischiofemoral ligaments and the hip flexor muscles are stretched at every step. On the other hand, even prone lying on a hard surface may provide only 160 degrees of hip extension. Unless a person stands and walks frequently each day, the normal reaction of fibrous connective tissue to shorten and fuse together results in progressive constriction of the iliotibial band, producing a flexion, abduction, and external rotation deformity (Kottke, 1971a). Therefore, bedridden patients tend to develop hip flexion deformities.

Knee flexion deformity develops due to shortening in the hamstring and gastrocnemius muscles and the posterior capsule of the joint if the knees are not stretched to full extension. The placing of pillows under the knees promotes knee flexion contractures. A parlytic patient who lies in bed without the support of a footboard tends to develop a dropfoot deformity.

Prevention of Loss of Normal Mobility: To prevent tightness is much easier than to correct it after it has developed. Immobilization seems to be the prime factor in the development of limitation of motion. Spasticity and poor positioning can also intensify the restriction of motion. Therefore, except in the case of necessary immobilization of a fracture or of an open draining wound, motion should begin immediately after surgery or trauma to insure that supple areolar connective tissue rather than dense scar develops at the site in which motion should occur (Kottke, 1971a).

Maintenance of normal mobility is based upon range of motion

exercises, good positioning, and proper splinting as well as the
control or elimination of other precipitating factors, such as pain
and inflammation. If for any reason the range is restricted, tightness develops and limits the arc of motion. Hills and Byrd (1973)
recently showed that immobilization of the forearm of a healthy
subject for thirty days resulted in a twenty percent decrease in
wrist flexion and extension and a forty-one percent decrease in
wrist abduction and adduction.

Range of Motion Exercises: Motion in joints and soft tissues
can be maintained by the normal movement of the part of the body,
including joint capsules, muscles, subcutaneous tissue, and ligaments through full range of motion (Lowenthal and Tobis, 1957;
Kottke, 1971a). At least twice daily, three times each, each joint
should be carried through full range of motion in each direction
that it can be moved. If the patient cannot move the extremity by
himself, the range of motion exercises are performed passively. If
he is moderately weak or has pain, exercises are carried out with
the assistance of a physical therapist. If the patient is able to
move his extremities freely, he should perform the range of motion
exercises actively. In the case of pain and inflammation, range
of motion exercises must be more gentle so that further injury
is not produced.

Bed Positioning: A post-traumatic patient who is unconscious,
paralyzed, or unable to move himself in bed should be placed on a
physiatric bed which consists of a firm, flat mattress on a three-
quarter inch bedboard, with short siderails and a footboard blocked
from the lower end of the mattress by a four-inch block leaving a
four-inch space to relieve the heels of pressure and allow the toes
between the board and the mattress when the patient is prone. An
alternating pressure bed may help to prevent bedsores. Under special
conditions powered rotating frames may also be required.

Positioning instructions and turning schedules are based on the
individual patient's needs in order to offer him a better chance to
prevent deformities and decubitus ulcers. Positioning techniques
are described in detail in several sources (Bergstrom and Coles;
1969; Ellwood, 1971).

The presence of conditions such as edema, tissue damage, pain,
or fracture may limit the length of time to be spent in each position and a variety of positions may be necessary. Positioning
should be checked more frequently if spasticity is present since it
may be changed by involuntary motions.

Splinting: When normal muscle balance and use are disturbed,
either temporarily or permanently, an orthosis may be required to
prevent deformity. For instance, in a post-traumatic hemiplegic

patient a proper cockup splint or a hand roll to support the wrist
at 200 degrees and the hand with the metacarpalphalangeal and interphalangeal joints flexed to 135 degrees is necessary.

In peripheral nerve injuries, the early fitting of static or
dynamic orthoses may help both to prevent deformity and to regain
fuction. When there is nerve or muscle injury, splinting should
support the part in the position of optimal function.

In burns, the principle of splinting is completely different.
Thermal injury is followed by an increase in vascular permeability
and edema. These lead to cellular invasion and proliferation of
collagen fibers which undergo characteristic contracture with
shrinking of the skin and subcutaneous tissue and keloid formation.
Splinting of the part in the "functional position" is not enough to
prevent deformity. The part must be put in a position opposite to
the anticipated deformity. For instance, if the dorsum of the hand
is burned fibrosis and contracture of the extensor structures will
occur causing extension of the wrist, hyperextension of the metacarpalphalangeal joints and flexion of the interphalangeal joints.
Therefore, the hand must be splinted with the wrist in ten-degree
volar flexion, the metacarpalphalangeal joints in acute flexion,
the interphalangeal joints in extension, and the thumb in extension
and abduction (Koepke et al, 1963; Koepke et al, 1970; Robitaille
et al).

Stretching to Increase Range of Motion: Connective tissue has
a very high tensile resistance to an applied tension of short duration. However, if it is placed under prolonged mild tension it shows
plastic elongation. On the basis of this plastic property of the
connective tissue, a method has been developed by Kottke et al (1966)
for stretching contractures in the hip flexors, knee flexors, and
ankle plantar flexors. With this technique significant restoration
of motion has been obtained within the limits of pain and without
evidence of tearing of tissues.

Recent studies have shown that the application of a sustained
load at a temperature of 45° C. produces a significant residual
increase in the length of the tendon (Lehmann et al, 1970; Warren
et al, 1971). This indicates that the combination of heat at the
therapeutic level plus prolonged stretch is more effective than
stretch at room temperature.

Only active motion should be used for mobilizing the contracture
of an elbow because stretch applied by the long lever of the forearm
to the relatively weak joint may result in overstretching and tearing
of the connective tissue (Kottke, 1971a). During active stretching
the discomfort experienced by the patient helps to protect against
overstretch.

For contractures which do not respond to stretching surgery is indicated. Some considerations about this surgical procedure are in the section concerned with spasticity in this chapter. Stretching of the flexor muscles of proximal joints of a spastic extremity may cause prolonged reflex flexion in that extremity, in the contralateral extremity and in the ipsilateral extremity, producing the crossed extension-flexion and long spinal reflexes. Therefore, when the hamstring muscles are transplanted to the femoral condyles to relieve flexor spasticity of the knee, if one or more muscle is left attached to the tibia, this reflex will cause knee flexion up to the limit of strength of that muscle. Complete hamstring transfer to the femoral condyle in children older than six years does not cause genu recurvatum unless the soleus also is tight so that the heel cord is short.

Heel cord lengthening is not always successful. Recurrence of spasticity and heel cord shortening up to thirty-three percent have been documented (Banks and Green, 1958; Sharrard and Bernstein, 1972). Based on the experiments of Buck (1953), confirmed by Lipscomb and Wakim (1961), we have found that simple complete percutaneous tenotomy at one level without suturing, followed by ambulation using a rigid ankle brace fixed at 90 degrees a few days after the surgery, and with the position protected with a splint at night, gives a satisfactory result. The muscular contraction induced by the activity applies a reasonable tension on the fibrin coagulum which helps to orient the collagen fibers in the long axis of the tendon. The brace restricts motion so that the tendon does not become overstretched. The ankle joint maintains mobility. The child's activity maintains coordination. The physiologically-regulated activity of the triceps surae determines the new length of the reformed tendon. Therefore, at the end of four months a new tendon has developed at a length appropriate to the neurophysiologic activity of the patient.

MUSCLE WEAKNESS

An important aspect of the effects of immobilization is muscle weakness and atrophy. Muscle weakness and atrophy may be the result of disuse or misuse, as well as central or peripheral nerve palsies or the lack of sensory feedback which helps the monitoring of the muscle force.

Strength may be defined as the maximal tensile force that can be exerted by a muscle during a contraction. Endurance is the ability of the muscle to contract and to exert tension for a prolonged period of time. Power is the rate of doing work or work done per unit of time. Tension during contraction of the muscle fiber is postulated to be the stimulus causing an increase in the strength of the muscle fiber (Kottke, 1971a).

When there is complete inactivity strength is lost at the rate of approximately five percent per day. The strength of seventeen muscle groups of a young student who was put on complete bedrest decreased on the average of 1.0 to 1.5 percent per day of the initial strength. The decrease of initial strength per day in the upper extremity of three subjects after immobilization of one arm in a plaster cast ranged from 1.3 to 5.5 percent of the initial strength (Muller, 1970). Immobilization of the hind limb of cats for twenty-two weeks was reflected by a decrease in weight to thirty percent of the normal (Cooper, 1972). After complete afferent and efferent denervation, the muscle weight declined to about sixty percent of normal in three weeks in cats (Eccles, 1941). The result was the same when tenotomy was performed in addition to denervation (Eccles, 1944). In denervated atrophying rat gastrocnemii, the loss in total isometric strength corresponded to the loss in weight (Fisher, 1940). Thirty days of immobilization of the forearm of a healthy man resulted in a forty-four percent decrease in grip strength (Hills and Byrd, 1973).

Strengthening: Work per unit time exceeding metabolic capacity was proposed by Hellebrandt (1945) to be the stimulus causing muscular hypertrophy, increased strength, and increased ability to recruit motor units. Hettinger and Muller (1953), based on their experiments, suggested that the stimulus for the increased maximal muscular strength is neither the degree of exertion of the muscle nor the daily metabolism of the muscle, but presumably the instantaneous maximal metabolic rate of the muscle fiber. This metabolic rate was considered the maximal stress tolerated. Therefore, they concluded that, when muscular exercise caused the metabolic rate of a single muscle fiber to exceed its oxygen or nutrient supply, hypertrophy and increasing strength occur, not necessarily related to its tensile stimulus.

A muscle contracting at two-thirds its maximal tension for six seconds once daily was reported to increase in strength as rapidly as when it was contracted for forty-five seconds once daily (Hettinger and Muller, 1953).

Maximal contraction of a muscle for six seconds five times daily resulted in greater limiting (eventual) strength than with maximal contraction for six seconds only once daily (Muller, 1970). Likewise, Liberson (1961) reported that isometric exercise of six seconds duration twenty times daily increased both the tensile strength and the endurance above the level achieved by exercising only once daily.

Regularly recurring exercises in which tension exceeded thirty-five percent of the strength of the muscle was reported to result in an increase in muscle strength (Muller, 1970). When the exerted

strength was between twenty and thirty-five percent of the maximal
tension, the strength of the muscle was maintained and when tension
did not exceed twenty percent each day the muscle strength started
to decrease.

Using a system of brief maximal exercises consisting of the contraction of a muscle against maximal resistance through the range
of motion with sustained contraction in the fully shortened position
for five seconds and increasing the resistance by a constant small
increment, Rose et al (1957) were able to increase strength regularly
up to a plateau after sixty days or longer and the strength could
be maintained near that plateau by exercising only once a week.
Likewise, in another experiment strength reached a plateau in twelve
to twenty weeks and was maintained near its peak level with a brief
isometric contraction once a week (Muller, 1959). DeLorme (1945)
proposed a system of exercises in which he determined the ten repetition maximal resistance by contracting the muscle against a light
load ten times and progressively increasing the load for bouts of
ten contractions each until the maximal load which can be lifted
ten times was reached. This was called the ten-repetition maximum.
The progressively-increasing, submaximal contractions were postulated
to exert a conditioning effect on the neuromuscular system to prepare it for maximal effort, which was considered to be the stimulus
causing the increasing muscular strength. Modifications of this
exercise, in which the number of bouts or number of repetitions in
each bout up to maximal resistance was decreased, were reported to
be equally effective in increasing strength (DeLorme and Watkins,
1948; McGovern and Luscombe, 1953; McMorris and Elkins, 1954).
Progressive resistive exercises, brief maximal exercises, and brief
maximal isometric exercises all have been used to increase muscle
strength.

Muscle endurance, which is the ability to carry on any exercise
or activity for a prolonged period of time, is related to the factors
of muscular strength, circulation, and muscular metabolism. Fatigue
might be considered to be the inverse of endurance. Mundale (1970)
showed that fatigue would occur in hand grip during a ten minute
exercise at an intermittent tension of five percent of maximal
strength. The amount of fatigue was increased by increasing the
exercise load. Therapeutic exercise to increase endurance has been
described as low resistance-high repetition exercise. The load on
the muscle should be between fifteen percent and forty percent of
the maximal strength and the muscle should be exercised until significant fatigue is produced. Hellebrandt and Houtz (1956) showed that
endurance exercises in which the muscles were loaded so that fatigue
would begin to occur at the end of a bout of twenty to thirty contractions increased the performance up to three-hundred percent of
the initial value by exercising ten or more bouts with short rest
periods between them three times weekly for eight weeks. Occupa-

tional therapy, including rowing, weaving with a loom, planing, woodworking, sanding, and many other activities, provides a variety of activities useful to developing endurance (Kottke, 1971a).

IMPAIRMENT OF COORDINATION

Coordination, which is the combination of the activities of a number of muscles into a smooth pattern, is monitored primarily through the feedback of sensory stimuli transmitted through proprioceptive pathways and partly by visual and tactile stimuli. The development of coordination depends upon the repetition of a precise pattern of an activity many times. During motor activity the brain is aware of the general performance rather than the precise motion of each muscle and joint. The monitoring of position and motion for a skilled motor pattern is largely automatic through the interaction between the cerebellum, basal ganglia, and premotor cortex. Maintenance of a high degree of coordination requires frequent performance of an activity under conditions in which the sensory perception of the motor performance can check it for accuracy and correct errors. Any trauma to the nervous system which causes an interruption at any part of the sensory-motor feedback loop results in incoordination. The spinal and supraspinal reflexes, which are not apparent in normal individuals because modification and inhibition suppress them, facilitate the normal motion pattern. They may become clearly evident as isolated phenomena after damage to the central nervous system. Because of improper feedback due to damage to the sensory-motor loop, ataxia, which is considered the overshooting of the target, occurs. Nonuse of volitional motion for a long period of time may also cause incoordination or decrease of coordination without any interruption in the sensory-feedback mechanism.

Coordination training depends mainly upon the re-establishment of patterns of activation of the muscles to be used and inhibition to prevent the irradiation of stimuli to muscles which should not contract. Increasing the capacity of the central nervous system to inhibit irradiation of stimuli and integration of multiple components into timed sequence of interrelated responses may be obtained by the repetition of precisely performed patterns of activity many thousands of times. The best way to develop precision and speed of coordination for a specific activity has been shown to be to practice that activity for prolonged periods, starting with minimal resistance at low speed, then increasing the speed to its maximum with maximal precision. For the coordination training of hands and upper extremities, occupational therapy is generally used since the constructive aspect of it aids in maintaining attention. For the lower extremities, Frenkel's exercises, which begin with simple movements with gravity eliminated and gradually progress to more complicated movement patterns against gravity, are utilized (Kottke, 1964, 1971a).

METABOLISM AND NUTRITION

Both the trauma itself and the immobilization which follows trauma may cause considerable changes in human metabolism. Cuthbertson (1930) revealed that a patient who has suffered a break in a long bone usually goes into a negative nitrogen balance for a period after the injury. In a series of nineteen cases of patients with fractures, fifteen lost nitrogen equivalent to three percent and four lost nitrogen equivalent to nine percent of their total bodyprotein (Cuthbertson, 1936). It has also been shown that in severe post-traumatic conditions, such as a thirty-five percent burn involving in considerable part the full thickness of skin, loss of body protein may reach 1.5 kilograms or twelve percent of the total body content of protein (Cuthbertson, 1964) and may even exceed that amount (Reiss et al, 1956). Besides the loss of nitrogen, injuries such as a long bone fracture may also cause a marked rise in the excretion of sulphur, phosphorus, potassium, and creatinine in the urine during the first ten days following an injury. A part of the protein loss is due to loss of tissue, hemmorrhage, or exudates, and the other part to true catabolic phase and disuse atrophy (Munro and Cuthbertson, 1943). For instance, a nitrogen loss of five to seven grams per day into the exudate is not uncommon in large burns (Artz and Moncrief, 1969).

In the experiments of Deitrich et al (1948) on healthy subjects it was shown that six or seven weeks of immobilization in a plaster cast caused a considerable amount of nitrogen, calcium, phosphorus, sulphur, and potassium loss as well as an increased tendency to postural hypotension and a decrease in exercise tolerance.

Blocker et al (1955) have advocated giving high amounts of protein, up to four grams per kilogram body weight, by force feeding in the early stage of burns. However, that has not attracted general agreement (Cuthbertson, 1964; Dolecek, 1969).

In rehabilitation wards a high caloric, high protein diet is given to all bedridden patients, particularly to those who have severe burns. If a burned patient loses fifteen percent of his body weight and fails to maintain an acceptable intake, tube feeding should be initiated (Artz and Moncrief, 1969). However, overfeeding is not desired, especially if the patient is expected to have problems in ambulation. An obese paraplegic, hemiplegic, or amputee may even require a reduction diet for successful ambulation.

Another problem may arise if the patient is not able to control deglutition, such as is seen in post-traumatic cerebral palsy patients. In these cases, special feeding techniques, force feeding, and even gastrostomy may be required.

OSTEOPOROSIS

Disuse osteoporosis is a frequent concomitant of post-traumatic conditions in which motor function becomes limited in an extremity or the whole body. Loss of considerable amounts of calcium and nitrogen during immobilization has been shown in experiments both on healthy subjects and on patients with bone fractures (Deitrich et al, 1948; Howard et al, 1945). There is a general tendency to attribute disuse osteoporosis to the reduction of stress and strain upon the bone secondary to lack of weight bearing (Stein et al, 1955). However, Abramson and Delagi (1961) believe that weight bearing alone exerts an inadequate stress to maintain normal bone strength and that the force of muscular contraction must be exerted against bone as well.

Localized osteoporosis seen after prolonged immobilization in fractures and in peripheral nerve paralysis of long duration may not need any specific treatment other than the usual heat and exercise therapy for the underlying conditions. On the other hand, rather generalized osteoporosis, such as that in paraplegics, may require treatment. Anabolic agents and early ambulation were suggested to treat that kind of osteoporosis (Abramson and Delagi, 1961). Recently, Jowsey et al (1972) showed that a therapeutic regimen of 50 milligrams of solium fluoride and at least 900 milligrams of calcium per day, and 50,000 units of Vitamin D twice weekly is capable of stopping osteoporosis and providing a modest increase in skeletal mass without undesirable effects on the skeleton.

Post-traumatic reflex sympathetic dystrophy also causes a considerable decalcification of bone accompanied by severe pain and sometimes sympathetic phenomena such as color and temperature changes in the skin of the affected extremity. In reflex sympathetic dystrophy treatment consists of blocking the sympathetic nerve ganglia with procaine, repeated as indicated, active motion, although it may be painful, and gentle massage. Since hyperemia is present heat is undesirable (Knapp, 1966). If the dystrophy persists in spite of the blocks sympathectomy may also be used (Barnes, 1953).

PAIN

Pain is one of the most common problems in post-traumatic conditions and is usually associated with inflammation. It generally occurs at the site of the trauma, such as is seen in traumatic arthritis, traumatic tenosynovitis, sprain, bursitis, fibrositis,

and injuries to muscles, tendons, ligaments, bones and joints.
However, referred pain or pain secondary to spasticity is not rare.

For relief of minor pain simple analgesics are effective. However, additional remedies may be required for more serious pain.
In acute and subacute conditions of traumatic origin, such as muscular contusions, ligamentous strains and sprains, tendonitis, and
synovitis, application of cold for the first twelve hours and immobilization of the part is advisable. Cold relieves the pain
(Grant, 1964) and decreases edema formation. Thereafter, intermittent application of mild heat and sedative massage may be indicated to decrease pain, muscle spasm, and inflammation (Gucker,
1965). For chronic painful conditions thermotherapy is the treatment of choice. A sedative massage following the heat is also helpful. The type and depth of heating depends on the locus of pain.
Corticosteroids and procaine injections in selected cases of tendonitis and bursitus may result in dramatic relief. Certainly, if there
is an underlying cause such as spasticity, this should also be
treated.

Sometimes pain is severe and persistent. Intractable pain may
be seen following an intervertebral disc surgery, an amputation or
a spinal cord injury. Various types of open (Spiller and Martin,
1912; Schwatz and L'Leary, 1941; Hyndman, 1942) and percutaneous
(Mullan et al, 1963) chordotomies have been performed for several
years to diminish the pain sensation by interrupting the pain tracts.
Recently, Shealy et al (1967, 1970) and Shealy (1972), based on the
Melzack and Walls (1965) gate theory of control of pain, have successfully used a dorsal column stimulator to relieve intractable
pain.

Another successful approach to the pain problem in intractable
cases has been the change of pain behavior of the patient using a
positive and negative rewarding system (Fordyce et al, 1968).

SPASTICITY

Injuries of the central nervous system, such as traumatic paraplegia, cerebral palsy, or brain injury, often result in spasticity
which limits the functional recovery of a patient. Spasticity may
be defined as an increased resistance to passive stretch, accompanied
by increased stretch reflexes. Any trauma which destroys the inhibitory area or inhibitory pathways along the central nervous system
causes spasticity, resulting in an augmentation in both monosynaptic
and polysynaptic stretch responses with an increase in gamma and alpha
motor neuron activities. Spasticity may cause the restriction of
join motions, facilitate contractures, and mask voluntary power.
Consequently, it may limit the activities of daily living of the

patient. In milder cases, discomfort is so slight that treatment
may not be indicated. In certain cases spasticity may assist the
ambulation of patients by reflex facilitation of knee and hip extension.

For pharmacological control of spasticity diazepam (chlordiaz-
epoxide) appears to be the most effective compound. Experiments
have shown that it diminishes spontaneous reflex and postural neuromuscular activity (Holt, 1964; Rushworth, 1964; Wilson and McKechnie, 1966). In humans, no loss of strength was detected in dyna-
meter measurements after twenty milligrams of diazepam was injected
intravenously (Cook and Nathan, 1967). Several other compounds,
such as mephenesin, meladrazine, barbiturates, and chlorpromazine,
have been tested to diminish spasticity (Pedersen, 1969). However,
none of them was as effective as diazepam, and relaxation could
only be obtained with higher doses which also cause sedation. Recently a new compound, dantrolene sodium, has been studied for its
ability to decrease spasticity (Herman et al, 1972; Basmajian and
Super, 1973); however, significant side effects such as muscle
weakness and dizziness have been seen in some patients (Basmajian
and Super, 1973).

Physical therapy may be used to modify spasticity. It has been
known for a long time that cutaneous stimulation under certain conditions can inhibit the gamma motor neurons of muscles and decrease
spasticity (Liddel and Sherrington, 1924; Ballif et al, 1952; Hunt
and Paintal, 1958). Cold applications to the skin also diminish
spasticity. Studies in cats indicate that progressive cooling of
motor nerves first blocks the gamma and later the alpha fibers without blocking the sensory nerves (Douglas and Malcom, 1955). It was
also reported that the frequency of the firing of the primary and
secondary sensory endings decreased with cooling and ceased at 20º C.
(Eldred et al, 1960). Local cold applications with ice packs (Hartviksen, 1962) and ice massage (Grant, 1964) may diminish spasticity
and pain for a period varying from a few minutes up to twenty-four
hours. Heat, with its well-known relaxing effect, is traditionally
a very valuable tool to reduce the muscle tone. In cerebral palsy,
placing the patient in a reflex-inhibiting position, which means
placing the hyperactive muscle in the stretched position so that
Golgi tendon reflex inhibition is induced, causes a considerable decrease in the abnormal reflexes, giving the opportunity to carry
out the necessary exercises and training (Bobath and Bobath, 1950).

Chemical interruption of the stretch reflex is also a valuable
adjunct. Intrathecal injection of phenol for the relief of spasticity has become popular since the late 1950's (Nathan, 1959; Kelly
and Gauthier-Smith, 1959). Intramuscular or perineural injection
of procaine caused a marked decrease in the spasticity and tendon
jerks of the experimental animals, and this was attributed to the

blocking effect of procaine on gamma motor neuron activities (Matthews and Rushworth, 1957; Rushworth, 1960). Khalili et al (1964)
have used peripheral nerve blocks with phenol solution, and Halpern
and Meelhuysen (1966, 1967) have used intramuscular neurolysis with
phenol solution for the control of spasticity. Intramuscular neurolysis has a duration from three to fourteen months. This procedure may also be repeated several times (Awad, 1972).

Surgical procedures may be indicated for the treatment of selected cases with severe muscle tightness and contractures. Tenotomies
for the correction of contractures have been used for centuries.
Soleus neurectomy (Eggers and Evans, 1963), open or subcutaneous
heel cord lengthening (Banks and Green, 1958; Frost, 1963; Conrad
and Frost, 1969), adductor myotomies with obturator neurectomy (Banks
and Green, 1960), gastrocnemius recession (Strayer, 1958), and the
complete transplantation of hamstring tendons to femoral condyles
with the division of the patellar retinacula, which is known as
Egger's procedure (Eggers, 1952), have been advocated and performed
for the treatment of the lower extremity spasticity and contractures.
Because of the fear of causing genu recurvatum, there is a tendency
to use the modification of Egger's original procedure in which transfer of one or more hamstring muscle is omitted. However, our experiences in rehabilitation practice have shown that these modifications
are not as effective as the original Egger's procedure because the
regional flexor reflexes of the secondary sensory endings of muscle
spindles induce persisting flexion in the remaining muscle (Kottke,
1972).

DECUBITUS ULCERS

Decubitus ulcers are the result of cellular necrosis of the skin
and subcutaneous tissue over weight-bearing body prominences, such
as the sacrum, trochanters, ischial tuberosities, heels, and malleoli
in debilitated patients who are unable to change the position of
their bodies frequently. After trauma a patient may not be able to
change his position, either because he is unconscious or paralyzed
or is not induced to move frequently because of analgesia. Kosiak
(1959, 1961) reported that decubitus ulcers arise from prolonged
tissue ischemia caused by pressure exceeding the normal capillary
pressure of the body. Microscopic changes were demonstrated after
the application of as little pressure as seventy millimeters of mercury for two hours. No appreciable differences were found between
normal and paralytic subjects in response to the amount or the duration of pressure. Dinsdale (1973) demonstrated that friction is also
an important factor in the formation of decubitus ulcers. Alterations in body metabolism following trauma, anemia, or edema with increased distance from the capillary to cells, are also considered
as factors contributing to decubitus ulcers.

When tissue is subjected to pressure for only short periods, the normal reactive hyperemic response partially compensates for the temporary ischemia with the result that the tissue does not undergo morphologic degeneration. The alternating relief of pressure at short intervals prevents pathologic changes even at pressures as high as 240 millimeters of mercury for three hours (Kosiak, 1961). Therefore, to place a patient who is unable to move on an automatically alternating pressure bed with a five-minute cycle or to change his position every thirty minutes appears to be a necessary measure to prevent bedsores. Experiments showed that cushions and padding of various forms were of little value in preventing the formation of decubitus ulcers during constant sitting. Pressures under the ischial tuberosities were generally in excess of 300 millimeters of mercury on a flat padded or unpadded surface and approximately 160 millimeters of mercury when subjects were seated on a firm flat surface padded with two inches of foam rubber (Kosiak et al, 1958).

Treatment of decubitus ulcers consists of (1) complete elimination of pressure on the ulcerated area, (2) improvement of the nutritional and anemic status, (3) cleaning the ulcer with normal saline, (4) ultraviolet radiation which is bacterial without being cytocidal to fibrocytes or epithelium, and (5) surgical repair of large ulcers.

DYSFUNCTION OF BLADDER AND BOWEL

Dysfunction of the bladder is common in spinal cord injuries, and mismanagement may shorten the life span of the patient. Dietrick and Russi (1958) found genitourinary disease in ninety percent of fifty-five paraplegics studied at autopsy, and concluded that ninety percent of all patients with spinal cord injuries die from renal failure. On the other hand, follow-up studies have shown that if damage to renal parenchyma by recurrent infection is avoided, glomerular filtration, tubular function, and renal circulation will not diminish within an eighteen-year period (Price et al, 1966; Price, 1972).

Bors (1957) classified bladder dysfunction as the (1) upper motor neuron type, in which the conus is undamaged and the reflex patterns via the autonomic and somatic (internal pudental) nerves are intact, and (2) lower motor neuron type, in which the conus and/or its peripheral nerves are damaged and the reflex patterns via the autonomic and somatic nerves are absent. In the first type, reflex activity of the baldder is preserved and the patient may be able to void reflexly. In the second type, the patient needs to increase the abdominal pressure by the Crede maneuver or abdominal straining in order to empty his bladder.

Immediate catheterization of the patient after the injury is necessary. Usually a Foley retention catheter is inserted. Gutt- mann (1954), Guttmann and Frankel (1966), and Bors (1967) have advocated intermittent catheterization as the method which provides drainage with the lowest incidence of infection and fastest recovery of reflex function. However, overdistention of the bladder, with its attendant overdistention paralysis, may occur during the inter- vals between catheterizations. Intermittent catheterization at the present time is possible only while the patient is at the hospital on a special urological service. However, Stober et al (1973) have reported that they have trained patients for a home intermittent catheterization program with good results. After the reflex activ- ity of the bladder returns, intermittent clamping of the catheter starts and a few days later the patient is given a trial of voiding. If the patient can void and the urine residual is less than 25 cubic centimeters, he may remain catheter-free. If this cannot be accom- plished after a few trials, the patient should continue to use an indwelling catheter. In patients with ureteral reflux, urinary di- version utilizing an isolated segment of ileum has been accepted as a satisfactory procedure to prevent kidney dysfunction (Cor- donnier, 1955).

Routine urinalysis, urine cultures and evaluation of bladder and kidney functions are important in the management of bladder dysfunction. Infection and urinary calculi are the two principal hazards of urinary management. A large urine volume helps to reduce the risk of both, so the daily fluid intake should be 3500 to 4000 cubic centimeters. Ascorbic acid is administered routinely to es- tablish an acid urine and mandelamine for its anti-bacterial effect. The catheter and bladder should be irrigated with ten percent Rena- cidin [R] solution twice daily to prevent or dissolve calcifications. Other complications are vesico-ureteral reflux, peri-urethral abscess formation, penoscrotal fistula, epididymitis, benign papilloma of the bladder and urethra, and prostatic diverticula (Comarr, 1959).

Bowel dysfunction can be managed easier than bladder dysfunction. Most patients regain the bowel habit in a short time using a stool softener, the gastrocolic reflex after a warm meal, digital reflex stimulation, a regular schedule, early ambulation, and sometimes Dulcolax suppositories. Enemas and laxitives are not recommended. For patients who cannot develop reflex bowel evacuation digital re- moval at two to three day intervals may be necessary.

COSMESIS

Trauma which causes deformities or facial disfigurement has a psychosocial impact on the patient. Burn scars, particularly those which leave facial disfigurement, are very difficult to manage.

Facial burn scars are repelling to social contacts, so the patient
may have difficulty in establishing friendships, attracting a marriage partner or even finding a job. Severe facial burn scars ob-
literate facial expression which is an important communication
mechanism to impress and to convince other people.

Seven years of experience in our Department has shown that good
results in diminution of keloids which have developed after second-
degree burns or after grafting for third-degree burns may be obtained using individually fitted elastic masks made of lycra, in
order to maintain adequate pressure over the edematous tissues of
the neck, chin, and fact (Robitaille et al).

COMMUNICATION DISORDERS

Brain injury to the dominant hemisphere may result in aphasia,
which is impairment of ability to comprehend spoken or written language or of linguistic expression by word-finding, syntax or phone-
mic production. Injuries to the brain stem or cerebellum result in
dysarthria in which the patient has intact language symbols and the
correct patterns of movement for articulating them, but has a difficulty secondary to paresis, paralysis, tremor, spasticity, or rigidity in motor function (Calvo, 1968). In bulbar disorders, the
speech is thick, nasal, and feeble. In cerebellar disorders, the
speech may be explosive and intermittent. In basal ganglia disorders, speech may be slurred and very feeble, reduced in volume,
and monotonous. Other problems seen in post-traumatic conditions
are voice disorders secondary to the tracheostomy, hearing impairment, apraxia, and agnosia. A patient who has an impairment in
communication skills should be referred to a speech pathologist as
soon as possible for evaluation and therapy.

FUNCTIONAL INDEPENDENCE

Each person who is physically independent in self care must be
able to perform more than seventy motor activities in the course of
each day. These basic activities include rising from bed, caring
for hygiene, eating, dressing, ambulating, climbing stairs, and per-
forming a wide variety of manual tasks. Any person who cannot accomplish all of these activities will be dependent on others for
help each day (Lawton, 1963; Kottke, 1971b). The lack of ability
to carry out some of these tasks is not uncommon in post-traumatic
cases. If the general condition of the patient indicates that he
may fail to perform any of the activities of daily living, the
ability of the patient to perform each of these activities should
be tested. Activities that cannot be performed should be analyzed,
proper exercises should be selected to increase his ability, and

these special techniques taught and practiced until he can perform
them. A prosthesis, orthesis, wheelchair, crutches or cane may be
necessary to make ambulation possible. A variety of assistive
equipment, such as utensils, feeders, holders, etcetera, are available to facilitate the training and achievement of the activities
of daily living (Rosenberg, 1968; Lowman and Rusk, 1962). Training
in homemaking activities may also be required for a housewife (Rusk
and Taylor, 1953; Sandler, 1971).

PROSTHETICS AND ORTHETICS

Temporary or permanent use of an orthesis may be required for
support or assistance when there is loss of muscle function following trauma. Ortheses are utilized for several purposes: (1) relief
of pain - such as a back brace to relieve the pain in a lumbar disc
syndrome; (2) rest - ischial weight-bearing brace with a pelvic
band in Legg-Perthes disease to provide rest to the hip joint;
(3) support of weakened muscles - a long leg brace to keep knee in
full extension and to prevent dropfoot in a paraplegic; (4) prevention of contractures - a short leg brace with dorsiflexion spring
for day and night use for the patient who has a tendency towards
heel cord shortening; (5) joint stability - a Boldrey brace to prevent the motion of the neck in a cervical vertebral fracture; (6)
prevention of unwanted motion - a head halter to diminish the unwanted head motion in an athetoid child; and (7) functional improvement - a dynamic splint designed to substitute for weak muscles in
peripheral nerve palsy. A great number of ortheses have been described for upper and lower extremities (Knapp, 1968a, b, c; Anderson,
1965). Interest in using externally-powered ortheses for the severe-
ly-handicapped, such as quadriplegics, has been growing during the
past decade (Reswick and Vodovnik, 1967; Long and Trombly, 1968;
Peizer, 1971; Patterson et al, 1971).

When amputation of an extremity has resulted from trauma, the
purpose of rehabilitation should be to fit a prosthesis which will
meet the patient's needs most appropriately according to his age,
occupation, general health, strength, coordination, and ability to
maintain balance in an erect position; to prepare the stump for
fitting by physical therapy, bandaging, and exercises; to provide
pre-prosthetic and prosthetic training; to relieve phantom pain if
it exists; and to adjust the patient to his disability. A variety
of sockets, joints, harnesses, and feet have been described and
used (Anderson et al, 1959; Friedmann, 1966; Wilson, 1970). Myoe-
lectrically-controlled upper extremity prostheses have also been
used for more than a decade (Reswick and Volovnik, 1967; Peizer,
1971). Immediate post surgical fitting of the lower extremity
prosthesis has been developed as a method which decreases complications and makes early ambulation possible. Because of these advan-

tages immediate post-surgical fitting is the method of choice whenever possible (Burgess et al, 1967).

PSYCHOSOCIAL AND VOCATIONAL PROBLEMS

Any trauma which causes a permanent or long-standing disability,
such as a spinal cord injury, hemiplegia, brain dysfunction, amputation, blindness, extensive burn scars, or deformity, may create
a considerable amount of emotional, social, and vocational problems.
In such a patient, first, a period of confusion and disorganization
generally occurs. Grieving and depression may accompany this condition. Most of the patients tend to deny their incapacities and
hold unrealistic hopes and ideas. This is not only true for the
patient, the family may also demonstrate the same behavior. Adjustment of the patient and his family to the disability and necessary
modifications in his behavior may require the collective efforts of
various rehabilitation disciplines. Unless the patient is motivated,
any goal for a rehabilitation program will be unrealistic, since
nothing can be achieved without the full cooperation of the patient
himself. When there are brain injuries, more complicated problems may
be faced, according to the type, location, and amount of deficit.
Reduction in intellectual function, perceptual deficits, impairment
of memory and learning ability, and changes in emotional responses
are the most important problems which interfere with a rehabilitation
program. Every severe disability, whether caused by trauma or not,
creates a social imbalance. According to the degree of the disability, changes in social life and social relations are expected. Some
of these changes affect the psychological condition and motivation
of the patient. Therefore, social work is considered an important
part of the rehabilitation process. Sometimes, a post-traumatic
patient is unable to return to his previous occupation because of
his disability. In this case, vocational guidance and training are
necessary to build a new life for the patient.

REFERENCES

1. Abercrombie, M., Flint, M. H., James, D. W.: Collagen forma-
        tion and wound contraction during repair of small excised
        wounds in the skin of rats. J. Embryol. Exp. Morphol.
        2:264-274, 1954.

2. Abercrombie, M., James, D. W.: Long term changes in the size
        and collagen content of scars in the skin of rats. J.
        Embryol. Exp. Morphol. 5:171-183, 1957.

3. Abramson, A. S., Delagi, E. F.: Influence of weight bearing
        and muscle contraction on disuse osteopororis. Arch.

Phys. Med. Rehabil. 42:147-151, 1961.

4. Anderson, M. H.: Upper Extremity Orthotics. Charles C. Thomas,
    Springfield, Illinois, 1965.

5. Anderson, M. H., Bechtol, C. O., Sollars, R. E.: Clinical
    Prosthetics for Physicians and Therapists. Charles C.
    Thomas, Springfield, Illinois, 1959.

6. Artz, C. P., Moncrief, J. A.: The Treatment of Burn. Second
    Edition. Saunders Company, Philadelphia, Pennsylvania,
    1969.

7. Awad, E. A.: Phenol block for control of hip flexors and ad-
    ductor spasticity. Arch. Phys. Med. Rehabil. 53:554-557,
    1972.

8. Ballif, L., Fulton, J. F., and Liddell, E. G. T.: Observation
    on spinal and decerebrate knee jerk with special reference
    to their inhibition by single break shocks. Proc. Roy.
    Soc. Lond. (Biol.) 98:589-607, 1925.

9. Banks, H. H., Green, W. T.: The correction of equinus deformity
    in cerebral palsy. J. Bone Joint Surg. (Am.) 40:1359-1370,
    1958.

10. Banks, H. H., Green, W. T.: Adductor myotomy and obturator neu-
    recomy for the correction of adduction contracture of the
    hip in cerebral palsy. J. Bone Joint Surg. (Am.) 42:111-
    126, 1960.

11. Barnes, R.: The role of sympathectomy in the treatment of caus-
    algia. J. Bone Joint Surg. (Br.) 35:172-180, 1953.

12. Basmajian, J. V., Super, G. A.: Dantrolene sodium in the treat-
    ment of spasticity. Arch. Phys. Med. Rehabil. 54:60-64,
    74, 1973.

13. Bergstrom, D., Coles, C. H.: Bed Positioning Procedures. Ameri-
    can Rehabilitation Foundation, Minneapolis, Minnesota, 1969.

14. Blocker, T. G., Levin, W. C., Nowinski, W. W., Lewis, R. R.,
    Blocer, V.: Nutritional studies in the severely burned.
    Ann. Surg. 141:589-597, 1955.

15. Bobath, K., Bobath, B.: Spastic paralysis-treatment of by the
    use of reflex inhibition. Brit. J. Phys. Med. 13:121-127,
    1950.

16. Bors, E.: Neurogenic bladder. Urol. Survey 7:177-250, 1957.

17. Bors, E.: Intermittent catheterization in paraplegic patients.
        Urol. Int. 22:236-249, 1967.

18. Buck, R. O.: Regeneration of tendon. J. Pathol. 66:1-18, 1953.

19. Boullough, W. S., Laurence, E. B.: The control of mitotic ac-
        tivity in the mouse. Proc. R. Soc. Lond. (Biol.) 151:
        517-536, 1960.

20. Burgess, E. M., Traub, J. E., Wilson, A. B., Jr.: Immediate
        Post-Surgical Prosthetics in the Management of Lower
        Extremity Amputees, TR 10-5. Prosthetic and Sensory Aid
        Service, Veterans Administration, Washington, D. C., 1967.

21. Calvo, R. J.: Rehabilitation of communication; in, Licht, S.:
        Rehabilitation and Medicine. Elizabeth Licht, New Haven,
        Connecticut, 1968. Pp. 104-128.

22. Commar, A. E.: The practical urological management of the pa-
        tient with spinal cord injury. Brit. J. Urol. 31:1-46,
        1959.

23. Conrad, J. A., Frost, H. U.: Evaluation of subcutaneous heel
        cord lengthening. Clin. Orthop. 64:121-127, 1969.

24. Cook, J. B., Nathan, P. S.: On the site of action of diazepam
        in spasticity in man. J. Neurol. Sci. 5:33-37, 1967.

25. Cooper, R. R.: Alterations during immobilization and regenera-
        tion of skeletal muscle in cats. J. Bone Joint Surg. (Am.)
        54:919-953, 1972.

26. Cordonnier, J. J.: Urinary diversion utilizing an isolated
        segment of ileum. J. Urol. 74:789-794, 1955.

27. Cuthbertson, D. P.: The disturbance of metabolism produced by
        bony and non-bony injury. Biochem. J. 24:1244-1262, 1930.

28. Cuthbertson, D. P.: Further observations on the disturbance of
        metabolism caused by injury, with particular reference to
        the dietary requirements of fracture cases. Br. J. Surg.
        23:505-520, 1936.

29. Cuthbertson, D. P.: Physical injury and its effects on protein
        metabolism; in, Munro, H. N., Allison, J. B.: Mammalian
        Protein Metabolism. Vol. II. Academic Press, New York,
        1964.

30. Deitrich, J. E., Whedon, D., Shorr, E.: Effect of immobiliza-
       tion upon various metabolic and physiologic functions of
       normal men. Am. J. Med. 4:3-36, 1948.

31. DeLorme, T. L.: Restoration of muscle power by heavy-resistance
       exercises. J. Bone Joint Surg. (Am.) 27:645-667, 1945.

32. DeLorme, T. L., Watkins, A. L.: Technics of progressive resis-
       tance exercises. Arch. Phys. Med. Rehabil. 29:263-273,
       1948.

33. Dietrick, R. B., Russi, S.: Tabulation and review of autopsy
       findings in fifty-five paraplegics. JAMA 166:41-44, 1958.

34. Dinsdale, S. M.: Decubitus ulcers in swine: Light and elec-
       tron microscopy study. Arch. Phys. Med. Rehabil. 54:51-
       56, 1973.

35. Dolecek, R.: Metabolic Response of the Burned Organism. Edited
       by Nowinski, W. W. Charles C. Thomas, Springfield, Illi-
       nois, 1969. Pp. 178-180.

36. Douglas, W. W., Malcom, J. L.: The effect of localized cooling
       on conduction in cat nerves. J. Physiol. (Lond.) 130:53-
       71, 1955.

37. Dunphy, J. E., Udupa, K. N.: Chemical and histochemical sequen-
       ces in the normal healing of wounds. N. Engl. J. Med. 253:
       847-851, 1955.

38. Eccles, J. C.: Disuse atrophy of skeletal muscle. Med. J. Aust.
       2:160-164, 1941.

39. Eccles, J. C.: Investigations on muscle atrophies arising from
       disuse and tenotomy. J. Physiol. (Lond.) 103:253-266,
       1944.

40. Eggers, W. N.: Transplantation of hamstring tendons to femoral
       condyles in order to improve hip extension and to decrease
       knee flexion in cerebral spastic paralysis. J. Bone Joint
       Surg. (Am.) 34:827-830, 1952.

41. Eggers, W. N., Evans, E. B.: Surgery in cerebral palsy. J.
       Bone Joint Surg. (Am.) 45:1275-1305, 1963.

42. Eldred, E., Lindsley, D. F., Buchwald, J. S.: The effects of
       cooling on mammalian muscle spindles. Exp. Neurol. 2:144-
       151, 1960.

43. Ellwood, P. M.: Bed positioning; in, Krusen, F. H., Kottke,

F. J., Ellwood, P. M.: <u>Handbook of Physical Medicine
and Rehabilitation</u>. Second Edition. W. B. Saunders Company, Philadelphia, Pennsylvania, 1971. Pp. 463-487.

44. Enneking, W. I., Horowitz, M.: The intra-articular effects of
immobilization on the human knee. J. Bone Joint Surg.
(Am.) 54:973-985, 1972.

45. Evans, E. B., Eggers, G. W. N., Butler, J. K., Blumel, J.:
Experimental immobilization and remobilization of rat
knee joints. J. Bone Joint Surg. (Am.) 42:737-758, 1960.

46. Fisher, E.: The relation between birefringence and contractile
power of normal, hypertrophied and atrophied skeletal
muscle. Am. J. Physiol. 131:156-164, 1940.

47. Fordyce, W. E., Fowler, R. S., Lehmann, J. F., DeLateur, B. J.:
Some implications of learning in problems of chronic pain.
J. Chronic Dis. 21:179-190, 1968.

48. Friedmann, L. W.: Rehabilitation of amputees; in Licht, S.:
<u>Rehabilitation and Medicine</u>. Elizabeth Licht, New Haven,
Connecticut, 1966. Pp. 296-389.

49. Frost, H. M.: Subcutaenous tendo achilles lengthening. Am.
J. Orthop. 5:256:257, 1963.

50. Grant, A. E.: Manage with ice (cryokinetics) in the treatment
of painful conditions of the musculoskeletal system.
Arch. Phy. Med. Rehabil. 45:233-238, 1964.

51. Grillo, H. C., Watts, G. T., Gross, J.: Studies in wound healing: I. Contraction and the wound contents. Ann. Surg.
148:145-152, 1958.

52. Gucker, T.: The use of heat and cold in orthopedics; in Licht,
S. <u>Therapeutic Heat and Cold</u>. Second Edition. Elizabeth
Licht, New Haven, Connecticut, 1965. Pp. 398-406.

53. Guttmann, L.: Statistical survey on one thousand paraplegics.
Initial treatment of traumatic paraplegia. Proc. R. Soc.
Med. 47:1099-1109, 1954.

54. Guttmann, L., Frankel, H.: Value of intermittent catheterization in early management of traumatic paraplegia and tetraplegia. Paraplegia 4:63-84, 1966.

55. Hall, M. C.: Cartilage changes after experimental immobilization of the knee joint of the young rat. J. Bone Joint
Surg. (Am.) 45:36-44, 1963.

56. Halpern, D., Meelhuysen, F. E.: Phenol motor point block in the management of muscular hypertonia. Arch. Phys. Med. Rehabil. 47:659-664, 1966.

57. Halpern, D., Meelhuysen, F. E.: Duration of relaxation after intramuscular neurolysis with phenol. JAMA 200:1152-1154, 1967.

58. Hartviksen, K.: Ice therapy in spasticity. Acta Neurol. Scand. (Suppl. 3) 38:79-84, 1962.

59. Hell, E. A., Cruickshank, C. N. D.: The effect of injury upon the uptake of H^3-thymidine by guinea pig epidermis. Exp. Cell. Res. 31:128-139, 1963.

60. Hellebrandt, F. A.: Application of the overload principle to muscle training in man. Am. J. Phys. Med. 37:278-283, 1945.

61. Hellebrandt, F. A., Houtz, S. J.: Mechanism of muscle training in man: Experimental demonstration of the overload principle. Phys. Ther. Rev. 36:371-383, 1956.

62. Herman, R., Mayer, N., Mecomber, S. A.: Clinical pharmacophysiology of dantrolene sodium. Am. J. Phys. Med. 51:296-311, 1972.

63. Hettinger, T., Muller, E. A.: Muskelleistung und muskeltraining. Arbeitsphysiologie 15:111-126, 1953.

64. Hills, W. L., Byrd, R. D.: Effects of immobilization in the human forearm. Arch. Phys. Med. Rehabil. 54:87-90, 1973.

65. Holt, K. S.: The use of diazepam in childhood cerebral palsy. Ann. Phys. Med. 8(Suppl.):16-24, 1964.

66. Howard, J. E., Parson, W., Bigham, R. S., Jr.: Studies on patients convalescent from fracture. Urinary excretion of calcium and phosphorus. Bull Johns Hopkins Hosp. 77:291-313, 1945.

67. Hunt, C. C., Paintal, A. S.: Spinal reflex regulation of fusimotor neurons. J. Physiol. 143:195-212, 1958.

68. Hyndman, O. R.: Lissauer's tract section: A contribution to chordotomy for the relief of pain. J. Internat. Coll. Surgeons 5:394-400, 1942.

69. Jowsey, J., Riggs, B. L.; Kelly, P. J., Hoffman, D. L.: Effect of combined therapy with sodium fluoride, vitamin D and

calcium in osteoporosis. Am. J. Med. 53:43-49, 1972.

70. Kelly, R. D., Gautier-Smith, P. C.: Intrathecal phenol in the treatment of reflex spasm and spasticity. Lancet 2:1102-1105, 1959.

71. Khalili, A. A., Harmel, M. H., Foster, S., Benton, J. G.: Management of spasticity by selective peripheral nerve block with dilute phenol solution in clinical rehabilitation. Arch. Phys. Med. Rehabil. 45:513-519, 1964.

72. Knapp, M. E.: Practical physical medicine and rehabilitation, late treatment of fractures and complications, Part 2. Postgrad Med. 40:A113-118, 1966.

73. Knapp, M. E.: Orthotics-Bracing the upper extremity. Postgrad Med 43:215-219, 1968a.

74. Knapp, M. C.: Orthotics-Bracing the lower extremity. Postgrad Med. 43:225-230, 1968b.

75. Knapp, M. C.: Orthotics (bracing). Postgrad Med. 43:241-246, 1968c.

76. Koepke, G. H.: The role of physical medicine in the treatment of burns. Surg. Clin. North Am. 50:1385-1399, 1970.

77. Koepke, G. H., Feallock, B., Feler, I.: Splinting the severely burned hand. Am. J. Occup. Ther. 17:147-150, 1963.

78. Kosiak, M.: Etiology and pathology of ischemic ulcers. Arch. Phys. Med. Rehabil. 40:62-69, 1959.

79. Kosiak, M.: Etiology of decubitus ulcers. Arch. Phys. Med. Rehabil. 42:19-29, 1961.

80. Kosiak, M., Kubicek, W., Olson, M., Dantz, J. N., Kottke, F. J.: Evaluation of pressure as a factor in the producing of ischial ulcers. Arch. Phys. Med. Rehabil. 39:623-629, 1958.

81. Kottke, F. J.: Recent advances in kinesiology. Proc. IV. Intol. Cong. Phys. Med., Paris, 6-11 Sept., 1964.

82. Kottke, F. J.: Secondary spindle reflexes from muscle spindle. Proc. VI Intl. Cong. Phys. Med., Barcelona, 2-6 July, 1972. (In Press)

83. Kottke, F. J.: Therapeutic exercise; in, Krusen, F. H.;

Kottke, F. J., Ellwood, P. M., Jr.: Handbook of Physical Medicine and Rehabilitation. Second Edition. W. B. Saunders Company, Philadelphia, Pennsylvania, 1971a. Pp. 385-428.

84. Kottke, F. J.: Training for functional independence; in, Krusen, F. H.; Kottke, F. J., Ellwood, P. M., Jr.: Handbook of Physical Medicine and Rehabilitation. Second Edition. W. B. Saunders Company, Philadelphia, Pennsylvania, 1971b. Pp. 429-451.

85. Kottke, F. J., Kubicek, W. G.: Relationship of the tilt of pelvis to stable position. Arch. Phys. Med. Rehabil. 37: 81-90, 1956.

86. Kottke, F. J., Pauley, D. L, Ptak, R. A.: The rationale for prolonged stretching for correction of shortening of connective tissue. Arch. Phys. Med. Rehabil. 47:345-352, 1966.

87. Lawton, E. B.: Activities of Daily Living for Physical Rehabilitation. McGraw-Hill, New York, 1963.

88. Lehmann, J. F., Masock, A. J., Warren, C. G., Koblanski, J. N.: Effect of therapeutic temperature on tendon-extensibility. Arch. Phys. Med. Rehabil. 51:481-487, 1970.

89. Liberson, W. T.: Brief isometric exercises; in Licht, S.: Therapeutic Exercise. Second Edition. Elizabeth Licht, New Haven, Connecticut, 1961. Pp. 307-326.

90. Liddell, E. G. T., Sherrington, C.: Reflexes in response to stretch (myotatic reflexes). Proc. R. Soc. Lond. (Biol.) 96:212-242, 1924.

91. Lipscomb, P. R., Wakim, K. G.: Regeneration of severed tendons. An experimental study. Proc. Staff Meet, Mayo Clin. 36: 271-276, 1961.

92. Long, C. II, Trombly, C. A.: Clinical applications of myoelectric control in upper extremity orthotics. Arch. Phys. Med. Rehabil. 49:661-664, 1968.

93. Lowenthal, M., Tobis, J. S.: Contractures in chronic neurologic disease. Arch. Phys. Med. Rehabil. 38:640-645, 1957.

94. Lowman, E. W., Rusk, H. A.: Self help devices. Rehabilitation Monograph XXI. The Institute of Physical Medicine and Rehabilitation, New York University Medical Center, 1962.

95. Matthews, P. B. C., Rushworth, G.: The selective effect of
     procaine on the stretch reflex and tendon jerk of soleus
     muscle when applied to its nerve. J. Physiol. 135:245-
     262, 1957.

96. McGovern, R. E., Luscombe, H. B.: Useful modifications of
     progressive resistive exercise technique. Arch. Phys.
     Med. Rehabil. 34:475-477, 1953.

97. McMinn, R. M. H.: Tissue Repair. Academic Press, New York,
     1969. P. 135.

98. McMorris, R. O., Elkins, E. C.: A study of production and
     evaluation of muscular hypertrophy. Arch. Phys. Med.
     Rehabil. 34:475-477, 1953.

99. Melzack, R., Wall, P. D.: Pain mechanism. Science 150:971-
     979, 1965.

100. Mullan, S., Harper, P. V., Hekmatpanah, J., Torres, H.,
     Dobbin, G.: Percutaneous interruption of spinal pain
     tracts by means of a strontium^{90} needle. J. Neurosurg.
     20:931-939, 1963.

101. Muller, E. A.: Training muscle strength. Ergonomics 2:216-
     222, 1959.

102. Muller, E. A.: Influence of training and of inactivity on
     muscle strength. Arch. Phys. Med. Rehabil. 51:449-462,
     1970.

103. Mundale, M. O.: The relationship of intermittent isometric
     exercise to fatigue of hand grip. Arch. Phys. Med.
     Rehabil. 51:532-539, 1970.

104. Mundale, M. O., Hislop, H. J., Rabideau, R. J., Kottke, F. J.:
     Evaluation of extension of the hip. Arch. Phys. Med.
     Rehabil. 37:75-80, 1956.

105. Munro, H. N., Cuthbertson, D. P.: The response of protein
     metabolism to injury. Biochem. J. 37:xii, 1943.

106. Nathan, P. W.: Intrathecal phenol to relieve spasticity in
     paraplegia. Lancet 2:1099-1102, 1959.

107. Patterson, R. P., Halpern, D., Kubicek, W. G.: A proportion-
     ally controlled externally powered hand splint. Arch.
     Phys. Med. Rehabil. 52:434-439, 1971.

108. Pedersen, E.: Spasticity: Mechanism, Measurement, Management. Charles C. Thomas, Springfield, Illinois 1969.

109. Peizer, E.: External power in prosthetics, orthotics and orthopedic aids. Prosth. Inter. 4:4-60, 1971.

110. Potenza, A. D.: Effect of associated trauma on healing of divided tendons. J. Trauma 2:175-184, 1962.

111. Price, M.: Renal function in patients with traumatic myelopathy. Arch. Phys. Med. Rehabil. 53:261-265, 1972.

112. Price, M., Tobin, J. A., Reiser, M., Olson, M. E., Kubicek, W. G., Boen, J., Kottke, F. J.: Renal function in patients with spinal cord injuries. Arch. Phys. Med. Rehabil. 47:406-411, 1966.

113. Reiss, E., Pearson, E., Artz, C. P.: The metabolic response to burns. J. Clin. Invest. 35:62-77, 1956.

114. Reswick, J. B., Vodovnik, L.: External power in prostheses and orthoses. Artif Limbs 11:3:5-21 (Autumn) 1967.

115. Robitaille, A., Halpern, D., Kottke, F. J., Burrill, C., Payne, L.: Correction of keloids and finger contractures in burn patients. Accepted for publication in Arch. Phys. Med. Rehabil.

116. Rose, D. L., Radzyminski, S. F., Beatty, R. R.: Effect of brief maximal exercise on the strength of the quadriceps femoris. Arch. Phys. Med. Rehabil. 38:157-164, 1957.

117. Rosenberg, C.: Assistive Devices for the Handicapped. American Rehabilitation Foundation, Minneapolis, Minnesota, 1968.

118. Rushworth, G.: Spasticity and rigidity. An experimental study and review. J. Neurol. Neurosurg. Psychiatry 23:99-118, 1960.

119. Rushworth, G.: Skeletal muscle spasm and some speculations on the mode of action of diazepam. Ann. Phys. Med. 8(Suppl.):1-2, 1964.

120. Rusk, H. A., Taylor, E. J.: Living with a Disability. The Blakiston Company, Inc., New York, 1953.

121. Sandler, B.: Training in homemaking activities; in, Krusen, F. H., Kottke, F. J., Ellwood, P. M., Jr.: Handbook of

Physical Medicine and Rehabilitation. W. B. Saunders
Company, Philadelphia, Pennsylvania, 1971. Pp. 488-509.

122. Schwartz, H. G., O'Leary, J. L.: Section of the spinothalamic
     tract in the medulla with observations on the pathway for
     pain. Surgery 9:183-193, 1941.

123. Sharrard, W. J. W., Bernstein, S.: Equinus deformity in cere-
     bral palsy. A comparison between elongation of the tendo
     calcaneus and gastrocnemius recession. J. Bone Joint
     Surg. (Br.) 54:272-276, 1972.

124. Shealy, C. N.: Current status of dorsal column stimulation
     for relief of pain; in, Somjen, G. G.: _Neurophysiology
     Studied in Man_. Proc. Symp. Faculte des Sciences, Paris,
     20-22 July, 1971.

125. Shealy, C. N., Mortimer, J. T., Reswick, J. B.: Electrical
     inhibition of pain by stimulation of the dorsal columns:
     Preliminary clinical report. Anesth. Analg. (Cleve.) 46:
     489-491, 1967.

126. Shealy, C. N., Mortimer, J. T., Hagfors, N. R.: Dorsal column
     electroanalgesia. J. Neurosurg. 32:560-564, 1970.

127. Spiller, W. G., Martin, E.: The treatment of persistent pain
     of organic origin in the lower part of the body by division
     of the anterolateral column of the spinal cord. JAMA 58:
     1489-1490, 1912.

128. Stein, I., Stein, R. O., Beller, M. L.: _Living Bone in Health
     and Disease_. J. B. Lippincott Company, Philadelphia, Pen-
     sylvania, 1955.

129. Stover, S. L., Miller, J. M., Nepomuceno, C. S.: Intermittent
     catheterization in patients previously on indwelling
     catheter drainage. Arch. Phys. Med. Rehabil. 54:25-30,
     1973.

130. Strayer, L. M.: Gastrocnemius recession. J. Bone Joint Surg.
     (Am.) 40:1019-1030, 1958.

131. Thaxter, T. H., Mann, R. A., Anderson, C. E.: Degeneration of
     immobilized knee joint in rats. Histological and autora-
     diographic study. J. Bone Joint Surg. (Am.) 47:567-585,
     1965.

132. Warren, C. G., Lehmann, J. I., Koblanski, J. N.: Elongation
     of rat tail tendon. Effect of load and temperature. Arch.

Phys. Med. Rehabil. 52:465-474, 1971.

133. Wilson, A. B.: Limb prosthetics. Artif. Limbs 14:1:1-52, Spring, 1970.

134. Wilson, L. A., McKechnie, A.: Oral diazepam in the treatment of spasticity in paraplegia. Scott Med. J. 11:46-51, 1966.

AIR DISASTER TRAUMA (Paper in Abstract)

William J. Reals

Consultant, Federal Aviation Administration

Washington, D. C.

The first investigation of air accident trauma occurred in 1908 at Ft. Myer, Virginia. Orville Wright and Lt. Thomas Selfridge were in an accident in which Orville Wright was killed by falling from the disabled aircraft and suffering a severe skull fracture. In the years between this accident and the beginning of World War II little effort was made by physicians to study air accident trauma. The Luftwagge in Germany, early in World War II, developed the first scientific approach to the pathological and medical investigation of aviation accident trauma.

In 1955, the pathological investigation of the British Comet aircraft disasters began a new era in the scientific study of crash injury and the correlation of trauma to the man-machine complex. Following this brilliant piece of work departments of aviation pathology were established at the Armed Forces Institute of Pathology in Washington, D. C. and at the RAF Institute of Pathology at Halton in the United Kingdom. These two Institutes and the Civil Aeromedical Institute of the Federal Aviation Administration have served as the principal centers for the investigation of aircraft trauma in the world. Other countries such as West Germany and Australia also have studied aircraft trauma. Until this effort began worldwide many accidents were considered to be "cause unknown," due to mechanical failure, weather or other unrelated causes but scientific investigation of trauma has now led to the solution of many air disasters.

Trauma is the major consideration in catastrophic accidents. A thorough scientific study of each victim, particularly the cockpit crew is necessary to find the cause of death. The autopsy

361

investigation of trauma in aircraft accidents should be very carefully done employing not only proper post mortem technique but
also biochemical and toxicological studies as well as photographs
and x-rays.

Trauma resulting from high-speed jet aircraft and multiengine
reciprocating engine aircraft usually results in disintegration of
bodies. Fire is often a complicating factor. Decapitation, dis-
memberment and loss of body integrity are commonly seen. Characteristic tears of the perineum are common. In general aviation
aircraft, (airplanes weighing less than 12,500 pounds), generally
showed different type injuries. Crashes occur at much lower terminal velocities and some of the accidents are survivable. In
light aircraft accidents the primary areas of lethal injuries are
face, legs, cardiovascular system and liver. In the cockpit area
the use of shoulder harness and helmets or hard hats will prevent
many, perhaps 50 percent of lethal brain injuries and crushing
injuries to the chest. Rupture of the cardiovascular system is a
complication usually caused by extreme compression of the heart
between the sternum and the vertebral column during crashes.
Flailing fractures of arms and legs may incapacitate the victim
so that he cannot escape the resulting fire, a complication of
many light aircraft accidents.

Helicopter accidents cause unique injuries in that many of the
creashes are uncontrolled, high-speed descents, usually the injuries are multiple rather than single with central nervous system
and the cardiovascular system commonly paired in severe injuries.
Crash injuries to the cardiovascular system are often caused by
stabbing due to fractures of the ribs. Glider accidents are very
low speed accidents and the pattern of injury is similar to light
aircraft accidents. Because of the increase of sport parachuting,
parchute accidents are increasingly common, often cause of death
is due to tangling of the parachute harness, causing asphyxiation.
Some parachutists may be accidently struck by the aircraft in
emergency escape and multiple severe injuries are caused by the
parachute failing on descent. In flight bird strikes have also
caused fatal accidents involving passenger airplanes and may be
the cause of severe injury to the pilot in general aviation
aircraft.

BLOOD BANK PROBLEMS IN TRAUMA MANAGEMENT

Howard F. Taswell

Mayo Clinic, Rochester

Blood bank problems arise in the management of trauma at three
levels: (1) those related specifically to trauma to blood or trauma
to the hemopoietic system or both, (2) those related to trauma to
an individual patient, and (3) those related to trauma to many persons within the community.

TRAUMA TO BLOOD AND HEMOPIETIC SYSTEM

Trauma may directly affect any component of the blood at a cellular level, resulting in cell lysis or agglutination from many agents
such as antibodies, drugs, toxins, bacteria, parasites, solutions
that are not isotonic or are excessively acidic or basic, extremes
of temperature, radiation, and mechanical effects related to abnormal
or artificial cardiac valves, tumors, abnormal vasculature, or extracorporeal devices. It is the responsibility of a well-run blood
bank and transfusion service to avoid these agents and to prevent
trauma to the blood.

Antibodies to erythrocyte antigens may be present in either the
donor's or the recipient's blood and should be detected by a routine
screening procedure in order to avoid an antigen-antibody reaction
that could lead to cell lysis, agglutination, or sequestration and
destruction of the erythrocyte. Most modern blood banks routinely
test all donor sera for antibodies by using several reagent red cells

selected to contain on their surface most of the common red cell antigens. Donor blood that contains antibodies either is not used at all or is used only after the removal of most of the plasma. Routine use of red blood cells instead of whole blood obviates most of the problems associated with transfusion of donor antibodies that may be found in whole blood. Similar testing procedures are used to detect antibodies in the serum of intended transfusion recipients. Recipient antibodies are harmless if only compatible blood that lacks the corresponding antigen is transfused. Careful compatibility tests or crossmatching procedures are necessary to be certain that no recipient antibodies are present that will react with the intended donor blood. Even when compatible blood is transfused, however, foreign antigens present in the donor blood may immunize the recipient. Generally, sensitization does not occur until the transfused donor erythrocytes have disappeared from the circulation of the recipient. However, in some instances, high titers of antibody may rapidly develop and may hemolyze transfused red cells that are still circulating.

The addition of drugs or medications to units of donor blood should always be avoided. Many drugs and solutions are incompatible with blood in terms of pH or osmolarity, and their mixture with donor blood will result in agglutination or hemolysis that may remain undetected unless the blood sediments and a color change is noted in the supernatant plasma. Solutions should never be connected in tandem with blood for intravenous administration. For similar reasons it is dangerous to use a Y-type set for the intravenous administration of solutions and blood.

Occasionally, hemolysis or a positive direct Coombs test (or both) related to a drug being taken by a patient is first noted in the blood bank during grouping or compatibility tests and may be of diagnostic value to the patient's physician.

Careful selection and phlebotomy of donors is necessary to prevent transfusion of blood containing bacteria (from contamination or donor disease) or containing parasies such as those causing malaria. All blood banks should carefully screen their donors by a brief medical history and physical examination, as outlined in the American Association of Blood Bank's Standards for Blood Banks and Transfusion Services.

Equal care must be taken during the handling and shipping of blood to avoid mechanical trauma and during storage to avoid hemolysis due to extremes of temperature. Blood must be stored only in carefully controlled refrigerators whose temperatures are maintained over a very narrow range. A change in temperature above or below this range must be promptly indicated by an alarm system.

TRAUMA TO THE PATIENT

Trauma to blood or the hemopietic system by the agents previously listed or trauma to the patient by various agents can result in anemia, leukopenia, thrombocytopenia, hypovolemia, hypoproteinemia, or problems of hemostasis. The subsequent treatment of the patient presents a second level of blood bank problems. These are problems related to the choice of proper component therapy -- therapy under emergency conditions -- and the proper identification of the donor unit and recipient. There are many other problems related to the available blood inventory, the age of the blood, the temperature of the blood and its rate of administration, compatibility testing, and the use of group-specific versus nongroup-specific blood.

Therapy with various selected blood components is rapidly replacing the previous indiscriminate use of whole blood. Component blood therapy became possible with the development and use of the flexible plastic bag in place of the rigid glass bottle as a container for blood. The usefulness of plastic bags as blood containers became particularly apparent during the Korean War. These containers greatly reduced the weight and space requirements of units of blood as well as of the empty containers. They were readily adaptable to rapid and safe transfusion since an air pressure cuff could be safely applied to the outer surface of the bag, as opposed to the older method in which air was pumped directly into the glass bottle, occasionally causing a fatal air embolism. Most important of all, however, plastic bags permitted easy centrifugation and manipulation of the unit of whole blood and resultant separation of the various blood fractions.

Whole blood should no longer be used except in exchange transfusions, during extracorporeal circulation, or to treat the anemia resulting from acute massive hemorrhage. Red blood cells, as opposed to whole blood, present a decreased load of potassium and metabolic products, provide twice as much hemoglobin per unit volume, and decrease the likelihood of circulatory overload as well as the incidence of other nonhemolytic transfusion reactions. In addition, more widespread use of red blood cells allows the production and utilization of many other fractions of blood that would otherwise be wasted if transfused in whole blood to a patient who does not need them.

Plasma products and crystalloid solutions which are hepatitis-free are readily avilable commercially and should be used for volume replacement in many situations of hypovolemia or hypotension or both. Serum albumin should be used for the treatment of hypoproteinemia due to severe burns or other causes. Increased use of such products would greatly reduce the incidence of post-transfusion hepatitis, which is the most common of the serious blood transfusion reactions.

Platelet concentrates or appropriate plasma protein fractions such as Factor VIII concentrate should be used for the treatment of thrombocytopenia or hemostatic problems that result from trauma (or from therapy for trauma). Platelet concentrates are indicated for the treatment of the thrombocytopenia that occurs after the trauma of intensive chemotherapy or ionizing radiation; after perfusion; after massive transfusion with stored blood; and in association with hypersplenism, diffuse intravascular coagulation, and many primary diseases.

Because the trauma patient frequently requires emergency treatment, he presents several special problems to the blood bank. Proper donor and recipient identification is critical in any blood tranfusion but becomes more difficult in the excitement of the emergency room, often with an unresponsive patient who may not yet have been issued a hospital identification number and who is critically ill and in need of prompt transfusion. This is a situation in which the physician and the technician – one ordering the blood and administering it and the other crossmatching and dispensing it – are both under extreme pressure, a situation under which errors are likely to occur. There is a definite association between hemolytic transfusion reactions and the indication for blood transfusion. Two-thirds of the hemolytic transfusion reactions occur in situations that are acute or require large amounts of blood (or both). There must be proper identification and linkage of the patient, his blood sample, and the donor blood units prepared for him – and this can be achieved with currently available donor-recipient wrist-band identification systems.

In an emergency situation, it is always advisable, and usually possible, to determine the patient's blood group and perform compatibility tests in order to provide the patient with group-specific compatible blood. Uncrossmatched group O, Rh-negative blood should rarely, if ever, be used. The use of group O or so-called universal donor blood carries the risk of hemolysis of the recipient's erythrocytes from high-titer donor hemolysins. Another type of catastrophe may occur from the hemolysis of group-specific blood that is administered after an earlier emergency transfusion of several units of group O blood. If group O blood is given in the emergency room, great care must be taken to record this information and to avoid subsequent early transfusion of group-specific blood which was crossmatched and compatible with the patient's original blood sample prior to transfusion of the group O blood. The use of universal group O whole blood should be avoided or, if group-specific blood is not available, only group O red blood cells should be used. In recent years, frozen, thawed, and washed group O red blood cells also have been used extensively as universal blood in several hospitals in the United States.

If the trauma patient requires large volumes of blood in a relatively brief period of time, the blood bank may face additional problems related to patient complications from circulatory overload, excess citrate or potassium, acidosis, hypothermia, or coagulation defects as well as the problem of the availab ility of blood of appropriate age and ABO or Rh type. Treatment for blood loss or anemia with large volumes of intravenously administered fluids and blood should be carefully monitored to avoid circulatory overload and congestive heart failure. Large volumes of rapidly transfused blood rarely, if ever, lead to continued bleeding because of excess citrate except in patients with liver disease or in hypothermia; however, complications due to excess potassium are not uncommon in patients with burns or crush injuries. Problems of hemostasis may occur from the transfusion of large quantities of blood that is deficient in viable platelets or in labile coagulation factors or as a complication of the primary injury or disease. As a general rule, transfusion of one relatively fresh unit of blood, every fourth or fifth unit transfused, will prevent this problem. Hypothermia may result from the rapid administration of large quantities of cold blood but can be avoided by warming the blood at the time of transfusion. This can be accomplished by the careful use of micro-wave blood-warming devices immediately prior to transfusion or by the use of a long blood-administration set coiled in a warm water bath. The latter method significantly slows the rate of trans-fusion, whereas the former causes, on rare occasions, hemloysis of donor erythrocytes overheated by microwave energy. Adequate sup-plies of blood of the proper blood group must be maintained to meet the needs of the individual trauma patient as well as the current and potential needs of all other patients. Volume expanders and blood substitutes frequently can be used until appropriate blood is available and crossmatched. The use of some blood substi-tutes such as dextran may alter the recipient's serum so that sub-sequently crossmatched blood appears to be incompatible.

TRAUMA TO THE POPULATION

A third level of problem arises for the blood bank in the man-agement of trauma involving many persons or even an entire community. Very large supplies of blood are indispensable for the treatment of the large number of patients who may be seen immediately after any disaster. The latter may be a natural disaster such as a tornado, earthquake, fire, or flood or may be a man-made disaster resulting from riot, arson, armed conflict, or transportation accidents. In such disasters, an urgent need for blood in excess of what is usually kept may pose a difficult problem. Long-term storage is not possi-ble because of the limited 21-day shelf-life of blood. Much depends on the type and scope of the disaster, which may be local, regional, or national.

All blood banks should have plans for emergency expansion of their blood drawing and processing capabilities. Alternative plans should be prepared to permit operation under shelter and in protected locations should disaster conditions require. If there is a tremendous demand for large quantities of blood on short notice, group O blood may be used as a universal blood group. The combined use of group O universal blood and long-term freeze-storage techniques has been successful in the treatment of the wounded in Southeast Asia. Under the worse conditions of a major disaster, the use of blood substitutes, blood derivatives, or balanced salt solutions may become necessary.

An effective response to disaster by a blood bank is contingent on the existence of trained personnel and a detailed plan covering four general areas: (1) a stockpile of disposable blood-collecting containers and related equipment; (2) a large blood donor population; (3) trained emergency laboratory personnel; and (4) a central registry of blood donors by blood type, of trained personnel, and of the quantity and location of blood-collecting equipment.

The plan must include a means of activating the system, expanding blood drawing and processing capabilities, establishing multiple alternate communication systems, notifying personnel, maintaining transportation routes, and meeting an immediate requirement for relocation of the blood bank.

Blood bank problems that arise in the prevention and management of trauma to blood and the hemopoietic system, to the individual patient, or to the community can be successfully handled only in an organized, disciplined blood bank with trained personnel and a detailed plan of action.

CONTRIBUTORS

Theodore M. Cole, M.D.
Associate Professor, Department
  of Physical Medicine and
  Rehabilitation
University of Minnesota Medical
  School
Minneapolis, Minnesota

Eugene P. Cronkite, M.D.
Chairman, Medical Department
Brookhaven National Laboratory,
  New York, and
Professor of Medicine
State University of New York at
  Stonybrook

Stacey B. Day, M.D., Ph.D., D.Sc.
Head, Biomedical Communications
  and Medical Education
Sloan Kettering Institute for
  Cancer Research, and
Professor of Biology, Sloan
  Kettering Division
Cornell University Medical
  College
New York City

W. Louis Fowlks, M.D.
Associate Professor, Department
  of Opthalmology
University of Minnesota Medical
  School
Minneapolis, Minnesota

Hugh C. Gilbert, M.D.
Professor of Surgery
University of Illinois Medical
  Center
Chicago, Illinois

Leon Goldman, M.D.
Professor and Chairman
Department of Dermatology, and
  Director, Laser Laboratory
Medical Center
University of Cincinnati
Cincinnati, Ohio

John R. Gordon, M.D.
Professor of Anesthesiology
University of Minnesota Medical
  School
Minneapolis, Minnesota

Robert M. Hardaway III, M.D., M.C.
Brigadier General
Medical Corps, U. S. Army, and
Commanding General
William Beaumont Army Medical
  Center
El Paso, Texas

Edward M. Humphrey, M.D., Ph.D.
Professor of Surgery
University of Minnesota Medical
  School, and
Surgeon-in-Chief
Veterans Administration Hospital
Minneapolis, Minnesota

Maynard E. Jacobson, M.D.
Associate Professor of Medicine
University of Minnesota Medical
  School
Minneapolis, Minnesota

Frederic J. Kottke, M.D., Ph.D.
Professor and Chairman
Department of Physical Medicine
  and Rehabilitation
University of Minnesota Medical
  School
Minneapolis, Minnesota

Gerald S. Moss, M. D.
Director, Division of Surgery
Cook County Hospital, and
Professor of Surgery
University of Illinois Medical
  Center
Chicago, Illinois

Charles A. Murray III, M.D.
Instructor, Department of
  Surgery
University of Minnesota Medical
  School
Minneapolis, Minnesota

M. J. Narasimhan, Jr., M.D.
Visiting Scientist, Pediatrics
  and
Fellow, Bell Museum of Patho-
  biology
University of Minnesota Medical
  School
Minneapolis, Minnesota

William F. Northrup, M.D.
Medical Fellow in Surgery
Department of Surgery
University of Minnesota Medical
  School
Minneapolis, Minnesota

A. Turkyilmaz Ozel, M.D.
Assistant Professor
Department of Physical Medicine
  and Rehabilitation
University of Minnesota Medical
  School
Minneapolis, Minnesota

William J. Reals, M.D.
Consultant to Federal Air
  Surgeon
Federal Aviation Administration
Washington, D. C.

Michael L. Schwartz, M.D.
Medical Fellow in Surgery
Department of Surgery
University of Minnesota Medical
  School
Minneapolis, Minnesota

Hans Selye, M.D., Ph.D., C. C.
Director, Institute of Experi-
  mental Medicine and Surgery
University of Montreal
Montreal, Canada

Quentin T. Smith, M.D.
Associate Professor, Oral Biology
School of Dentistry
University of Minnesota
Minneapolis, Minnesota

Howard F. Taswell, M.D.
Director, Blood Bank and
  Transfusion Service
Mayo Clinic
Rochester, Minnesota

Robert W. ten Bensel, M.D.
Associate Professor
Pediatrics and Public Health
University of Minnesota Medical
    School, and
Medical Director, Ambulatory
    Services
Hennepin County General Hospital
Minneapolis, Minnesota

John A. Washington II, M.D.
Director and Professor
Department of Microbiology
Mayo Clinic
Rochester, Minnesota

Walter Zingg, M.D., F.R.C.S.(C)
Associate Professor of Surgery
University of Toronto
    and
Senior Scientist, Research
    Institute for Sick Children
Toronto, Canada

SUBJECT INDEX